Engineering Aspects of Cereal and Cereal-Based Products

Contemporary Food Engineering

Series Editor

Professor Da-Wen Sun, Director
Food Refrigeration & Computerized Food Technology
National University of Ireland, Dublin
(University College Dublin)
Dublin, Ireland
http://www.ucd.ie/sun/

Innovation in Food Engineering: New Techniques and Products, *edited by Maria Laura Passos and Claudio P. Ribeiro* (2009)

Processing Effects on Safety and Quality of Foods, *edited by Enrique Ortega-Rivas* (2009)

Engineering Aspects of Thermal Food Processing, *edited by Ricardo Simpson* (2009)

Ultraviolet Light in Food Technology: Principles and Applications, *Tatiana N. Koutchma, Larry J. Forney, and Carmen I. Moraru* (2009)

Advances in Deep-Fat Frying of Foods, *edited by Serpil Sahin and Servet Gülüm Sumnu* (2009)

Extracting Bioactive Compounds for Food Products: Theory and Applications, *edited by M. Angela A. Meireles* (2009)

Advances in Food Dehydration, *edited by Cristina Ratti* (2009)

Optimization in Food Engineering, *edited by Ferruh Erdoğdu* (2009)

Optical Monitoring of Fresh and Processed Agricultural Crops, *edited by Manuela Zude* (2009)

Food Engineering Aspects of Baking Sweet Goods, *edited by Servet Gülüm Sumnu and Serpil Sahin* (2008)

Computational Fluid Dynamics in Food Processing, *edited by Da-Wen Sun* (2007)

Engineering Aspects of Cereal and Cereal-Based Products

Edited by
Raquel de Pinho Ferreira Guiné
Paula Maria dos Reis Correia

CRC Press
Taylor & Francis Group
Boca Raton London New York

CRC Press is an imprint of the
Taylor & Francis Group, an **informa** business

Cover Image: Designed by Patricia Correia Rodrigues.

CRC Press
Taylor & Francis Group
6000 Broken Sound Parkway NW, Suite 300
Boca Raton, FL 33487-2742

First issued in paperback 2016

© 2014 by Taylor & Francis Group, LLC
CRC Press is an imprint of Taylor & Francis Group, an Informa business

No claim to original U.S. Government works

Version Date: 20130315

ISBN 13: 978-1-138-19973-6 (pbk)
ISBN 13: 978-1-4398-8702-8 (hbk)

Visit the Taylor & Francis Web site at
http://www.taylorandfrancis.com

and the CRC Press Web site at
http://www.crcpress.com

Contents

Series Preface

Contemporary Food Engineering

Food engineering is the multidisciplinary field of applied physical sciences combined with the knowledge of product properties. Food engineers provide the technological knowledge transfer essential to the cost-effective production and commercialization of food products and services. In particular, food engineers develop and design processes and equipment to convert raw agricultural materials and ingredients into safe, convenient, and nutritious consumer food products. However, food engineering topics are continuously undergoing changes to meet diverse consumer demands, and the subject is being rapidly developed to reflect market needs.

In the development of food engineering, one of the many challenges is to employ modern tools and knowledge, such as computational materials science and nanotechnology, to develop new products and processes. Simultaneously, improving food quality, safety, and security continues to be a critical issue in food engineering study. New packaging materials and techniques are being developed to provide more protection to foods, and novel preservation technologies are emerging to enhance food security and defense. Additionally, process control and automation regularly appear among the top priorities identified in food engineering. Advanced monitoring and control systems are developed to facilitate automation and flexible food manufacturing. Furthermore, energy saving and minimization of environmental problems continue to be important food engineering issues, and significant progress is being made in waste management, efficient utilization of energy, and reduction of effluents and emissions in food production.

The *Contemporary Food Engineering Series*, consisting of edited books, attempts to address some of the recent developments in food engineering. The series covers advances in classical unit operations in engineering applied to food manufacturing as well as such topics as progress in the transport and storage of liquid and solid foods; heating, chilling, and freezing of foods; mass transfer in foods; chemical and biochemical aspects of food engineering and the use of kinetic analysis; dehydration, thermal processing, nonthermal processing, extrusion, liquid food concentration, membrane processes, and applications of membranes in food processing; shelf-life and electronic indicators in inventory management; sustainable technologies in food processing; and packaging, cleaning, and sanitation. These books are aimed at professional food scientists, academics researching food engineering problems, and graduate-level students.

The editors of these books are leading engineers and scientists from many parts of the world. All the editors were asked to present their books to address the market's need and pinpoint the cutting-edge technologies in food engineering.

All the contributions have been written by internationally renowned experts who have both academic and professional credentials. All the authors have attempted to provide critical, comprehensive, and readily accessible information on the art and science of a relevant topic in each chapter, with reference lists for further information. Therefore, each book can serve as an essential reference source to students and researchers in universities and research institutions.

Da-Wen Sun
Series Editor

Preface

The cereal technology and baked foods industry is a major scientific and technological area, one that is in constant development and is of crucial importance for today's food industry and nutrition.

The field of cereal science and engineering is very wide, and further developments have been achieved in the last few years, whose effects will be decisive in today's cereal products technologies. This book intends to give an updated contribution to provide food science professionals and students the most recent information available.

Cereals include a vast number of biochemical entities, very diverse in composition and properties, as well as technological abilities. Their production varies according to the type of cereal, cultivar, or place of growing, as well as cultural practices. Their quality is of the utmost importance, not only at harvest, but also mostly after storage, when they are used for transformation. Technological operations such as milling, either wet or dry, and extrusion have been important in the past and are still a main area of concern in cereal technology. Some particular processing operations, like bread or confectionary baking, are subject to improvements nowadays, and the nutritional and functional properties of cereal products are more and more important from the consumer's point of view. For these reasons, all the aspects referred to, as well as others, are included in the present book, showing some recent advances in the cereal technology and baked foods science.

Raquel de Pinho Ferreira Guiné
Paula Maria dos Reis Correia

Series Editor

Professor Da-Wen Sun, PhD, is a world authority on food engineering research and education; he is a member of the Royal Irish Academy, which is the highest academic honor in Ireland; he is also a member of Academia Europaea (The Academy of Europe) and a fellow of International Academy of Food Science and Technology. His main research activities include cooling, drying, and refrigeration processes and systems; quality and safety of food products; bioprocess simulation and optimization; and computer vision and spectral imaging technologies.

In particular, his many scholarly works have become standard reference materials for researchers in the areas of computer vision, computational fluid dynamics modeling, vacuum cooling, etc. Results of his work have been published in more than 600 papers, including over 250 peer-reviewed journal papers (Web of Science h-index = 41; Google Scholar h-index = 47). He has also edited 13 authoritative books. According to Thomson Reuters's *Essential Science Indicators*[SM] updated as of July 1, 2010, based on data derived over a period of ten years and four months (January 1, 2000–April 30, 2010) from the ISI Web of Science, a total of 2554 scientists are among the top 1% of the most cited scientists in the category of agriculture sciences, and Professor Sun is listed at the top with a ranking of 31.

Dr. Sun received his first class BSc honors and his MSc in mechanical engineering, and his PhD in chemical engineering in China before working at various universities in Europe. He became the first Chinese national to be permanently employed in an Irish university when he was appointed a college lecturer at the National University of Ireland, Dublin (University College Dublin [UCD]), in 1995. He was then continuously promoted in the shortest possible time to the position of senior lecturer, associate professor, and full professor. Dr. Sun is now a professor of food and biosystems engineering and director of the Food Refrigeration and Computerized Food Technology Research Group at UCD.

As a leading educator in food engineering, Dr. Sun has contributed significantly to the field of food engineering. He has guided many PhD students who have made their own contributions to the industry and academia. He has also, on a regular basis, given lectures on the advances in food engineering at international academic institutions and delivered keynote speeches at international conferences. As a recognized authority in food engineering, Dr. Sun has been conferred adjunct/visiting/consulting professorships by over ten top universities in China, including Zhejiang University, Shanghai Jiaotong University, Harbin Institute of Technology, China Agricultural University, South China University of Technology, and Jiangnan University. In recognition of his significant contribution to food engineering worldwide, and for his outstanding leadership in the field, the International Commission of Agricultural and Biosystems Engineering (CIGR) awarded him the CIGR Merit Award in 2000 and again in 2006; the UK-based Institution of Mechanical Engineers named him Food Engineer of the Year 2004; in 2008, he was awarded the CIGR Recognition Award in recognition of his distinguished achievements as the top 1% of agricultural engineering scientists around the world; in 2007, he was presented with the only AFST(I) Fellow Award in that year by the Association of Food Scientists and Technologists (India); in 2010, he was presented with the CIGR Fellow Award (the title of "Fellow" is the highest honor in CIGR and is conferred upon individuals who have made sustained, outstanding contributions worldwide); and in 2013, he was awarded by the International Association of Food Protection (IAFP) with the Frozen Food Foundation Freezing Research Award for his preeminence and outstanding contributions in research that impact food safety attributes of freezing.

Dr. Sun is a fellow of the Institution of Agricultural Engineers and a fellow of Engineers Ireland (the Institution of Engineers of Ireland). He has also received numerous awards for teaching and research excellence, including the President's Research Fellowship, and has received the President's Research Award from UCD on two occasions. He is also the editor in chief of *Food and Bioprocess Technology—An International Journal* (Springer) (2011 Impact Factor = 3.703, ranked at the fourth position among 128 ISI-listed food science and technology journals); series editor of the Contemporary Food Engineering Series (CRC Press/Taylor & Francis Group); former editor of *Journal of Food Engineering* (Elsevier); and an Editorial Board Member for a number of international journals including *Journal of Food Process Engineering, Sensing and Instrumentation for Food Quality and Safety, Polish Journal of Food and Nutritional Sciences, etc.* Dr. Sun is also a chartered engineer.

On May 28, 2010, he was awarded membership to the Royal Irish Academy (RIA), which is the highest honor that can be attained by scholars and scientists working in Ireland. At the 51st CIGR General Assembly held during the CIGR World Congress in Québec City, Canada, in June 2010, he was elected as incoming president of CIGR and will become CIGR president in 2013 to 2014. The term of the presidency is six years—two years each for serving as incoming president, president, and past president. On September 20, 2011, he was elected to Academia Europaea (The Academy of Europe), which is functioning as European Academy of Humanities, Letters and Sciences and is one of the most prestigious academies in the world; election to the Academia Europaea represents the highest academic distinction.

Editors

Raquel de Pinho Ferreira Guiné has a license in Chemical Engineering, MSc in Engineering Science specializing in industrial engineering, and a PhD in Chemical Engineering specializing in Unit Operations and Transfer Phenomena, all at the Faculty of Sciences, Coimbra University, Coimbra, Portugal. She is a coordinator–professor at the CI&DETS Research Center and Food Engineering Department, Polytechnic Institute of Viseu, Portugal.

Paula Maria dos Reis Correia has a license in Agro-Industrial Engineering, MSc in Food Science and Technology, and a PhD in Food Engineering, all at the Institute of Agronomy, Technical University of Lisbon, Lisboa, Portugal. She is an adjunct professor at the CI&DETS Research Center and Food Engineering Department, Polytechnic Institute of Viseu, Portugal.

Contributors

El-Sayed M. Abdel-Aal
Guelph Food Research Centre
Agriculture and Agri-Food Canada
Guelph, Ontario, Canada

Vítor D. Alves
CEER-Biosystems Engineering
Institute of Agronomy
Technical University of Lisbon
Lisbon, Portugal

Franco Antoniazzi
Department of Food Science
University of Parma
Parma, Italy

Satish Bal
Agricultural and Food Engineering
    Department
Indian Institute of Technology
Kharagpur, India

Rintu Banerjee
Agricultural and Food Engineering
    Department
Indian Institute of Technology
Kharagpur, India

Maria João Barroca
CERNAS and Department of Biological
    and Chemical Engineering
Coimbra Institute of Engineering
Coimbra, Portugal

Zoltán Bedő
Agricultural Institute
Centre for Agricultural Research
Hungarian Academy of Sciences
Martonvásár, Hungary

Eleonora Carini
Università Telematica San Raffaele
    Roma
Rome, Italy
and
Department of Food Science
University of Parma
Parma, Italy

Concha Collar
Food Science Department
Instituto de Agroquímica y Tecnología
    de Alimentos
Consejo Superior de Investigaciones
    Científicas
Paterna, Spain

Paula Maria dos Reis Correia
CI&DETS/Department of
    Food Industry
Agrarian School
Polytechnic Institute of Viseu
Quinta da Alagoa, Viseu, Portugal

Duška Ćurić
Faculty of Food Technology and
    Biotechnology
University of Zagreb
Zagreb, Croatia

Elena Curti
Siteia.Parma Interdepartmental Centre
University of Parma
and
Department of Food Science
University of Parma
Parma, Italy

Luísa Beirão da Costa
CEER-Biosystems Engineering
Institute of Agronomy
Technical University of Lisbon
Lisbon, Portugal

Sara Beirão da Costa
CEER-Biosystems Engineering
Institute of Agronomy
Technical University of Lisbon
Lisbon, Portugal

Mithu Das
Agricultural and Food Engineering
    Department
Indian Institute of Technology
Kharagpur, India

Dardo M. De Greef
Instituto de Tecnología de Alimentos
Facultad de Ingeniería Química
Universidad Nacional del Litoral
Santa Fe, Argentina

Ozge Sakiyan Demirkol
Department of Food Engineering
Ankara University
Ankara, Turkey

Silvina R. Drago
Instituto de Tecnología de Alimentos
Facultad de Ingeniería Química
Universidad Nacional del Litoral
Santa Fe, Argentina

Rolando J. González
Instituto de Tecnología de Alimentos
Facultad de Ingeniería Química
Universidad Nacional del Litoral
Santa Fe, Argentina

Luís F. Guido
REQUIMTE
Departamento de Química e
    Bioquímica da Faculdade de
    Ciências da Universidade do Porto
Porto, Portugal

Raquel de Pinho Ferreira Guiné
CI&DETS/Department of
    Food Industry
Agrarian School
Polytechnic Institute of Viseu
Quinta da Alagoa, Viseu, Portugal

László Láng
Agricultural Institute
Centre for Agricultural Research
Hungarian Academy of Sciences
Martonvásár, Hungary

Michele Minucciani
Department of Food Science
University of Parma
Parma, Italy

Manuela M. Moreira
REQUIMTE
Departamento de Química e
    Bioquímica da Faculdade de
    Ciências da Universidade do Porto
Porto, Portugal

Dubravka Novotni
Faculty of Food Technology and
    Biotechnology
University of Zagreb
Zagreb, Croatia

Maria Papageorgiou
Department of Food Technology
Alexander Technological Educational
    Institute of Thessaloniki
Thessaloniki, Greece

Sanaa Ragaee
Department of Food Science
University of Guelph
Guelph, Ontario, Canada

Mariann Rakszegi
Agricultural Institute
Centre for Agricultural Research
Hungarian Academy of Sciences
Martonvásár, Hungary

Koushik Seethraman
Department of Food Science
University of Guelph
Guelph, Ontario, Canada

Sergio O. Serna Saldivar
Department of Biotechnology and
    Food Engineering
Tecnologico de Monterrey
Monterrey, Nuevo León, Mexico

Anshu Singh
Agricultural and Food Engineering
    Department
Indian Institute of Technology
Kharagpur, India

Adriana Skendi
Technological Educational Institute of
    Kavala
Department of Oenology and Beverage
    Technology
Drama Branch
Drama, Greece

Bojana Smerdel
Faculty of Food Technology and
    Biotechnology
University of Zagreb
Zagreb, Croatia

Servet Gulum Sumnu
Department of Food Engineering
Middle East Technical University
Ankara, Turkey

Roberto L. Torres
Instituto de Tecnología de Alimentos
Facultad de Ingeniería Química
Universidad Nacional del Litoral
Santa Fe, Argentina

Elena Vittadini
Department of Food Science
University of Parma
Parma, Italy

1 Cereal Production and its Characteristics

László Láng, Mariann Rakszegi, and Zoltán Bedő

CONTENTS

1.1 INTRODUCTION

Cereals play a decisive and irreplaceable role both in agricultural production and in feeding the world population. Its widespread distribution was made possible by the great variability in the species belonging to the cereal group (e.g., wheat, rice, maize, barley, rye, triticale, oats, sorghum, buckwheat, millet) and, within each species, by the great choice of varieties capable of adapting to extremely different climatic conditions. All in all, these species occupy 45% of the arable land in the world. This percentage differs from one continent or climatic region to another (FAOSTAT 2010), but at least one representative of this group can be grown in all regions inhabited by humans.

The three most important cereal crops, rice, wheat, and maize, are the staple foods of mankind. The proportions used for human consumption are 85% for rice, 72% for wheat, and 19% for maize (Maclean et al. 2002), and these three plant species directly supply 44% of all calories consumed by the entire human population (Braun et al. 2010). Rice provides 20% of the global per capita energy and 13% of the per capita protein of humans, whereas wheat provides 19% of the food calories and 20% of the protein to the world population. Approximately 5% of mankind's calorie requirements and 4% of our protein requirements are met by maize (Braun et al. 2010). In addition to the ability of these species to be grown under a wide range of conditions and their excellent nutritional properties, their use is promoted by the fact that cereal grain is easy to store and transport, making it the primary component of food reserves and the basis of the food trade.

1

1.2 ECONOMIC AND PRODUCTION ASPECTS

Land suitable for cultivation is only available in limited quantities, so there is constant competition for land between food crops, fodder crops, and plants grown for industrial or bioenergy purposes. The relative competitiveness of different species is determined by the geographical location, population level, price conditions, and to a decisive extent, by political decisions. As basic foodstuffs and fodder crops, cereals can be regarded as strategic products, so every country endeavors to become self-supplying with respect to cereals. This is reflected in the crop structure: over the last 30 to 35 years, the global area sown to cereals has fluctuated between 670 and 720 million ha, with a stable trend. Within the cereal group, however, there have been considerable changes in the sowing area of different species in recent decades. The current geographical distribution of cereal species is the result of historical processes developing due to complex interactions between climatic conditions in individual regions, the climatic requirements of the crops, and the population, social, and economic aspects of the given region. Due to the extremely diverse levels of production intensity and yields, the sowing area does not always give a true reflection of the production potential of the region but, nevertheless, reliably indicates both the geographical limits within which a given plant species can be cultivated and the sowing structure, which for biological and economic reasons, realistically can only be changed slowly and to a limited extent. The temperature conditions of the growing site (cold and heat), the rainfall conditions or prospects for irrigation, the relief conditions, and the soil status determine the range of crops that can be successfully cultivated in a given area.

Among the cereals, wheat has the most stable sowing area on a global scale. Over the last half-century, there has been approximately a 30% increase in the growing area of rice and a 50% increase in that of maize, whereas the area sown to triticale (*Triticosecale* Wittm.), a new crop species developed in the 1970s, has now reached 3 million ha. The sowing areas of all of the other cereal species have declined over the same period. This decline was greatest for oats and rye. Simultaneously, with the reduction in the number of horses, the sowing area of oats has decreased from nearly 40 million ha to less than 10 million ha, whereas rye is currently grown on 5 to 6 million ha, compared with 30 million ha in earlier decades. The reduction in the rye-growing area can be attributed to the spread of wheat to traditionally rye-producing areas, changes in eating habits and, to some extent, to the spread of the rival crop triticale. In the case of millet, sorghum, and buckwheat, decreases in the sowing area of 20%, 10% to 12%, and 40%, respectively, were observed in recent decades.

The geographical distribution of cereal production is shown in Table 1.1. In each of the last 50 years, wheat has occupied 31% to 33% of the total cereal sowing area, being grown on 204 to 239 million ha (FAOSTAT). It is grown in the highest proportions between the latitudes of 30° and 60°N, and 27° and 40°S, primarily in South Asia, East Asia, North America, and Eastern Europe. The countries with the largest wheat-growing areas are India, China, Russia, the United States, Australia, and Kazakhstan. Its widespread distribution is facilitated partly by the fact that several wheat species are cultivated, and partly by the fact that most of the species consist of winter, facultative, and spring variants. A decisive proportion of the area is occupied

TABLE 1.1

Geographical Distribution of Cereal Production in 2010 (million ha)

	Wheat	Maize	Rice	Barley	Sorghum	Millet	Oats	Rye	Triticale	Buckwheat
Eastern Africa	2.025.290	14.348.297	2.571.816	1.230.233	4.828.124	1.776.847	28.088	0	0	9.500
Middle Africa	14.628	4.042.696	714.074	1.200	1.526.888	1.281.448	0	0	0	0
Northern Africa	6.834.130	1.226.729	473.465	3.378.452	5.763.877	2.022.000	122.400	27.000	9.900	0
Southern Africa	573.893	3.023.540	1.135	84.764	186.221	280.300	17.239	3.700	0	530
Western Africa	53.624	8.177.044	5.289.835	500	12.459.546	15.746.809	50	0	0	0
Africa Total	9.501.565	30.818.306	9.050.325	4.695.149	24.764.656	21.107.404	167.777	30.700	9.900	10.030
Northern America	27.546.900	34.163.300	1.462.950	3.384.760	1.945.750	146.900	1.352.520	196.340	22.200	80.100
Central America	686.250	9.167.825	332.168	268.768	1.993.451	1.900	66.756	42	723	0
Caribbean		564.527	423.324		90.838	0	0	0	0	0
South America	8.150.346	19.203.052	5.090.149	1.195.889	1.913.864	6.675	555.376	28.584	67.565	45.900
Americas Total	36.383.496	63.098.704	7.308.591	4.849.417	5.943.903	155.475	1.974.652	224.966	90.488	126.000
Central Asia	16.125.566	232.234	241.254	1.682.720	4.501	29.804	167.568	45.817	0	63.873
Eastern Asia	24.798.767	33.038.498	33.206.974	782.533	566.670	768.500	208.550	229.800	200.000	749.806
Southern Asia	48.301.207	9.610.118	54.441.967	2.757.529	7.891.340	12.372.763	800	0	0	3.700
Southeastern Asia	96.500	9.861.547	48.511.763	11.000	28.246	210.500	0	0	0	0

(continued)

TABLE 1.1 (Continued)
Geographical Distribution of Cereal Production in 2010 (million ha)

	Wheat	Maize	Rice	Barley	Sorghum	Millet	Oats	Rye	Triticale	Buckwheat
Western Asia	12.136.440	968.914	148.580	5.659.333	635.378	150.007	97.880	141.064	26.844	90
Asia Total	101.458.480	53.711.311	136.550.538	10.893.115	9.126.135	13.531.574	474.798	416.681	226.844	817.469
Eastern Europe	36.717.243	8.432.689	256.503	12.752.577	54.555	278.364	3.581.672	3.494.443	2.050.180	863.781
Northern Europe	4.418.661	7.800	0	2.988.619	0	0	868.731	211.530	216.300	26.800
Southern Europe	5.239.180	3.360.257	437.425	3.516.190	50.160	659	852.816	188.392	139.550	1.348
Western Europe	9.495.065	2.310.956	23.800	3.701.012	52.100	16.300	275.360	703.355	858.616	36.900
Europe Total	55.870.149	14.111.702	717.728	22.958.398	156.815	295.323	5.578.579	4.597.720	3.264.646	928.829
Australia and New Zealand	13.561.762	76.548	19.000	4.140.340	516.000	38.200	858.966	57.400	334.200	0
Melanesia	10	4.610	4.310	0	1.080	0	0	0	0	0
Micronesia	0	70	90	0	10	0	0	0	0	0
Oceania	13.561.772	81.228	23.400	4.140.340	517.090	38.200	858.966	57.400	334.200	0
World	216.775.462	161.821.251	153.650.582	47.536.419	40.508.600	35.127.976	9.054.772	5.327.467	3.926.078	1.882.328

Source: FAOSTAT 2010. FAO Database. Food and Agriculture Organization of the United Nations. Available at http://faostat.fao.org.

by bread wheat. The winter form has the greatest yield potential because the long vegetation period allows it to develop a deep root system, which contributes greatly to its ability to survive dry periods and to take up nutrients efficiently. The transition from the vegetative to the reproductive stage requires several weeks of cold weather (2°C–5°C), that is, it has a clearly defined vernalization requirement. Its frost resistance and winter hardiness, which vary among cultivars, enables it to survive cold periods in winter. It can be grown in areas with a Continental climate, where the winter is cold enough for vernalization, but not cold enough for the plants to die of cold during the winter. Spring wheat is grown either on areas with an extremely cold winter, or in Mediterranean and subtropical regions. Spring wheat may also have a short vernalization requirement. In areas with a mild winter, the higher temperatures (6°C–10°C) mean that vernalization takes place very slowly, so varieties with a vernalization requirement or which are sensitive to the day length can be sown in autumn without the risk of the plants becoming reproductive before spring arrives, which could cause the developing spikes to be damaged by cold or become sterile. Spring wheat sown in autumn also develops a deep root system, improving its yield potential and yield stability compared with spring-sown spring wheat.

Durum wheat (*Triticum durum* Desf.) has greater drought and heat tolerance than bread wheat, so it can be grown more successfully than the latter in areas with a hot climate. The widespread production of its winter variant is inhibited by its poor winter hardiness compared with *Triticum aestivum* wheats. The durum wheat area makes up approximately 6% of the total wheat-sowing area (12–14 million ha; USDA 2010). The countries with the largest durum-growing areas are those of the European Union, Canada, the United States, and Turkey, and the Mediterranean states in North Africa.

Spelt wheat (*Triticum spelta* L.) is a cereal crop grown in cool mountainous regions of Europe, but it is also grown as an alternative crop in the United States and Australia. Its entire growing area is only a fraction of that of durum.

Although the global sowing area of wheat is stable, a significant regrouping of the major growing regions has been observed in recent decades. Among the continents, the wheat-growing area in Asia has greatly increased over the last 40 years, particularly in the southern and southeastern regions, whereas in Southern Europe, wheat is now grown on only half of the previous areas, and this has only been partially balanced by increases in the wheat-growing areas in Western Europe and Northern Europe. The size of the growing area in the two large export-oriented wheat-producing regions, North America and Australia, fluctuates greatly in response to market demands and economic conditions. In developed countries with favorable climates, where domestic demand does not require the size of wheat-growing areas to be maintained, the sowing area depends on how profitable wheat production is compared with that of maize, soybeans, or oil crops.

Traditionally, the crop grown on the second largest area was rice. Due to its high heat and water requirements, the geographical regions where it can be grown are more limited than in the case of wheat, but its importance is enhanced by the fact that its growing regions coincide with the areas with the densest population, as almost 90% of the total rice-growing area is to be found in Asia, particularly in eastern, southern, and southeastern Asia. Between 1960 and 1980, there was a linear increase in its growing area, with nearly 1.5 million ha of new land planted to rice each year. The growing area is continuing to increase, but the rate of growth has now slowed to 0.7 million ha a year. The countries that produce the largest amounts of rice are India, China, Indonesia, Bangladesh, Thailand, Myanmar, and Vietnam. The availability of satisfactory water supplies is a critical factor for the global future of rice production. Approximately 3700 to 4000 L of water is required for each kilogram of irrigated rice. The reduction in water reserves, combined with climate change, is likely to limit the quantity of water available for rice irrigation in many regions, thus endangering the maintenance of current production levels. To counteract these unfavorable trends, it will be necessary to breed varieties with a lower water requirement and apply water-saving technologies (Facon 2000).

Compared with the two staples, wheat and rice, maize is characterized by a much wider range of end uses. Its importance as a foodstuff is greatest in Africa and in Central and South America, whereas in other regions, it is used chiefly as animal feed, increasing the energy content of feed mixes. It is also the raw material for silage production and is increasingly used for industrial purposes (biofuel, invert sugar, and bioplastics). Of all the cereals, maize is currently the most rapidly spreading crop, due partly to the increase in meat consumption and due partly to the increasing demand for bioethanol. Over the last 50 years, its growing area has increased from 100 million to 160 million ha, with the most intensive territorial increase (3 million ha year^{-1}) in the first years of the twenty-first century. In 2007, maize overtook rice as the crop grown on the second largest area in the world. Because of its great heat requirement, it can only be grown in geographical regions where the cumulative heat sum during the vegetation period satisfies the minimum requirements of the species. The main maize-producing countries are the United States, China, Brazil, India, Mexico, Nigeria, Argentina, and Ukraine. Intensive increases in the growing area can be observed primarily in the United States, South America, and Africa. The super-early maize hybrids developed by breeders have increasingly allowed maize to be grown in more northerly areas, so it is gradually becoming competitive with small grain cereals, especially in northeastern Europe.

Although wheat is grown in every country in the temperate zone, rye is a typically European crop, with 85% of its growing area in this continent, mostly in the eastern and northern regions. Rye is the most frost-resistant cereal, allowing it to be grown reliably in northern regions where the production of wheat is risky. It is an undemanding crop, capable of surviving on sandy or poorly fertile soils with a shallow humus layer, and under such conditions, it gives a larger, more stable yield than wheat or triticale. On better soils, however, its yield potential is poorer than that of wheat, so it is not competitive under such conditions. The substantial decline in the sowing area has several causes. Apart from changes in consumption habits and market demand, it has been ousted primarily due to the development of more tolerant wheat (and to some extent triticale) varieties with better adaptability, which are able to take the place of rye on transitional areas.

The agronomic traits of the allopolyploid species triticale, produced by crossing wheat and rye, are closer to those of wheat than of rye, so it is chiefly grown on poor wheat soils. Approximately 90% of its almost 4 million ha growing area is to be found in Eastern and Western Europe. In these regions winter types are grown almost exclusively, whereas the spring/facultative type has spread in the Southern Hemisphere. Despite the fact that bread can be made from some triticale varieties, the grain is mostly used as fodder, making it a market rival of barley and maize, rather than of wheat.

Barley is the fourth most widely cultivated cereal after wheat, maize, and rice. It has been able to spread so widely thanks to the fact that the winter and spring types are capable of adapting to very diverse climatic conditions, and its good heat and drought tolerance allow it to be successfully grown in regions that are unfavorable for other cereals. Barley is an important feed grain in many areas of the world that are not typically suited for maize production. This is especially true of northern climates, such as those experienced in Northern and Eastern Europe. Barley is the principal feed grain in Canada, Europe, and the northern United States. Barley varieties can be divided not only into winter and spring groups, but also into four subtypes in each group: six-row malting varieties, six-row feed varieties, two-row malting varieties, and two-row feed varieties. The most widespread of these are the winter six-row feed varieties and the spring two-row malting barleys. It is primarily the latter that is used for human consumption, in beer production.

Sorghum is able to grow in the warmer, tropical regions of the world. In Africa, it is the cereal grown on the second largest area, and 60% of the world sowing area is located on this continent. Southern Asia is the second most important sorghum-producing region, with around 20% of the growing area. Its use for human consumption is restricted exclusively to these two regions and is stagnating even here, so all in all, the majority of the yield is now used as fodder (Anglani 1998).

Oats are grown in regions with a cooler climate, with 80% of the growing area in Europe and North America. Russia has the largest area sown to oats (2.2 million ha), but Australia, Canada, Spain, Poland, and the United States also grow oats on areas of over 0.5 million ha. It is a spring-sown cereal, although winter types do exist. However, their poor winter hardiness means that they can only be cultivated sporadically. Oats have traditionally been used to feed horses and cattle, but the global use of oats as feed has been steadily declining, whereas its food use is on the rise. In the course of 40 years, the feed use of oats has decreased from 90% of the total oat production to just over 70% in 2007 (Strychar 2011).

Millet is a heat-requiring plant with a short vegetation period that can be grown with limited water supplies and as a second crop. Two regions, West Africa and South Asia, contain 80% of the global sowing area (35 million ha), although it is also grown on considerable areas in other parts of Africa. In developing countries, 90% of the grain is used for human consumption in the form of porridge, whereas in developed countries, the growing area is small and the crop is used mainly as feed. The least widespread cereal crop is buckwheat, which is grown on a total area of less than 2 million ha, primarily in Eastern Europe and East Asia. Its main advantage is its extremely short vegetation period, so it can be grown as a main or second crop in areas where other cereals are unable to grow to maturity.

Over the last 50 years, cereal yields have increased 2.7 times, as seen in Figure 1.1. Approximately 94% of the yield increase could be attributed to the increase in average yield per hectare, and only 6% of the increase came from an increase in the growing area. Although the average yield increase per hectare per year was 44 kg,

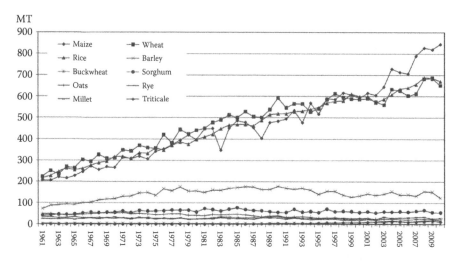

FIGURE 1.1 World cereal production (1961–2010). (From FAOSTAT 2010. FAO Database. Food and Agriculture Organization of the United Nations. Available at http://faostat.fao.org.)

averaged over all the cereals, this figure was 65 kg for maize, 53 kg for rice, and 40 kg for wheat. The yield increases recorded for barley, rye, and oats were substantially lower (23, 27, and 14 kg, respectively). The yield levels of millet and sorghum, which are typically grown in developing countries, have stagnated or increased only slightly, whereas that of buckwheat has declined.

The increase in the yield level was made possible by a combination of improved cultivation techniques and the breeding of new plant cultivars. On the one hand, cultivars adapted to new management practices or growing conditions are developed; on the other hand, the properties of these new cultivars facilitate the further improvement of cultivation techniques, within the limits represented by the environmental conditions. The three decades starting in 1960 were a period of rapid development, involving intensive mechanization and the improvement of management practices. Fertilizer use increased from 50 kg ha^{-1} to more than 200 kg ha^{-1} (FAOSTAT), and there were similar increases in the application of fungicides and insecticides.

The increasing rates of mineral fertilizer and the greater mechanization required the cultivation of cultivars that could be harvested without lodging and which responded with higher yields to the larger quantities of inputs. The identification of dwarf and semidwarf genes in rice and wheat led to the development of nonlodging cultivars with good fertilizer responses, the cultivation of which, combined with higher levels of fertilization, resulted in dramatic yield increases in Southeast Asia and Mexico. This "Green Revolution" contributed decisively to improving world food supplies and fighting famine.

The different rates of development achieved for individual crops, geographical regions, and periods clearly indicate the difficulties that can be expected when endeavoring to achieve further increases in yield. The annual yield increment of 3.6% achieved during the period when wheat production underwent the most intensive development (1966–1979) soon dropped, first to 2.8% (1984–1994) and then to 1.1% (1995–2005). A similar decline in yield increments was observed in rice production. Averaged over 70 years, the growth rate in maize production in the United States was 115 kg ha^{-1} year^{-1}, 50% to 60% of which could be attributed to the hybrids and 40% to 50% to improvements in management practices. The genetic contribution to wheat yield gains was estimated at 28%, whereas increases in the use of nitrogen fertilizer alone accounted for 48% of the total gain (Bell et al. 1995).

The substantial differences in the rate of yield increase in geographical regions with diverse climatic and soil conditions is well illustrated by the fact that during the same period (1959–2008), the estimated genetic gain for wheat in the United States was 1.1% year^{-1} in the Southern Great Plains and only 0.79% year^{-1} in the Northern Great Plains (Graybosh and Peterson 2010). Similar differences could be detected in other regions. The average annual genetic gain in grain yield in different provinces of China was found to range from 32.1 to 72.1 kg ha^{-1} year^{-1}, or from 0.48% to 1.23% (Zhou et al. 2007).

Achieving further improvements in productivity in the countries (regions) with the highest yield levels is complicated by limiting environmental factors, the decreasing returns from surplus investments and environment protection concerns. Until 1985, the increase in productivity in cereal production exceeded the population growth rate, so the per capita cereal production increased from 286 kg year^{-1} in 1961 to 375 kg year^{-1} in 1985. By 2002, however, this figure had declined to 323 kg year^{-1}. The global per capita cereal production is illustrated in Figure 1.2. The current improvements in the per capita cereal production have allowed a return to the cereal supply levels recorded 20 to 30 years ago, but it has proved impossible to exceed these levels. As an annual population growth rate of 1.14% is predicted, further increases will be needed in the future and, as in the past, this will have to be based chiefly on the enhancement of yields per hectare.

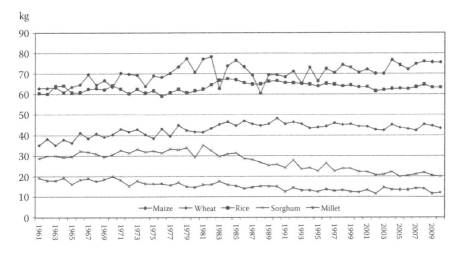

FIGURE 1.2 Per capita cereal production (1961–2010). (From FAOSTAT 2010. FAO Database. Food and Agriculture Organization of the United Nations. Available at http://faostat.fao.org.)

1.3   TYPES OF CEREALS AND INDUSTRIAL USES

One common property of cereals is that they have seeds with a floury texture due to the rich starch content of the endosperm, making them primary energy sources for both human nutrition and animal feeding. Depending on the species, the seeds also contain substantial quantities of protein, lipid, fiber, vitamins, and minerals. Due to the differing availability of various cereal species, differences in the level of development, and the consequent differences in eating habits in various geographical regions, there is considerable variability in the human consumption of individual cereal species and in the food processing capacity supplying these needs. At the same time, economic developments, urbanization, and the expansion of the cereal trade have been accompanied by changes in these traditions, leading to a reduction in cereal consumption in more developed regions and to changes in its composition. Cereals are raw materials with a wide range of uses, and hundreds of products are manufactured from them, in addition to the major use characteristics of each species. A summary will be given here of the main types available for each of the major cereal species and of how they are used.

The cereal species used in the greatest volume by the food industry is wheat. The types cultivated include hard (H) and soft (S) grained, red (R) and white (W) grained, and winter (W) and spring (S) sown variants. On the basis of these properties, a distinction is made between HRW, HRS, SRW, SRS, HWW, HWS, SWW, and SWS wheat types. Durum wheat forms a further category. In the United States, Canada, and Australia, various wheat production zones are also distinguished. This means that wheat varieties in various quality categories are grown in separate areas, thus making it impossible for different quality types to become mixed. In the United States, for example, six quality groups have been set up, among which the hard red winter type, having high protein content and strong gluten, is cultivated in the central regions of the country, whereas durum is grown in the north, and the soft white type, suitable for the manufacture of biscuits and pastry, is grown in the northwest.

Grain color is primarily of traditional and aesthetic importance because the nutritional composition of red and white wheat is the same. However, whole wheat and high-extraction flour made from white wheat is more attractive. On the other hand, white-grained wheat has the disadvantage that it tends to germinate in the spike, leading to a decrease in the falling number. A low falling number has a negative influence on the bread-making quality. The flour absorbs less water, which affects the bread yield; the crust strength and crumb texture are inferior, the shelf life is shorter, and the product may be sticky or gummy.

The suitability of wheat for industrial use is determined not only by the grain hardness but also by the protein content and gluten quality. Nevertheless, the primary basis for discriminating different end uses is not the protein content, but the grain hardness. The endosperm texture is the single most important and defining quality characteristic, as it facilitates wheat classification and affects milling, baking, and end-use quality. Among the genotypes of common wheat (*T. aestivum*), soft-grained types give fine, powdery flour when milled, whereas the flour from hard-grained types is coarser (Biffen 1908). The difference between the two types is caused principally by the fact that the kernels of hard-grained cultivars split along the cell walls during milling, whereas in the case of soft grains, the cells tend to disintegrate (Hoseney et al. 1988). The particle size of the flour is greater for hard-grained than for soft-grained wheats and they undergo greater mechanical injury. The greater extent of starch injury results in greater water uptake ability. Consequently, the wheat endosperm structure is decisive for the technological quality of a given genotype and is an important indicator of milling and bread-making quality, as well as being a reliable predictor of the qualitative and quantitative traits of the resulting flour. For processing reasons, a hard endosperm structure is correlated with high flour yield, greater starch injury, and higher water uptake, loaf volume, bread quality parameters (Pomeranz and Williams 1990), and protein content (Bedő et al. 1998).

Possible end uses for wheat, based on various combinations of grain hardness and protein content, are illustrated in Figure 1.3.

Most hard-grained wheats are used to make bread. Different types of bread are made in various regions or countries, however, and these require different technological parameters. Some wheat flour is used in the pasta industry, where the requirements in terms of wet gluten content and rheological properties are not as strict as for bread-making, but where the milling technique plays a much greater role in achieving the quality required for pasta production. The least frequent special use of wheat is for the manufacture of wheat beer, which can only be made from wheat varieties with low protein content and low falling number. As soft wheat flours absorb less water, these are typically used for pastry, biscuits, and cakes (Morris and Rose 1996).

The flour obtained by milling wheat kernels contains 70% to 80% starch, 10% to 15% protein, and 1% to 2% lipids, among which the protein has the most important role in determining functional properties. In the course of kneading, the storage proteins, most of which are soluble in alcohol, form a continuous protein matrix around the starch granules. This then forms a spatial network in response to water and mechanical mixing, resulting in the production of gluten. Wheat (and rye) is

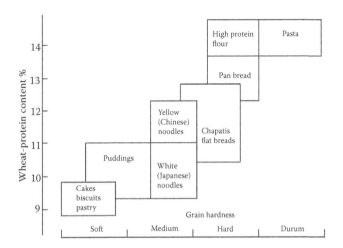

FIGURE 1.3 Possible end-uses for wheat, based on various combinations of grain hardness and protein content. (From Moss, H. J., *J. Aust. Inst. Agric. Sci.* 39: 109–115, 1973.)

unique among cereals in the ability of its flour proteins to form gluten. It is gluten that enables the dough to retain the gases produced by yeast and thus to give well-risen loaves. The bakery products produced using wheat flour thus have a light texture and are easily digestible (Gianibelli et al. 2001). It was by virtue of the flour's ability to form dough that wheat became one of the most important raw materials for human nutrition.

Wheat quality is influenced by the quantity, quality, and relative proportions of the gluten proteins. An increase in the low molecular weight gliadins results in a reduction in the dough development time (MT), the dough resistance (PR), the maximum resistance to stretching (Rmax), and the loaf volume, whereas the dough softening accelerates and the extensibility increases (Uthayakumaran et al. 2001; Fido et al. 1997). Like the gliadins, the low molecular weight glutenins also determine the extensibility of the dough, whereas the high molecular weight glutenin subunits are closely correlated with dough strength (Shewry et al. 2003; Uthayakumaran et al. 2002). Due to this close, well-known correlation between high molecular weight glutenin composition and bread-making quality, the production of raw materials with optimum quality for the baking of various products can be best achieved through the integrated cultivation of wheat varieties with known high molecular weight composition.

Although the main determinants of wheat quality are the gluten proteins, the starch quality may also have a decisive role in the manufacture of certain products (Oda et al. 1980; Iriki et al. 1987). Starch makes up 65% to 70% of the dry matter content of wheat kernels and consists of high molecular weight amylopectin molecules and amylose, which has a lower molecular weight. Of the two polymers, amylose makes up 18% to 35%, whereas amylopectin makes up 68% to 75%, of the starch in hexaploid wheat and durum. The waxy type of wheat contains practically no amylose (Nakamura et al. 1995). Numerous factors contribute to the final characteristics of wheat starch. The genotype, environmental effects, and processing techniques, such as milling, cooking, heat treatment, storage period, and temperature, all have a fundamental influence on the properties of starch, as they may lead to changes in granule size, extent of starch damage, water absorption of the flour, rheological traits, gelling and swelling of the starch, and the viscosity and quantity of resistant starch. The functional properties of starch are also influenced by the nature and ratio of the types of starch granules in the dry endosperm, by the starch content, by the size distribution and shape of the starch granules, by the amylopectin structure and the amylose/amylopectin ratio, by the kernel hardness, by the presence of lipids, and even by the characteristics of other components surrounding the starch matrix. The types of pasta consumed in Asian countries require wheat varieties with special composition and quality. The quality of these products primarily depends not on the gluten but on the starch composition. The manufacture of Japanese noodles, for example, requires the breeding and production of waxy wheat containing no amylose.

Durum wheat has harder kernels than common wheat and has a high protein content. Traditionally, it is mainly used to make pasta. Milling leads to a semolina fraction with high granule size, which forms an attractive yellow type of pasta after the addition of water. The main parameters determining durum quality are kernel hardness, protein content, kernel vitreousness, semolina yield, and yellow pigment content. Spring durum usually has better quality than winter durum, but the yield is substantially lower. In the course of industrial use, the two types are often mixed.

Rice has long been a staple food in many parts of the world and is principally consumed as a whole grain that is threshed and polished after harvesting. In traditional rice-consuming countries, such as Thailand, it has been estimated that 8.79% of the milled rice is used in processing industries for making crackers, rice flour, rice vermicelli, and rice starch (Isvilanonda 2006), whereas in the United States, for example, approximately 58% of the rice is consumed as table rice, with 21% being used in processed foods, 11% in beer, and 10% in pet food (U.S. Rice Federation 2009). Rice is divided into two types—japonica and indica. The characteristics and forms of these two types of rice differ. The grains of japonica rice are round and do not easily crack or break. When cooked, this rice is sticky and moist. The rice produced in Japan is mostly japonica. The grains of indica rice are long and tend to break easily. When cooked, the rice is fluffy and does not stick together. Most of the rice produced in Southern Asia, including India, Thailand, Vietnam, and Southern China, is indica rice. Both the japonica and indica types of rice include nonglutinous and glutinous forms. Each type of rice has its own special characteristics and each has its own place in rice cooking. Nonglutinous rice is popularly used in general rice cooking. This rice is somewhat transparent and when cooked is less sticky than glutinous rice. It is usually cooked in water and served plain. Glutinous rice tends to be white and opaque and is very sticky when cooked. It is commonly used to make rice cakes and various kinds of desserts, or processed to make rice snacks.

Most rice varieties that are planted and consumed throughout the world have a white pericarp, but rice can also produce grains with red, black, or purple pericarps. The color is visible when the grains are dehulled, but it can be removed by polishing to reveal the white endosperm. The cultivation and consumption of colored varieties is limited in Western countries, but in some growing areas of Asia, traditional varieties with a colored pericarp are particularly valued in local markets (Finocchiaro et al. 2007). The quality of rice affects consumer acceptance and market value. The quality traits encompass physical appearance, cooking and eating properties and, more recently, nutritional value (Fitzgerald et al. 2009). Milled rice contains 7.3% to 8.3% protein, around 90% starch, and 0.4% to 0.6% lipid, of which starch is the main determinant of cooking and eating quality (Lásztity 1999). As in wheat, the amylose/amylopectin ratio in the starch varies over a wide range, depending on the variety. The amylose content of waxy rice is only 0.8% to 1.3%, whereas nonwaxy varieties contain 8% to 37%. Volume expansion and water absorption during cooking are positively correlated with the amylose content. The cooking time is directly affected by the gelatinization temperature and the protein content. The eating quality of nonglutinous rice can be improved by a decrease in the amylose and protein concentrations in the grain (Inatsu 1988). Rice also has a wide range of industrial applications. After malting, it serves as the raw material for beer and sake. It can also be used to make puffed rice, whereas its flour is an ingredient in baby food, rice pasta, starch, crispies, cereals, snacks, and coatings.

Despite maize being the third most important staple food, it is mainly used for livestock feeding. Its industrial use is also increasing due to the rapid spread of bioethanol manufacture. In 2009, one-third of the maize produced in the United States (104 million tonnes) was used for this purpose (Oladosu et al. 2011). Maize is also used directly as food or is processed to make food and feed ingredients (such as high-fructose corn syrup and corn starch). Its use for direct human consumption is mostly in the form of polenta and tortillas. Corn flakes are also widely consumed in developed countries, whereas sweet corn can be regarded more as a vegetable.

Maize grain can be classified on the basis of kernel shape, color, and hardness. With regard to shape, maize kernels may be flint (round) or dent (tooth shaped), whereas the color may be white, yellow, or colored. In North America, dent is the principal type grown, whereas most flint corn is grown in Central America and South America. The maize grown for food processing is mostly white (corn flakes, tortillas, and corn flour). Dent corn has a relatively soft, inner starchy layer that is easily ground to a powder, whereas flint corn has a very hard starchy interior, making its flour grittier. The kernels of waxy maize have a waxy appearance when cut and contain only branched-chain starches. This starch consists of more than 99% amylopectin, whereas regular types contain 72% to 76% of amylopectin and 24% to 28% of amylase (Alexander and Creech 1977). This type is used for industrial purposes. Unless industry raises special quality requirements (high protein, high lysine, high oil, opaque, etc.), the determination of kernel type and dry matter content is an adequate way of estimating the value of the crop, unlike the situation for other cereals. The greater the dry matter content, the higher the starch yield and the better the storage stability.

As a raw material for the food industry, barley is used to the greatest extent for malting, a process that develops the enzyme systems of the barley grain. The sugars produced are dissolved out during the mashing process to form the wort, which is fermented by yeast into beer (Atherton 1984). Barley grains of uniform size and germination are required for malting. The grains developing in the spikelets of six-rowed barley vary in both size and shape, whereas the grains of two-rowed barley are 25% to 50% larger and are of uniform size. Larger grains are more favorable for malting as they have a smaller husk ratio and a greater extract content. The protein content is decisive for malting quality. A low protein content of 9% to 11.5% is important not only due to the correspondingly higher carbohydrate content but also because a high protein content has a negative influence on beer quality and causes technological problems when the beer is filtered.

Due to both the greater quantity of sugar that can be extracted from the starch and to the lower protein content, two-rowed spring barley is the most popular raw material for malting. For economic reasons (higher yield potential), two-rowed winter barley is increasingly malted, although six-rowed cultivars may be used to make special products. Although the choice of cultivars with satisfactory quality is important for the industrial processing of all plant species, this is particularly true of malting barley. Due to the strict technological criteria for beer making and the need to achieve constant, uniform quality, large-scale processors only use raw material from carefully chosen varieties. Apart from the grain size and protein content, other traits important for beer quality (husk structure, β-glucan content, enzyme activity, and friability) are also cultivar-dependent; therefore, true-to-cultivar raw material and the strict quality control of purchased lots are essential to ensure a satisfactory product.

In Africa, sorghum is processed into a wide variety of attractive and nutritious traditional foods, such as semileavened bread, couscous, dumplings, and fermented or nonfermented porridges. Millet is consumed in the form of leavened or unleavened bread, porridge, boiled or steamed foods, and beverages. In the Sahel and elsewhere in northern Africa, pearl millet is an important ingredient in couscous. In the United States, Australia, and Europe, millet is also grown to feed cattle and birds.

1.4   CEREAL BREEDING

Over the course of the centuries, continuous production in various regions resulted in the natural selection of characteristic landraces, which are morphologically uniform and can be regarded as a mixture of pure progeny lines. These landraces then spread as the result of human activity to regions they would never have reached naturally. The diverse soil and climatic conditions in these completely new environments led to the development of a wide variety of new landraces. With the spread of improved cultivars, the area sown to landraces rapidly decreased throughout the world from the first half of the twentieth century, and they are now grown only in the poorest regions with extreme environmental conditions. Landraces are cultivated on approximately 10% of the world's wheat growing area (Heisey et al. 1999). In the case of spring wheat, 3.5% of the sowing area in developing countries is occupied by landraces or varieties of unknown pedigree, whereas this ratio is somewhat higher for spring durum and for winter/facultative wheat.

For each climatic region, breeders aim to develop new plant varieties that are adapted to the given climatic conditions, have greater productivity than earlier cultivars, can be reliably grown, and whose grain quality is acceptable to end-users and the market. In the case of winter cereal species, the size of the area over which new cultivars can potentially spread is fairly limited, whereas for maize, it may be considerably larger. Due to the cultivar × environment interaction, regional breeding centers or regional experimental networks have been established to provide farmers with a constant supply of new varieties.

An increase in the production level, or even the maintenance of the existing level, requires a regular change of variety. There have been several periods in the history of breeding when a major change in plant habit or genetic background has led to a spectacular improvement in productivity. The identification of dwarf and semidwarf genes in rice and wheat made it possible to develop nonlodging cultivars with high yield in response to fertilization. The greater yield potential of dwarf plants can be attributed to an increase in the kernel number per unit area and to a great improvement in the harvest index. In the case of wheat, the incorporation of the 1B/1R rye translocation into a large number of modern varieties led to improved productivity and an increase in stress tolerance. For several plant species (maize, rice, sorghum, millet, and rye) the replacement of varieties by hybrids led to a rapid increase in yields. Nowadays, breeding is responsible for a gradual but continual increase in yield potential, amounting to 1% to 2% a year. If more rapid genetic gain is to be achieved in the future, it will be necessary to improve the efficiency of nutrient and water use or the efficiency of photosynthesis using conventional or biotechnological methods. The cultivation of new varieties is able to improve the yield level without additional costs, so a change of variety is the most important way of increasing yields, especially in regions where the standard of farming has stagnated.

Two clearly distinguishable groups of risk factors endanger the development of the yield. The adaptability of plants in the face of abiotic stress (cold tolerance, heat tolerance, drought tolerance, salt tolerance, etc.) can be improved by breeding. The simultaneous cultivation of varieties suited to the local conditions but differing in development dynamics, vegetation period, height, etc., helps to mitigate the yield-reducing effect of unfavorable weather conditions and the extent of yield fluctuation.

Among the biotic stress factors, fungal diseases, viruses, and insect pests may cause considerable reductions in yield. Disease-resistant plants possess resistance gene(s) that are effective against the fungus races present in the given growing region. The plants remain symptom-free and healthy until a new pathogen race appears to which the variety is not resistant. It is often said in such cases that the variety has deteriorated, when in fact it is not the variety that has changed, but the fungal pathogen. The appearance of new, virulent fungus races makes regular changes of variety essential. In the case of wheat, a survey conducted in 26 countries revealed that rust resistance may be effective for a period of 1 to 15 years, depending on the cultivar, the sowing area on which it is grown, the country, and the type of rust (Kilpatrick 1975). In recent decades, breeders have achieved resounding success in breeding for resistance to stem rust. For many years, the major stem rust resistance genes incorporated into wheat effectively protected the plants from a pathogen capable of causing enormous damage. However, a new virulent race then appeared, first in Africa and then in several regions of Asia. The spread of the Ug99 stem rust race is a timely warning that resistance breeding cannot stand still, as constant efforts are required not only to improve resistance but also to maintain the resistance level already achieved. In the case of insect pests, the results achieved with conventional breeding methods have been far from spectacular. Because effective resistance genes are not available for all of the insects attacking cultivated species, the development of genetically modified plants is a possible way to improve insect resistance (Bt maize, Bt rice).

In both developed and developing countries, an increase in the number of varieties cultivated and a reduction in the areas sown to leading varieties can be observed. Farmers are chiefly interested in obtaining higher yields with little fluctuation in quality, so the varieties available are not all grown or not in the same proportions. In some places, the choice of variety is greatly limited by the demands of the processing industry or the criteria laid down in quality standards. A reduction in production risks would be better served by an increase in the number of varieties, whereas the homogeneity of the quality would be improved if fewer varieties were grown. This irreconcilable contradiction can be observed to a different extent for different regions and plant species.

Breeding has always been carried out in the framework of national or regional programs within distinct ecological regions, and although the international exchange of breeding material has led to demonstrable relationships between the cultivars developed in various parts of the world, breeding programs have nevertheless retained their individual characters. The appearance of multinational breeding companies has tended to strengthen the decline in genotypic differences and has resulted in a reduction in the number of breeding workshops with a distinctive character.

The cereal species of most importance in human nutrition are self-fertile plants, meaning that 8 to 10 years are required to develop a new variety. Breeding can be divided into three consecutive stages. The first involves the creation of populations with very wide genetic variability; in most cases, by crossing parental varieties with diverse genetic backgrounds, although genetic variability can also be increased through mutations or genetic modification. The landraces, old varieties, and genetic material preserved in the world's major gene banks and the genotypes, cultivated varieties, and breeding lines collected by individual breeders can all be used for crossing. An increasing role is played in the development of biotic and abiotic resistance by wild species related to cultivated plants, from which the desired trait can be transferred to cultivated species by conventional crossing.

As the result of crossing, the various traits of the parents are recombined in the progeny, resulting in populations with great genetic variability, from which stable, homozygous lines can be selected over a number of years. With few exceptions, the screening of available diversity and selection is based on phenotypic evaluation. The phenotype is the result of the expression of the genotype in a given environment. In the early years, selection is performed for monogenic or oligogenic traits (resistance, plant height, vegetation period), whereas the selection of traits with polygenic inheritance and high $G \times E$ interaction (e.g., yield, quality) is only effective in later stages of breeding. With the development of new molecular techniques, the genotype can be directly studied for an increasing number of monogenic traits. Marker-assisted selection will be of particular significance in the future in resistance breeding, for the pyramiding of resistance genes.

The third stage of breeding is testing. Lines capable of giving significantly higher yields than earlier varieties, with an acceptable level of yield fluctuation and quality satisfying the criteria raised by end-users, can only be identified after several years of experiments in a range of environments.

When breeding hybrid plants, selection involves the development of inbred lines, from which experimental hybrids are produced. In the case of hybrids, the third stage of breeding involves testing the performance and adaptability of repeatedly developed F1 hybrids.

In addition to facilitating yield increases and improved yield stability, plant breeding also makes a major contribution to improvements in food quality and food safety. Resistance breeding makes it possible to reduce the quantity of pesticide applied in crop production, resulting in foodstuffs free of toxins and chemical residues. By changing the protein, starch, fiber, fat, oil, and amino acid compositions of cereals, optimum raw materials can be produced for various end-uses. It is important to note, however, that the plant variety only represents the possibility of achieving high yields and the desired quality. Due to the substantial modifying effect of the year, the environment, and the cultivation technology, fluctuations occur in both the quantity and quality of the yield, and reducing the extent of this fluctuation is the task and responsibility of the farmer.

1.5   HARVEST TO PRESERVE GRAIN QUALITY

The genetically determined traits of the variety, the location, and the year all combine to determine the quantity and quality of yield that can be achieved in a given field. The production of satisfactory industrial raw material thus begins with the sowing of high-quality seed from a wisely chosen variety, followed by the optimum nutrition of the plant stand and its protection against diseases and pathogens. The yield is formed in the last phase of the vegetation period, during grain filling, and any effects the crop is exposed to during this stage may have a decisive influence on the quality. As the end of grain filling approaches, the rate of assimilate accumulation slows and the kernels lose water more rapidly. Hot weather or other unfavorable environmental effects at the milk stage may damage the plants, inhibiting the translocation of assimilates from the stem and other plant organs and resulting in the development of shrunken grains, that is, a loss of yield and quality.

If the quantity and quality of yield are to be preserved, cereals should be harvested immediately after they mature. In most growing areas, the kernels of small grain cereal species dry in the ear sufficiently for them to be harvested and stored without damage. In the case of mechanical harvesting, the harvest can be begun at a grain moisture content of 14% without the need to dry the crop after harvest. If the harvesting capacity is limited or in regions with a wet climate, mechanical harvesting can begin at 16% to 18% grain moisture content, but the drying costs make production less profitable (Láng 1976). Two-stage harvesting, which was previously the norm, has lost its significance in regions where agriculture has been mechanized. This method of harvesting, in which the plants are cut in the late dough stage, left to dry in windrows, and then threshed later, had a favorable effect on quality.

On a large part of the maize-growing area, the environmental conditions do not allow the kernels to dry to the moisture content required for storage, especially in the case of hybrids with longer vegetation periods. The crop may be harvested as whole ears or as kernels, using a combine-harvester. The harvesting of ears, either manually or mechanically, can begin at a grain moisture content of 30% to 35%. The husks are then removed and the ears are stored in well-ventilated granaries where they dry without becoming moldy or musty. The hot-air drying of ears is regularly used in seed production. Combine-harvesting can begin at a grain moisture content of 22% to 23% if the kernels are hard enough for neither the kernels nor the germs to be damaged in the course of threshing. The threshed kernels must be dried immediately after harvest.

After harvesting (threshing) and, if necessary, drying, but before storage, the grain must be cleaned by sifting and winnowing to remove contaminants, dust, stalk and leaf residues, and husks. During the cleaning process, most of the damaged and infected kernels are also removed, making it easier to preserve the quality of the grain lot during storage.

If the harvest is delayed, there is a reduction in the test weight and thus in the yield quantity, but there is also the risk of quality deterioration. The hardness and vitreousness of the grain and the protein and gluten contents are all the more favorable if the harvest is carried out at the optimum date, as soon after maturity as possible.

REFERENCES

Alexander, D.E. and R.G. Creech. 1977. Breeding special industrial and nutritional types. In: *Corn and Corn Improvement*, ed. G.F. Sprague, 363–390. American Society of Agronomy, Madison, Wisconsin.

Anglani, C. 1998. Sorghum for human food—a review. *Plant Foods for Human Nutrition* 52: 85–95.

Atherton, M.J. 1984. Quality requirements: malting barley. In: *Cereal Production*, ed. E.J. Gallagher, 119–130. Butterworth, in association with Royal Dublin Society, London, Boston, Durba, Singapore, Sydney, Toronto, Wellington.

Bedő, Z., Gy. Vida, L. Láng, and I. Karsai. 1998. Breeding for breadmaking quality using old Hungarian wheat varieties. *Euphytica* 100: 179–182.

Bell, M.A., R.A. Fischer, D. Byerlee, and K. Sayre. 1995. Genetic and agronomic contributions to yield gains: a case study for wheat. *Field Crops Research* 44: 55–65.

Biffen, R.H. 1908. On the inheritance of "strength" in wheat. *Journal of Agricultural Science* 3: 86–101.

Braun, H.J., G. Atlin, and T. Payne. 2010. Multi-location testing as a tool to identify plant response to global climate change. In: *Climate Change and Crop Production*, ed. M.P. Reynolds, 115–138. CABI, London.

Facon, T. 2000. Water management in rice in Asia: some issues for the future. In: *Bridging the Rice Yield Gap in the Asia-Pacific Region*, eds. M.K. Papademetriou, F.J. Dent and E.M. Herath. RAP Publication (FAO), no. 2000/16.

FAOSTAT 2010, FAO Database. Food and Agriculture Organization of the United Nations. Available at: http://faostat.fao.org.

Fido, R.J., F. Békés, P.W. Gras, and A.S. Tatham. 1997. Effects of alpha-, beta-, gamma-, and omega-gliadins on the dough mixing properties of wheat flour. *Journal of Cereal Science* 26: 271–277.

Finocchiaro, F., B. Ferrari, A. Gianinetti, C. Dall'asta, G. Galaverna, F. Scazzina, and N. Pellegrini. 2007. Characterization of antioxidant compounds of red and white rice and changes in total antioxidant capacity during processing. *Molecular Nutrition and Food Research* 51: 1006–1019.

Fitzgerald, M.A., S.R. McCouch, and R.D. Hall. 2009. Not just grain of rice: the quest for quality. *Trends in Plant Science* 14: 133–139.

Gianibelli, M.C., O.R. Larroque, F. MacRitchie, and C.W. Wrigley. 2001. Biochemical, genetic, and molecular characterization of wheat glutenin and its component subunits. *Cereal Chemistry* 78: 635–646.

Graybosh, R.A. and C.J. Peterson. 2010. Genetic improvement in winter wheat yields in the Great Plains of North America, 1959–2008. *Crop Science* 50: 1882–1890.

Heisey, P.W., M.A. Lantican, and H.J. Dubin. 1999. Assessing the benefits of international wheat breeding research. An overview of the global wheat impacts study. In: *CIMMYT 1998/99 World Wheat Facts and Trends*, ed. P. Pingali. 19–26. CIMMYT, Mexico, D.F.

Hoseney, C.R., P. Wade, and J.W. Finley. 1988. Soft wheat products. In: *Wheat Chemistry and technology, part II*, ed. Y. Pomeranz, 407–456. Am. Assoc. Cereal Chemists, St. Paul, MN.

Inatsu, O. 1988. Studies on improving the eating quality of Hokkaido rice. *Report of Hokkaido Prefectural Agricultural Experiment Stations* 66: 1–89.

Iriki, N., F. Yamauchi, H. Takada, and T. Kuwabara. 1987. Evaluation of flour quality and screening method for Japanese noodles in wheat breeding. *Research Bulletin of the Hokkaido National Agricultural Experiment Station* 148: 85–94.

Isvilanonda, S. 2006. Rice consumption in Thailand: the slackening demand. http://www.mae.eco.ku.ac.th/documents/Somporn_Thailand%20Rice%20consumption%20November%2029.pdf.

Kilpatrick, R.A. 1975. New wheat cultivars and longevity of rust resistance, 1971-75. US ARS-NE-64.

Láng, G. 1976. *Szántóföldi Növénytermesztés*. Mezőgazdasági Kiadó, Budapest.

Lásztity, R. 1999. *Cereal Chemistry*. Akadémiai Kiadó, Budapest.

Maclean, J.L., D.C. Dawe, B. Hardy, and G.P. Hettel, eds. 2002. *Rice Almanac: Source Book for the Most Important Economic Activity on Earth*. Third edition. IRRI, Los Baños, Philippines.

Morris, C.F. and S.P. Rose. 1996. Wheat. In: *Cereal Grain Quality*, eds. R.J. Henry and P.S. Kettwell, 3–54. Chapman and Hall, New York.

Moss, H.J. 1973. Quality standards for wheat varieties. *Journal of the Australian Institute of Agricultural Science* 39: 109–115.

Nakamura, T., M. Yamamori, H. Hirano, S. Hidaka, and T. Nagamine. 1995. Production of waxy (amylose free) wheats. *Molecular and General Genetics* 248: 253–259.

Oda, M., Y. Yasuda, S. Okazaki, Y. Yamauchi, and Y. Yokoyama. 1980. A method of flour quality assessment for Japanese noodles. *Cereal Chemistry* 57: 253–254.

Oladosu, G., K. Kline, R. Uria-Martinez, and L. Eaton. 2011. Sources of corn for ethanol production in the United States: a decomposition analysis of the empirical data. *Biofuels, Bioproducts and Biorefining* 5: 640–653.

Pomeranz, Y. and P.C. Williams. 1990. Wheat hardness: its genetic, structural and biochemical background, measurements and significance. In: *Advances in Cereal Science and Technology*, ed. Y. Pomeranz, 471–544. Am. Assoc. Cereal Chemists, St. Paul, MN.

Shewry, P.R., N.G. Halford, A.S. Tatham, Y. Popineau, D. Lafiandra, and P. Belton. 2003. The high molecular weight subunits of wheat glutenin and their role in determining wheat processing properties. *Advances in Food and Nutrition Research* 45: 219–302.

Strychar, R. 2011. World oat production, trade and usage. In *Oats: Chemistry and Technology*, eds. F.H. Webster and P.J. Wood. Am. Assoc. Cereal Chemists, St. Paul, MN.

U.S. Rice Federation. 2009. U.S. Rice Domestic Usage Report, Milling Year 2008–2009. http://www.usarice.com/doclib/188/231/4677.pdf.

USDA Foreign Agricultural Service, Commodity intelligence Report, December 22, 2010.

Uthayakumaran, S., S. Tömösközi, A.S. Tatham, A.W.J. Savage, M.C. Gianibelli, F.L. Stoppard, and F. Békés. 2001. Effects of gliadin fractions on functional properties of dough depending on molecular size and hydrophobicity. *Cereal Chemistry* 78: 138–141.

Uthayakumaran, S., H.L. Beasley, F.L. Stoppard, M. Keentok, N. Phan-Thien, R.I. Tanner, and F. Békés. 2002. Synergistic and additive effects of three high molecular weight glutenin subunit loci. I. Effects on wheat dough rheology. *Cereal Chemistry* 79: 294–300.

Zhou, Y., Z.H. He, X.X. Sui, X.C. Xia, X.K. Zhang, and G.S. Zhang. 2007. Genetic improvement of grain yield and associated traits in the Northern China winter wheat region from 1960 to 2000. *Crop Science* 47: 245–253.

2 Transportation and Storage of Cereals

Paula Maria dos Reis Correia and
Raquel de Pinho Ferreira Guiné

CONTENTS

2.1 INTRODUCTION

Cereals are a versatile and reliable source of foods. They are relatively easy to store and may be used to produce several food products, and are thus essential in the food production chain.

The postharvest steps are fundamental in the supply, which can be understood as a set of measures aimed at guaranteeing that foods reach the tables of consumers with the possible or necessary quality and at costs compatible with the financial capacity of the population.

The objective of adequate grain transportation and storage is to maintain, for as long as possible, the biological, chemical, and physical qualities that grains have immediately after harvest. Therefore, adequate grain storage is fundamental to provide the supply in between harvests and to avoid losses, therefore increasing competitiveness. Nowadays, producing grains of the highest quality is a priority; however, that quality has to be preserved until the moment of final consumption. In this way, it is necessary to rethink the transport and storage of these grains so as to reduce costs, but without ever having to compromise on quality. The techniques used in the storage of cereals are aimed at preserving their quality for the time necessary toward their later use as foods or as food components. The postharvest activities and techniques, which deal with living biological products such as cereals, start with the transportation of the agricultural crops, followed by reception in the preprocessing or storing units, and finishing with conditioning or safe storage. These activities are complemented by treatments with the finality of controlling both deterioration processes and plagues. Finally, cereals are moved inside the storing system for expedition, having in mind their commercialization or industrialization.

Grain cereals are constituted by independent granular units, its total volume being constituted by 40% of air, thus giving them the behavior of a fluid. This structure allows ventilation and facilitates the removal of heat and moisture, as well as enabling it to modify the composition of the interstitial atmosphere (CO_2, O_2, N_2), thus making it possible to use modified atmosphere for improved conservation.

With regard to the thermal properties of granular cereals, both thermal conductivity and specific heat are very low. Consequently, natural convection is insufficient to promote cooling and, for that reason, important increases in temperature may occur in the absence of forced ventilation, which may result in some conservation as well as safety issues.

2.2   TRANSPORTATION OF CEREALS

The transportation of cereal grains is an important step in the food chain and there are different types of vehicles to consider: trucks, railway (in railcars), or ships (in holds; Figure 2.1). Generally, cereals are transported in bulk, as shown in Figure 2.1. The vehicles have to be well designed and constructed, and they should be driven or operated by trained people with knowledge regarding food safety, hygiene, and security. Furthermore, those engaged in cereal transportation must keep a high level of hygiene, similar to the practices with other food products. Efforts have to be done to assist the food transport industry in preventing problems related to food safety and quality during transportation.

Persons that are engaged in food transport must concentrate their efforts on some preventive controls to preserve the cereals' quality and safety between the producer's field and the storage facilities or between the field/storage and the first transformation enterprise. Thus, there are some recommended requirements to provide good cereal transportation:

- Appropriate temperature control of cereals
- Good sanitation of transportation containers, including pest control and the guarantee of effective sanitation monitoring and procedures

FIGURE 2.1 Types of cereal transportation: (a and b) truck, (c) railway, and (d) ship.

- Good communications between transporter and receiver
- Appropriate training and awareness of employees

It is also important not to forget the following aspects:

- Cereals should not be transported if they do not present the proper conditions for the trip and subsequent storing period at the destination, despite their natural resistance.
- Specify relative humidity or other special conditions or treatments if they are requested.
- Select cargo, dunnage, blocking, and bracing so that the load covers the entire floor of the container.
- The container company should inspect and clean each container before each trip, to avoid cross-contamination.
- Avoid transporting mixed loads that increase the risk for cross-contamination.
- The loading practices and conditions should be carried out properly, including proper sanitation of loading equipment. The same care must be given to the unloading of the cereals, which should not be left on loading docks for hours.
- Improve holding practices for cereals, for example, when they are awaiting shipment or inspection, including unattended products, delayed holding, shipping of products while in quarantine, and rotation and throughput.
- Provide a preventive maintenance for transportation units (or storage facilities used during transport). If this should not be accomplished, it could result in roof leaks, gaps in doors, and dripping condensation.

2.3   EVALUATION OF CEREALS ARRIVING AT THE STORAGE FACILITY

When a lot of grain cereal arrives at a storage facility, its sanitary state needs to be evaluated. This evaluation is complex, and involves different areas, such as the physical, chemical, biochemical, microbiological, as well as quality attributes in its various perspectives. Table 2.1 shows the different evaluations that cereals can be submitted to, according to their nature. With regard to the evaluation of enzymatic activity, this should be complemented with an assessment of the results it produces because this leads to the consumption of substrates and the appearance of new products. With regard to the quality, this involves the quality of the product as a food to be ingested by people, as well as its technological abilities when the cereal is aimed at later transformation.

TABLE 2.1
Evaluation of Cereal Status on Arrival at Storage Facility

Physical State of the Grain

- Weight
- Moisture content
- Dispersion of individual grain moistures
- Temperature
- Impurities

Infestation Status

- Insects (weight losses, number of perforated grains, detection of hidden infestation by x-rays, acoustic measurements, or CO_2 release)
- Microorganisms (qualitative and quantitative analyses)
- Content and concentration of specific metabolites (resulting from certain microorganisms)

Biochemical Status

- Enzymatic activity (amylolytic, proteolytic, lipolytic, etc.)

Quality—Food Point of View

- Innocuous (dosage of mycotoxins and pesticide residues)
- Nutritional value (*in vitro* and *in vivo*, if applicable)
- Sensorial characteristics (attention must be given because a deficient taste or aroma will compromise not only the cereal but also all the products derived from it)

Quality—Technological Point of View

- Aptitude for use in transformation industries, in particular:
- Bread and pastry (specific evaluations using alveograph, farinograph, extensograph, and amylograph)
- Pasta (evaluations made with viscoelastomer)
- Germination capacity of the grains (example: barley for malting)

The sanitary state of the grain is evolutionary. Its current state represents the value of the immediate usage potential for a given technology, which greatly depends on the history of the grain. The prospective sanitary state represents the potential risk of alteration for a given storage period, with the view toward a certain usage, which is dependent on the present state of the grain, on factors that may lead to alteration and that are present in the grain, such as the microflora, and finally, on the characteristics of environmental factors, such as moisture or temperature.

2.4   STORAGE PROCESS: EQUIPMENT AND PHYSICAL REQUIREMENTS

In view of the quality of granular cereals, its conservation is directly associated with storage, which represents one of the principal reasons for waste, both in terms of quantitative and qualitative losses due mostly to inadequate treatment or storage systems (Bragatto and Barrella 2001).

Storage aims at preserving the physical, sanitary, and nutritional qualities of the grains after harvest. During storage, the factors that influence the preservation of the grains are temperature and relative humidity of the intergranular air, and the temperature and moisture content of the grains. Besides these factors, the structural characteristics as well as hygienic conditions of the facilities are pivotal for good storage practices.

The storage conditions of the grains are established according to spatial and timely factors. Concerning the first factor, it is important to consider the location, facility of flow, climate, exposition, soil resistance and costs associated with energy and labor. As to the latter factor, they include the duration of storage, future marketing conditions, maintenance techniques, and amortization. The probable duration of storage is established in view of the storage conditions (moisture, temperature, and atmosphere, among others), the present state of the lot, previous usages of the grain, and finally, the quality standards demanded.

2.4.1   TYPES OF STORAGE

The storage of cereals has been practiced since ancient times, mainly because they are seasonal crops that are consumed during the whole year. Decentralization and spreading of grain storage over a wide area is recommended, both for strategic reasons and for minimizing transportation, by using the grain, preferably, close to where it is produced (De Martini et al. 2009).

A grain storage facility can be defined as a system designed, structured, and equipped for receiving, cleaning, drying, storing, and dispatching grains or oilseeds. To perform these operations, equipment and structures need to be linked in a logical sequence. The coordination between different operations may involve complex decision-making processes due to the stochastic nature of receiving and dispatch operations (Silva 2002; Silva et al. 2012).

Storage began as a simple procedure as it was done in rural areas with traditional houses, inside of which the cereal was piled (Figure 2.2a). Nowadays, in some rural areas, farmers still use this type of storage for cereals for their own consumption.

FIGURE 2.2   Different types of cereal storage: (a) traditional, (b) cellar storage, and (c) silos.

Cellars are also used for the storage of grains, which are simply piled on the ground (Figure 2.2b). For industrial purposes, storage procedures are far more complex. Generally, storage is undertaken in bins or silos with a round shape (most commonly, but they can also be square) built from metal or cement, with different capacities, having sophisticated and well-equipped structures (Figure 2.2c). Metal silos are cylindrical structures that are hermetically sealed and made from galvanized iron sheets. The metal silo technology has proven to be effective in protecting harvested grains from the attacks of storage insects and rodent pests. The metal silo is airtight (with no oxygen inside) and, consequently, any insect pest that may get inside will die. In addition, it completely locks out any pest or pathogen that may invade the grains (Terefa et al. 2011).

Simulation models for grain storage are very complex because realistic solutions have to consider the equations of heat, mass, and momentum transport as well as the boundary conditions that simulate the environmental temperature. In the storage of grains in silos, heat and mass transfer phenomena are coupled and occur simultaneously. Therefore, an optimum design for the conservation of cereals necessarily requires an understanding of the transport phenomena that are present in the storage of grains, the environmental conditions of the stream flows, and the temperature and concentration profiles in the silo. Carrera-Rodríguez et al. (2011) presented a numerical model about the transient heat and mass convection of grain storage in a cylindrical silo. Temperature gradients were induced by the heat of respiration, and thermal gradients were generated by variations of the temperature surrounding the cavity. The model was developed by using the equations of heat, mass, and momentum transport for multiphase media.

Abalone et al. (2011a,b) developed a model to simulate the evolution of gas concentration in a silo-bag based on the more relevant variables, which included grain respiration rate and permeability of the silo-bag. The respiration rate depends on many biological factors, the variations of which make it difficult to develop accurate models for the prediction of gas concentration in the silo-bag. Furthermore, the permeability of the silo-bag depends on the thickness of the plastic and the material's composition.

The manipulation and storage of grain cereals in bulk is a universal tendency, being generalized in developed countries where it is integrated immediately after the harvest. Basically, deposits aimed at the storage of grains are classified into vertical or horizontal silos according to the shape of the storage structure. Vertical silos are deposits in which the height is more extensive than the corresponding diameter and, in this case, can be designated as a bin, upright, or vertical storage. Horizontal silos or grain warehouses have a height that is smaller than its base and, in this case, is called a horizontal or flat storage (D'Arce 2006). Silos can be further classified in relation to their capacity and location as collection center (small capacity, 100–500 tons), collection silo (500–2000 tons), center in a cereal region (2000–20,000 tons), intermediate silo (100,000 tons), and port silo (200,000 tons).

Upon arriving at the storage facility, the product, which is usually transported via trucks or wagons, is weighed. Later, it is unloaded and transported through several mechanized systems of elevators and horizontal transporters to the storing cells (Bragatto and Barrella 2001). When leaving the facility, the cereal goes through a horizontal transporter located underneath the facility, so that this procedure is done by gravity all the way from the silo to the truck or wagon for expedition. Also, upon leaving the facility, the vehicle carrying the cereal is weighed before proceeding to the primary transforming industries. If it is necessary to perform cleaning and drying operations, the cereal is received into one of the cells of the storage unit in which those operations are then carried out.

For the medium-capacity and high-capacity units, the automatic functions are done through a control panel where the different operations are followed with lights. This process minimizes the manual labor, and a single worker, by way of buttons and keys, receives, weighs, cleans, dries, and stores the products in the silos. The operations are adjusted to move the mass with the minimum of interruptions, and to ensure that the proper equipment controls the flow of the grains. The visualization of the whole is facilitated through flowcharts graphically representing all the operations and sequences of the movements of the material (D'Arce 2006).

Storage in silos can represent an important investment. However, this can be compensated by the low maintenance costs, long life of the facilities, and the guarantee of quality for the stored product. Furthermore, it allows a fast, economic, and versatile manipulation of the products, and enables the storage of different varieties of grains at the same time, without any direct contact.

The principal factors to consider when choosing the type of storage facility that should be implemented for a given situation are the type(s) of product(s) to be stored and the finality of the unit, technical aspects, economic factors (investment and operating costs), and location (D'Arce 2006).

2.4.2 CEREAL DRYING

The drying operation is an important phase of cereal processing and is performed before storage. This technique has been used since the beginning of civilization to preserve all kinds of foods. Cereal grains are generally suitable for storage for relatively long periods of time. They are usually harvested during the hot season, and therefore present low moisture contents at harvest. When they are harvested during this season and, when stored conveniently, with protection against insects and other predators, they can be kept for long periods of time (years or even decades). In wet harvest years, however, it is recommended that cereal grains must be further dried. In Great Britain, for example, grain is often harvested at moisture contents of around

16% to 20%, whereas in exceptionally late seasons in northern regions, grain may be harvested at a moisture content of approximately 25%. In this case, for safe storage, the grain has to be dried to a moisture content within the range of 13% to 15% (Evans 2001).

Some of the drying methods for grains will be described in the following sections. Certainly, there are more possibilities based on process fundamentals and creativity to conceive different and innovative driers.

The options for drying granular cereals include natural drying and artificial drying. The natural process, if the climate allows, is preferable for several reasons ranging from product quality to economic savings. This drying method uses solar radiation and low atmospheric relative humidity to dry the grains, thus minimizing grain damage. However, there is one important disadvantage to this method, which is the great dependence on climatic conditions (De Martini et al. 2009). Artificial drying consists of two types of drying systems: near-ambient air temperatures (low temperatures) or high-temperature driers.

The driers can be classified as continuous or discontinuous functioning systems. The first dries the grain in a single step, achieving the desired final moisture content. In the second, the grain circulates through the equipment several times to produce the same result. Sometimes, and according to the needs, the same equipment can function in intermittent or continuous modes.

Temperature is a pivotal factor when discussing cereal drying, and with respect to this operating condition, the process can be carried out at a low temperature or at a high temperature. The first is cheaper but also longer, when compared with the latter. Next, these two types of drying systems will be discussed in further detail.

2.4.2.1   Low-Temperature Drying

Low-temperature drying uses air which is not heated above ambient conditions. This process is a relatively slow process, which constitutes an important disadvantage, whereby the grain is stored in bins or on the floor and is dried by forcing ambient or slightly warmer air to flow through the grain (Figure 2.3). To guarantee acceptable

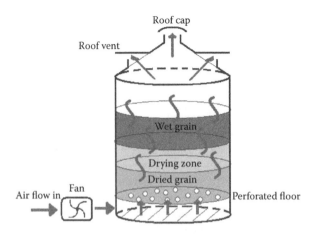

FIGURE 2.3   Bin for the low-temperature drying of grains.

drying efficiency, the air used has to present low relative humidity. The advantages of this process are that it is relatively energy efficient (it only requires energy to force the air through the grain), the grain is not damaged and is not submitted to heat (Hoseney 1990; Evans 2001). Drying in this type of equipment lasts for a period of between 15 and 30 days, depending on the temperature, relative humidity, and air flow. These three process variables must be adequately defined so as to promote full drying and prevent deterioration of the grain (Silva 2005).

2.4.2.2 High-Temperature Drying

When the volume of grain is large and the commercial transactions are high, the drying process can be speeded up by heating the air, which removes the water from the gain faster due to an increased driving force (temperature difference between the grain and the surrounding air). This is the major advantage of this process. High-temperature batch and continuous-flow driers operate with air temperatures between 40°C and 120°C, but in this case, it is necessary to cool the grain before storage (Evans 2001). Despite being faster, this drying method presents some drawbacks (Bragatto and Barrella 2001). In fact, the use of high drying temperatures could cause damage in the grains, such as increasing brittleness and susceptibility to breakage, discoloration, and loss of germination ability. These could be particularly problematic if the grains were destined for milling, malting, or for seed (Hoseney 1990).

For the artificial drying of grains, several types of driers can be used: fluidized bed; cross flow, cocurrent flow; countercurrent flow; mixed flow, or cascade drier. The choice of the most adequate method depends on several factors, among which are the level of technological instruction and enterprising capacity of the producer, the volume of cereal to dry, rate of harvest, and grain destination.

2.4.2.3 Aeration

Aeration, which is the forced movement of ambient air through a grain bulk, is conducted to preserve or improve the physical condition of the product (Lopes et al. 2008). Aeration uses relatively low airflow rates to cool grain as well as eliminate temperature and moisture differences in the storage bin. Aeration may be used with field-dried grain or with grain that is harvested damp (Khatchatourian and Oliveira 2006), then dried and cooled in a heated air drier. In both cases, temperature and moisture content variations may exist in the grain, or the grain may be too warm to be stored safely. Variations in grain temperature are also caused by changes in the outside air temperature after the grain is stored.

Airflow distribution in grain bulks depends on several factors, such as the filling method and grain bulk porosity; the depth of the grain; the grain morphology, configuration, and size of the interstitial space in the mass; the air velocity; and the form and size of any extraneous impurity in the mass, among others (Khatchatourian et al. 2009). Khatchatourian and Binelo (2008) developed a mathematical model and software for the three-dimensional simulation of airflow through high-capacity grain storage bins by considering the nonuniformity of the seed mass.

Warm air rising in the center of the bin cools when it reaches the cold grain near the surface. This results in an increase in moisture content near the surface, which can lead to rapid spoilage. Crusting on the surface of stored grain is a common symptom of moisture migration. Significant migration can occur in cereal grains at moisture contents as low as 12%, if they are placed into storage at high temperature and are not properly cooled.

When the cereal grains are stored in a bin where temperature changes are large, this can lead to nonuniform temperatures in the grains and moisture migration, resulting in moisture accumulation at particular points in the cereal mass. Aeration corrects this condition, as the movements through the grains make the temperature more uniform and decrease the moisture accumulation.

Aeration aims at optimizing the conditions of the storage atmosphere, and this is achieved through the removal of the interstitial air (Silva 2011). In this way, a new microclimate is generated, improving the preservation of the grain quality in bulk storage containers, equilibrating the bulk temperature, minimizing fungal activity, diminishing the respiration rate of the product, and when possible, even reducing the temperature of the grain.

Regarding the usage of ambient air for this operation, the most adequate conditions are not always met in terms of relative humidity and temperature during the time necessary for aeration. This may lead to proceeding with the aeration under improper conditions, thus implying great weight losses or undesirable wetting of the product and consequent fermentations, with considerable economical losses.

Under the structural point of view, the aeration system should include the following elements: (1) a fan, which provides the air with increased speed and pressure to flow through the grain column; (2) main pipes, to carry the air flux to the application points; and (3) distribution pipes, equipped with perforated plates through which the air flows, thus penetrating the intragranular spaces.

It is important to consider that bulk cereal is constituted by independent granular units, conferring to the whole mass the behavior of a fluid. On average, approximately 30% of the bulk volume is constituted by air (Hoseney 1990). Table 2.2 shows the density for some cereal grains and Table 2.3 shows the porosity of different grains at different moisture contents. This porous structure allows the easy elimination of heat and moisture as well as modifying the composition of the interstitial atmosphere (CO_2, O_2, N_2), eventually leading to conservation under a modified atmosphere. Modified atmospheres can be used for insect control in unsealed grain stores as an alternative to pesticide administration. This technology requires the alteration of the natural proportions of the atmospheric gases (nitrogen, oxygen, and carbon dioxide) to produce an atmosphere that is lethal to pests. This type of pest control is frequently used because consumer concerns about the effects of pesticide residues has increased the demand for pesticide-free products (Bibby and Conyers 1998).

According to Silva (2005), for aeration, a medium flux of 1 to 3 m^3/min of air per ton of product is recommended for product temperature reduction and for the removal of 2% to 3% of grain moisture content.

Temperature, dry matter loss, and moisture content of stored grains can be predicted by simulation models. Generally, these models are used to evaluate the efficacy of ambient air ventilation, to estimate the maximum safe storage period of grains, and therefore, to predict the necessary aeration time. These evaluations allow for the analysis of the viability of aeration and the optimization of control strategies (Lopes et al. 2006).

TABLE 2.2
Bulk Specific Mass for Some Cereal Grains

Cereal Grain	Moisture Content (%, wet basis)	Specific Mass (kg/m³)
Barley	7.9	585
	13.3	593
	19.5	569
Corn	7.3	753
	16.2	721
	24.9	656
Sorghum	6.8	753
	14.3	753
	22.1	721
Rice	12.0	586
	16.0	605
	18.0	615
Wheat	7.3	790
	14.1	756
	19.3	703

Source: Meneguello, E.L. *Projeto e automação de fábricas de ração.* Erechim RS, Brazil: URIAUM, 2006.

TABLE 2.3
Bulk Porosity for Some Cereal Grains

Cereal Grain	Moisture Content (%, wet basis)	Porosity (%)
Rice	12.0	59.6
	14.0	59.3
	16.0	57.9
Sorghum	14.3	42.0
	18.6	42.0
	22.1	45.5
Corn	13.4	40.1
	14.9	39.6
	16.8	40.5
Barley	10.3	55.5
Wheat	9.8	42.6

Source: Meneguello, E.L. *Projeto e automação de fábricas de ração.* Erechim RS, Brazil: URIAUM, 2006.

2.5   CONSERVATION OF THE GRAIN

The grain bulk is a collection of living organisms, and in a cereal lot, the most important organism is the grain itself (Brod 2005; Meneguello 2006). This is actually composed of millions of small, living plants. Although in a dormancy stage, it possesses all the characteristics of a living organism (Dunkel 1995). In this way, the lot constitutes an ecosystem (thermodynamic system grouping different interacting organisms and their environment) comprising a microbiotype characterized by a set of specific biological, physical, and chemical properties (Figure 2.4). The grains are living organisms with a slow metabolism, thus acting as a natural way of preservation. They have moisture contents between 5% and 40%, contrasting with the 75% to 97% of other living organisms.

In cereal storage, two types of variables are to be considered: abiotic and biotic variables. The first includes temperature, atmospheric composition (CO_2 and O_2, making the conservation better in the absence of oxygen), moisture, and organic compounds, which are by-products resulting from biological activity (De Martini et al. 2009). As to the latter, they include microorganisms (yeasts, fungi, and bacteria), insects, arthropods, and vertebrates (rats and birds). Regarding the grain bacteria, they are essentially eubacteria. As to the fungi, mostly resulting from production, they include *Alternaria, Cladosporium, Fusarium, Mucor, Rhizopus, Penicillium,* and *Aspergillus.*

Mycotoxins are poisonous metabolites produced by certain fungus and can infect crops when still in the field or later during storage. The conditions that favor the development of mycotoxins in cereals before and after harvest are not fully understood, but are of particular importance to maintain high standards for product quality. Some factors that may affect mycotoxin formation are moisture, temperature, time, invertebrate vectors, damage to the seed, O_2 and CO_2 levels, inoculum load, composition of substrate, fungal infection level, prevalence of toxigenic strains, and microbiological interactions. Crop spoilage, fungal growth, and mycotoxin formation

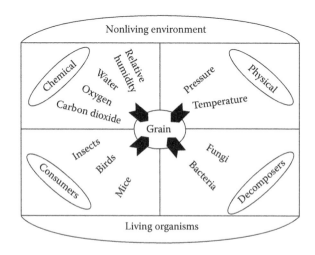

FIGURE 2.4   Cereal ecosystem.

result from the interaction of several of these factors in stored products (Abramson et al. 1999).

An alteration process arises from one of the following cases: (1) the presence in the grain of one or more causes for change, and these can be of biological, physical, or chemical natures; and (2) the existence of environmental factors favorable to activity causing these changes. In an ideal storage system, the grain and microorganisms are in a dormant state and insects, rats, or other pests are absent (Meneguello 2006). On the contrary, the abiotic environment is always present and is changeable, although the levels of relative humidity, atmospheric pressure, temperature, CO_2, and O_2 may slowly go up and down. An abnormal variation in any of these factors may create conditions favorable to the development and multiplication of those dormant organisms (D'Arce 2006).

Among the various causes for alteration in a lot of cereals, the following are highlighted:

1. Chemical degrading reactions, such as Maillard reactions, which occur due to the increase in temperature and long storage periods, denaturation of proteins and nucleic acids, nonenzymatic modifications in starch, destruction of vitamins, and nonenzymatic oxidations due to the presence of lipids.
2. Enzymatic alterations.
3. Biological alterations: the microorganisms, rats, and insects (by eating the grain) leave behind enzymes on the grains, thus constituting a contamination focus. The existence of spores does not constitute a problem as long as the moisture and temperature are not favorable to the spores' growth.
4. Mechanical or physical alterations: damaged or broken grains are more susceptible to deterioration.

Several ambient factors, such as time, temperature, moisture, and oxygen or carbon dioxide contents could influence the cereal's conservation. Time influences the rates of reaction, which also depend on concentration according to the classic chemical kinetic laws. Temperature is also pretty important because, by increasing temperature, the rate of reaction is generally increased. By increasing the molecular agitation, the kinetic energy of the molecules increases and the probability of them meeting each other for the reaction also increases. The degree of hydration strongly influences storage, as previously mentioned. For a moisture content that is between 14% and 15%, the corresponding water activity is 0.65, and for these conditions, there will be no microbial changes and stability is guaranteed. However, and in a general way, there is an inverse relation between water activity and conservation time. The amounts of O_2 and CO_2 influence the type of metabolism of the microorganisms (aerobic or anaerobic) and of living cells, also influencing the enzymatic oxidations and other enzymatic reactions.

2.5.1 BIOTIC FACTORS

Insects and microorganisms, as well as rodents and birds, are the destroyers of stored grains. Insects consume grains and contaminate them with feces, webbing, body parts, and microorganisms. Beetles and moths are the most ruinous of the grain insects and some of them can completely destroy a grain store.

Grains harvested from the fields carry part of the microflora that occurs naturally with the growing plants, water, and soil. The amount of microorganisms present and their growing rate will depend on the type of organisms, climatic conditions, commodity, harvesting, and handling practices.

2.5.1.1 Living Organisms

Microorganisms are undesirable because of their danger to public health and also because they destroy and contaminate cereal grains. Fungi are the major microorganisms responsible for the greatest contamination and damage to stored grains. Some of them produce secondary metabolites such as mycotoxins that are unsuitable for human or animal consumption.

Cereal grains are susceptible to attack by fungi during growth, maturation, and postharvest. During storage, attack from insects and rodents may occur, which (together with fungi) can lead to diminished weight, rancidification, fermentation, and other processes that alter the structures of the grain and the germination of the seed. Fungi are currently the primary microorganisms responsible for most of the damage to agricultural products, both in the field and in storage. Fungi in the field attack seeds, growing grains, and maturing grains (moisture content >25%). The fungi at storage develop in seeds and grains with moisture contents of less than 17% (Table 2.4). The losses caused by fungi during inappropriate storage can be as much as 100%.

The moisture content of the grain, temperature, storage period, contaminations, levels of impurity, attack by insects, oxygen concentration, and physical damage during harvest are the principal factors determining the level of susceptibility of the grains and seeds being attacked by fungi. The principal damages caused include diminishing of the germination power, loss of color, production of mycotoxins, heating, biochemical

TABLE 2.4
Temperature and Relative Humidity for the Development of Fungi in Storage

| Fungi | Temperature (°C) | | | Relative Humidity (%) |
	Minimum	Optimal	Maximum	Minimum
Aspergillus halophilicus				68
Aspergillus restrictus	5–10	30–35	40–45	70
Aspergillus glaucus	0–5	30–35	40–45	73
Aspergillus candidus	10–15	45–50	50–55	80
Aspergillus ochraceus				80
Aspergillus flavus	10–15	40–45	45–50	85
Penicillium sp.	−5 to 0	20–25	35–40	80–90

transformations, cellular modifications, appearance of yeasts, and rotting. In high dosages, mycotoxins are lethal to humans and animals. The most common and potentially toxic are aflatoxins, produced by the fungus *Aspergilus flavus* and *Aspergillus parasiticus*. *A. flavus* is common and widespread in nature and is most often found when certain grains are grown under stressful conditions. This mold is found widely on inadequately dried food and feed grains in subtropical and tropical climates throughout the world.

Most of the bacteria associated with grain are represented by the families *Pseudomonadaceae*, *Bacillaceae*, *Micrococcaceae*, *Lactobacillaceae*, and *Enterobacteriaceae*. Their growth in grain depends largely on moisture content and temperature. Besides these microorganisms, there are other microorganisms commonly associated with grains, mainly actinomycetes, myxomycetes, and yeasts. Graves et al. (1967) mentioned that the majority of actinomycetes in grain are *Streptomyces*. All three of these microorganism groups are of minor significance in grain because they grow only when conditions are extreme and when other organisms, mainly fungi, predominate. Olstorpe et al. (2010) studied storage systems in farms to determine how far the spontaneous character of the microbial fermentation during storage results in varying qualities of the resultant feed. They identified lactic acid bacteria and yeasts, in particular, to determine which species were present and likely to be associated with fermentation and storage stability. These authors found that the microbial flora in airtight moist barley grain storage varied considerably among farms. The spontaneous process did not result in a consistent microflora toward the end of storage. Conservation of the barley did not depend on low pH, but probably on low concentrations of O_2 and competition for nutrients between the microorganisms on the barley. A large number of lactic acid bacteria at one farm resulted in low numbers of *Enterobacteriaceae*, molds, and yeasts, although the pH was not different from barley storage on other farms. *Enterobacteriaceae* grew in barley and exceeded guideline values for silage, even in material that had relatively low moisture content. Obviously, low moisture content was not sufficient to suppress the growth of these bacteria. For example, yeasts and molds were found in barley, but in different amounts and with varying species composition. This study shows that the microbial population in airtight stored moist barley is highly diverse and not predictable.

Besides microorganisms, there are other factors that could cause damage to stored grain. These include insects, rodents, mites, birds, and the metabolic activity of the grain itself (respiration). In tropical conditions, insects are a major concern for grain storage because the mass of grain represents their ideal habitat. Attack by insects could result in important losses in weight, reduction in the germination power of the seeds, contamination, and devaluation of the product.

Insects can be classified into two categories: primary insects, which can perforate whole healthy grain; and secondary or associated insects, which feed from the debris and fungi but could still compromise the appearance and quality of the product. Among these are the beetles and mites, commonly found in stores of grain and flour. As to the primary insects, these can be distinguished as internal or external. The primary internal insects have developed mandibles to perforate the grain so that they can feed from inside the product. These include weevils and moths. Primary external insects feed from the grains' outside, or attack the internal parts, opening a way for other pests. These insects are not able to penetrate the grains and attack those grains previously opened or damaged by primary insects (Table 2.5).

TABLE 2.5

Most Common Insects in Cereal Storage

Common Name	Scientific Name	Visual Aspect	Some Important Aspects
Rice weevil	*Sitophilus oryzae* (Linnaeus)		Primary grain pest. Principal plague in stored cereals and finished products (flours, biscuits, etc.). It is winged and may occasionally fly.
Granary weevil	*Sitophilus granarius* (Linnaeus)		Primary grain pest. A serious pest of stored grains, especially of stored wheat and barley, in elevators, mills, and bulk storage. It is similar to the rice weevil but it cannot fly, and field infestations do not occur.
Lesser grain borer	*Rhizopertha dominaca* (Fabricius)		They chew grain voraciously causing damage, which may facilitate infestation by a secondary pest. It is a strong flyer and may rapidly migrate from infested grain to begin new infestations elsewhere.
Red flour beetle	*Tribolium castaneum* (Herbst)		Secondary pest. It is frequently found on farms, attacks mostly the embryos when they are already broken. The red flour beetle may impart a bad odor that affects the taste of infested products.
Saw-toothed grain beetle	*Oryzaephilus surinamensis* (Linnaeus)		Secondary pest. In grain bins, it feeds on broken kernels and grain residues. It is a common pest not only in grain bins, but also in elevators, mills, and processing plants. It frequently hides in cracks and crevices of buildings and machinery.
Flat grain beetle	*Cryptolestes ferrugineus* (Stephens)		Secondary pest. Attacks grains with breaks in the pericarp.
	Trogoderma granarium (Everts)		Secondary pest. Highly disseminated, attacking cereals and by-products.

(continued)

TABLE 2.5 (Continued)
Most Common Insects in Cereal Storage

Common Name	Scientific Name	Visual Aspect	Some Important Aspects
Angoumois grain moth	*Sitotroga cerealella* (Olivier)		Primary pest. Lives on the surface of stored bulk grain or bagged grain. Adults are not capable of penetrating deeply into the grain, but the larvae penetrate them and cause weight reduction. It can impart an unpleasant smell and taste to the cereal.
Indian meal moth	*Plodia interpunctella* (Hübner)		Secondary pest. It is a cereal plague in warehouses and transformation units. Is frequent in hot climates. The attack is confined to surface layers of stored shelled corn and small grains. It possesses a high capacity for producing webs infesting the surface of the cereal.
Warehouse moth	*Ephestia (Anagasta) kuehniella* (Zeller)		Secondary pest. The larvae move rapidly and with great capacity for production of webs forming compact masses, which can lead to obstruction of pipes in transforming units.
Flour mite or grain mite	*Acarus siro* (Linnaeus)		Mites are a serious pest for stored flour, bran, and other stored products in temperate regions.

Insects associated with raw cereals and processed food cause quantitative and qualitative losses. Infestations can occur just before harvest, during storage in a variety of structures (such as cribs and metal or concrete bins), or during transportation. Therefore, preventing economic losses caused by stored-product pests is important from the farmer's field to the consumer's table.

Rodents are probably the most damaging animals known. They consume and contaminate great quantities of cereal grains. Rodents adapt to every variation of ambient condition, are extremely agile, and possess a huge reproductive capacity.

2.5.1.2 Controls

To ensure the quality and safety of cereal grains, lots of money has been spent and efforts have been made in the construction and maintenance of facilities, transportation, and application of various methods for controlling grain pests (e.g., sanitation, drying, refrigeration, and chemical treatments as modified atmospheres, oxygen depletion, and pheromones). In this way, during storage, some operations can be performed for the adequate conservation of the product, namely the following (Bragatto and Barrella 2001):

- Aeration: as previously noted, it consists of the forced movement of air through the mass of grain, aimed at diminishing and standardizing temperature, creating favorable conditions for conservation of quality, preventing the migration of moisture and the formation of heat spots.
- New silage: moving of the grain mass, allowing diminishing and standardization of temperature.
- Thermometry: sets of sensors distributed symmetrically inside the silo to periodically measure the temperature of the mass of grain.
- Phytosanitary treatment: aims at preventing the appearance of insects or eliminate them when already present. Like other operations, systematic monitoring produces an efficient control from the beginning of the infestation.
- Hygienization of the warehouse: prevents the formation of a focus of infestation by insects or rodents.

Several tools are available for managing insects associated with raw grain and processed food, such us the use of pesticides and its alternatives. Insects may present some resistance to insecticides, thus compromising the effectiveness of chemical controls and increasing the costs of application and risks during storage. These treatments usually require a thorough understanding of the pest's ecology, and their application is only justified when pest populations exceed acceptable levels. At the same time, it will be necessary to manage an evaluation of risks, costs, and benefits.

Specifically, to fight rodents, two methods can be applied: direct combat and indirect combat. The first comprises the elimination of high vegetation, garbage, or rubbish near the storage facilities or warehouses, the use of constructions proper to avoid rodents, and the recourse to repelling agents. As to the latter, it comprises poisons, traps, and biological control. Killing of rats and mice, whether by baits, traps, or otherwise, is only effective over short time spans. In fact, the answer to rodent control is rodent-proofing of buildings and good sanitation.

2.5.2 ABIOTIC FACTORS

As mentioned previously, there are two main factors that affect, in a special way, the quality of cereal grains: high levels of moisture and inadequate harvesting (De Martini et al. 2009). Considering that this chapter is focused on transportation and storage, the attention will be centered on moisture (the main abiotic factor for a safe storage of grains), as well as on temperature. As previously seen, microorganisms, specially certain species of fungi, are the major causes of grain deterioration. Moisture can be used to control fungal growth on cereals. At moisture contents lower than 14%, fungi will not grow. However, when the moisture increases to values between 14% and 20%, great changes in the rate of fungal growth are produced along with changes in the species that are developed (Hoseney 1990). The same author stated that the maximum moisture contents for a safe storage are generally accepted to be 13% for corn, barley, oats and sorghum, 14% for wheat, and 12% to 13% for rice.

It is well known that cereal grains generally present different moisture contents, even if they come from the same field, and also vary from kernel to kernel. Thus, when the grains come from different production sources, it is expected that the final moisture will also vary. Grain, like any other hygroscopic material, keeps equilibrium with the air's relative humidity for a given temperature. In this way, they absorb or free moisture according to the humidity of the air in the bulk mass of grain until equilibrium is reached. At equilibrium, the water vapor pressure inside the grain will be equivalent to the vapor pressure of the air, and moisture transfer will cease (D'Arce 2006). For this reason, grain stored in permeable containers, such as bags, will suffer frequent oscillations in moisture according to the humidity of the surrounding air. On the other hand, for hermetically closed recipients, this will not happen. The safe storage of grains is very much dependent on hygroscopicity and therefore the moisture profiles in the grain are usually accessed using the water sorption isotherms.

Cereal grains have several layers of different cells, with different characteristics, and therefore produce a nonuniform behavior toward water. Thus, there is a heterogeneity in the moisture distribution inside the grain because the absorption of moisture by the different constituent fractions is different. Besides, there is also heterogeneity between grains in the same lot, and with different sizes. Effectively, all grains absorb water as long as the ambient relative humidity is high, but the way they do it is different. The water absorbance will continue until equilibrium is reached, in a very low rate process.

In grain storage, transfer of moisture and heat occur and if a spot of heat appears, this can lead to alterations that can spread to the whole cereal. However, these transfers can be positively used as a means to fight the problem through forced circulation of air (hot or cold). Also, thermal conductivity is important in these cases. Considering the porous mass of grains, heat is propagated from one point to another by conduction and convection: by conduction from one grain to its neighbors and by microconvection through the air flux in the pores of the bulk mass.

Heat and moisture transfers are important physical factors in the conservation of grain. The transfer can be natural or could originate from forced ventilation (by hot or cold air):

1. Transfer caused by forced convection (of dried or hot air): In this case, the transfers are established between the grains and the surrounding air until reaching equilibrium (moisture and temperature). The drying curves for cereals are slow and very energy demanding because the initial moisture content is already low and thus the limiting factor of drying rate is not the vaporization of the liquid water at the surface of the product, but the desorption of the adsorbed water and the internal resistance to moisture diffusion of the water vapor. Based on this knowledge, the technique called "slow differed drying" for corn, in which the rest phase corresponds to a period of diffusion of water vapor inside the grain from the center (still wet) to the periphery (already dehydrated), is improved.

2. Natural transfer of heat or water vapor (or both): This may occur in an unpredicted or predicted way, with severe consequences on the quality of the grain. They can be caused by simultaneous heat and moisture gradients, by temperature gradients (without moisture gradients), or by moisture gradients (without temperature changes). Moisture gradients may arise from the mixture of more or less wet grains (in this case, the transfer will be from grain to grain), or from the existence of layers with different moisture gradients (in this case, the process will be over a front and slower). Regardless of the mode, the tendency is to establish equilibrium but the mechanisms are different.

When there is just a moisture gradient, two situations can arise: (*a*) mixture—a representative sample presents two populations that tend to homogenize and equilibrium will be achieved in 1 week; and (*b*) overlap— a slower process, in which successive equilibriums are established, that is, there are layers with different moisture levels therefore initiating transfer between layers (Figure 2.5).

When there is only a temperature gradient, resulting from a hot or cold spot, the low specific heat and thermal conductivity of cereals produces a very slow transfer, therefore initiating localized elevations of temperature. Table 2.6 presents some thermal properties of cereal grains.

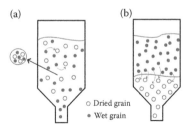

FIGURE 2.5   Moisture transfer in moisture gradient conditions: (a) mixture and (b) overlap.

TABLE 2.6
Thermal Properties of Some Cereal Grains

Cereal Grain	Moisture Content (%, wet basis)	Specific Heat (kJ/kg K)	Thermal Conductivity (W/m K)	Thermal Diffusivity (m²/s)
Corn	5.1	1.691	0.1466	9.83×10^{-8}
	20.1	2.223	0.1636	8.67×10^{-8}
	30.2	2.462	0.1724	9.25×10^{-8}
Wheat	9.2	1.549	0.1402	3.17×10^{-8}

Source: Meneguello, E.L. *Projeto e automação de fábricas de ração.* Erechim RS, Brazil: URIAUM, 2006.

Intergranular air is continuously moving though natural convection currents produced by differences in density of hot and cold air. When air moves from hotter to colder zones, the hot air presents a higher relative humidity, which is transferred in part to the grains until equilibrium. This moisture migration happens naturally, even though the grains initially all presented the same moisture. This migration occurs if the partial vapor pressure on the surface of the grain (P_s) is higher than in the air (P_a) causing moisture desorption. On the other hand, when $P_a > P_s$, moisture is absorbed/condensed. These localized accumulations of moisture may propitiate conditions favorable to the development of organisms responsible for degradation (D'Arce 2006).

2.5.2.1 Cold Wall Effect

The mass of grain deposited in a silo usually presents temperature differences. The layers closer to the silo walls and at the surface acquire a higher or lower temperature compared with the rest of the mass as a result of external temperatures. The cold wall effect is due to alternating daily and nightly temperatures, temperature differences between the two faces of a silo (one facing light and the other facing shadow), or differences in temperature during shipment between the grains and the wall of the ship (under water). Also, in these cases, the transfer occurs from the hotter to the colder grains, dehydrating the first and humidifying the latter (Figure 2.6).

In the case of the sun–shadow effect, due to contact with the hotter wall, the grain on one side gets hotter compared with the grain on the shadow side of the wall. The transfer goes to the cold zone, thus creating condensation with a consequent increase in moisture.

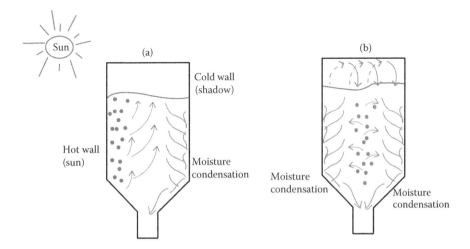

FIGURE 2.6 Cold wall effect: (a) sun–shadow effect, and (b) night cold effect.

In the case of the night cold effect, the walls cool during the night (and being colder than the grains), therefore cool the grains closer to the wall. The gradient is from the center to the periphery; and from above, a mist is created and condensation appears at the wall. If the moisture levels achieve critical values, it may lead to the development of fungi, and eventually, bacteria and insects. If so, the products resulting from metabolism will release more heat and may compromise the preservation of the grain.

The cold wall effect depends greatly on the type of soil and construction techniques. Metallic silos are more affected than concrete silos, and isolation of the walls or construction under a shelter or underground limits the risk of cold points.

2.5.2.2 Hot Spot Effect

The hot spot effect is formed when there is a localized generation of heat resulting from biological activity (like respiration, insects, or microorganisms). During the cold periods, moisture is moved from the hot grains in the center to the grains at the surface. During the hot period, the effect is reversed, and the critical region is then the bottom of the silo. This movement of moisture may occur even if the grain was stored with the correct moisture content (D'Arce 2006). There is an ascending movement of heat and water vapor from the hotter point to the drier mass of grain situated in the higher layers, thus humidifying them. The humidified and heated zone can also be the place where biological heat appears (respiration), producing a new transfer of heat and vapor in a higher point, which may propagate (Figure 2.7).

Furthermore, Dunkel (1995) stated that the use of sealed (airtight), underground storage at a sufficient depth below the surface prevents the major causes of loss in grain by maintaining a uniform low temperature, establishing and maintaining a low oxygen atmosphere, and minimizing accessibility to the grain. This author also mentioned that stored grain losses are minimized by these key biological factors without the use of residual or fumigant pesticides.

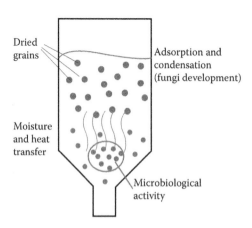

FIGURE 2.7   Hot spot effect.

2.5.3 Losses during Storage

The set of physical, chemical, biological, and sensorial characteristics of the cereal are established at harvest, and are related to the harvest procedures and cleaning techniques. This will define the capacity of the grain to resist the attacks of insects and microorganisms. During harvest, cleaning, drying, and transportation, the grains are subjected to mechanical impacts that may cause cracking and breaking, and these will serve as entrance for invasion by fungi and insects.

Intact grains have a better preservation capacity and deterioration is faster when the grain is prepared for extraction. Under adequate conditions, the grain can be stored for many years with a minimum rate of deterioration. However, if the conditions are unfavorable, the grain suffers considerable damage in just a few days.

The rate of deterioration depends on the activity of the biotic variables, and these are mostly affected by abiotic factors such as temperature and moisture. As stated previously, the principal agents affecting grain quality and quantity are rodents, insects, birds, and fungi. These deteriorating agents use the nutrients present in the grain to their growth and reproduction. Respiration of the grain, although in a smaller scale, could also contribute to the loss of dry matter during storage (De Martini et al. 2009).

When grains are stored under temperature and moisture conditions adequate for the maintenance of equilibrium, grains will keep their properties for long periods. In these conditions, if temperature is kept between 14°C and 20°C, the weight will be maintained and pests and fungi will not be able to develop and cause damage (Filho et al. 2007). For moisture contents between 14% and 16%, cereals need cooling to less than 20°C to be stored for up to 1 month. If a longer storage period is needed, moisture will need to be further reduced. At 12.5% moisture, most cereals can be stored for years with grain aeration-cooling only. At less than 12.5%, robust cereals such as wheat can be stored without cooling in most areas (Banks 1999).

Stored grains are usually dried to about 12.5% moisture, treated against insects, and the grain conditions at the container are monitored by sensorial evaluation, moisture content, and temperature changes. It is good to emphasize that effective storage of dried grain eliminates microorganism activity. Properly protected and drained sheds and bins also minimize bird and rodent damage. However, insect damage to stored grain is minimized primarily by use of chemicals.

Grain quality is largely determined during the growing season, and once the grain has been harvested, it is difficult to improve quality. However, grain quality can largely be affected by conditions during harvest and subsequent drying and storage (Evans 2001).

According to Bragatto and Barrella (2001), losses during storage are irreversible and may occur due to one of the following factors:

1. Unloading of the product in the warehouse or storage facility—this operation leads to the appearance of impurities of different sizes, from small to large, thus creating preferential pathways to aeration and opening a means for deterioration.
2. Nonhomogeneity of the grains with impurities—foreign particles and impurities, under the same conditions of moisture and relative humidity as

the grain, present moisture contents higher than the product, initiating the formation of a compact moist mass favorable to the development of micro-organisms, thus accelerating deterioration.

3. Nonhomogeneous moisture—the higher the temperature or the moisture content of the grain, the more intense is the respiratory process leading to heat and moisturizing of the higher layers, a higher incidence of insects, and the appearance of fungi and foreign odors resulting from deterioration.

4. Appearance of hot spots—these are formed due to the low thermal con-ductivity of the grain and represent one of the most important causes for damage under bulk storage, leading to the formation of convection currents of moisture and temperature.

5. Insect infestation—leads to the heating of the grain mass and reduction in weight and nutritive value.

6. Proliferation of fungi—results in the heating of the grain mass, loss of ger-minating capacity, and diminishing of nutritive value (besides being harm-ful to humans and animals in case of mycotoxin formation).

7. Mechanical damage—leads to the appearance of broken, scratched, and damaged grains, and this presents conditions favorable to the attack of insects and the development of fungi. Also, because of an increase in the surface area, the oxidation of lipids is increased.

8. Drying of the grain—this involves hot air and causes damage in a certain percentage of grains during the rapid cooling phase. In drying, moisture is initially removed from the external layers of the product, initiating a gradi-ent from inside to the periphery. When this gradient is too high, the internal tensions cause cracking, thus affecting storage and quality.

9. Aeration—even if well performed, its efficiency is limited, depending on relative humidity and temperature conditions.

To evaluate the losses during grain storage, two classifications are considered: (1) weight loss or physical damage, and (2) quality changes. The first occurs due to damage caused mainly by insects and, to a lesser extent, by rodents or birds. The second refers to the intrinsic properties of the product that are altered, mostly by fungi, which cause fermentation, organoleptic modifications (mainly in aroma and/or taste), and loss of nutritive value. Contamination by foreign bodies and other dam-ages that affect the quality of the raw matter for agro-industry are also included in this category.

2.6   DUST EXPLOSIONS IN GRAIN STORAGE

The principal accidents that occur in grain storage facilities include the following: (1) drowning and suffocation in the grain mass. In the first, the victim is pushed by the mass of cereal, and in the latter, is covered by it. (2) Intoxication caused by gases. This occurs in confined environments in which toxic gases accumulate (CO, CO_2, NO_2, CH_4, among others), affecting the respiratory processes of the victim. (3) Explosions. Due to the movement of huge masses of grain in the different phases that precede the cargo of the cells/silos, great quantities of dust accumulate inside the silo. This dust, aside from being a pollutant that compromises the workers' respiratory function, constitutes an increased danger because it is susceptible to exploding. For these reasons, it must be removed from the inside of the silo through exhaust machines and other equipment.

Explosions of dusts may occur wherever mineral or organic goods are processed, handled, or stored, and when an appropriate concentration of inflammable dust is airborne and ignited by the slightest heat. The explosions in storage facilities occur due to the mixture of the substances present with the atmospheric air (in particular its oxygen). The particles that explode correspond to impurities that accompany the mass of grain plus the ones resulting from the mechanical damage of the grain.

Since the beginning of the twentieth century, the frequency of dust explosions has increased, first in coal industries and later in the storage of agricultural products, grains in particular. In the United States, for example, 400 silos (3% of storage equipment) were destroyed from 1960 to 1980, killing 200 persons. In France, awareness of the problem only developed after the first major explosion in a malthouse that killed 12 people in 1982. Fifteen years later, in 1997, a similar accident killed 11 people in Blaye, north of Bordeaux. Many factors of gravity are usually combined in explosions of silos, and often result in catastrophe (Ribereau-Gayon 2000).

Pedersen and Eckhoff (1987) investigated a comprehensive series of single-impact ignition experiments, by generating explosible dust clouds of dried maize starch, in the region of tangential impact between a moving body and a stagnant horizontal anvil. They observed that steel sparks produced by single impacts of net energies of up to 20 J, between steel and concrete, steel, or rusty steel, are unlikely to ignite clouds of dried maize starch. Therefore, it seems unlikely that dust explosions involving dusts of grain, feed, or flour can be initiated by heat from accidental single impact, unless the net impact energies are much higher than 20 J.

Dust explosions happen if a certain number of conditions are met:

1. The dust is flammable (like in the case of dust freed by cereal grains and flours).
2. The quantity of oxygen and concentration of dust are sufficient, and this will depend on the type of material. In a general way, a reduction for levels under 10% of O_2 in the atmosphere will eliminate the risk of explosion.
3. The ignition source has to possess the necessary energy. The most common ignition sources are heat and sparks. Regarding the first, the level is defined by the temperature at which the material is flammable and the latter has to possess a minimum level of energy, also dependent on the material, but which usually is from 10 to 100 mJ.

After ignition, the evolution and development of an explosion varies according to the configuration of the space where it occurs, and these are basically of two types: propagation in a spherical space (or critical) and propagation in a long enclosure. In the first case, the propagation happens through successive spherical layers. The rate of propagation results from the combustion velocity and the expansion of the burned gases. The explosion caused by combustion moves faster than the front flame—deflagration. When the explosion happens in a long space, the rate of propagation of flames increases as the explosion develops (due to unidirectional expansion of burnt grains). The pressure is not regular along the walls and may achieve critical levels—detonation.

Explosions are affected by many factors, such as the nature of the dust, the concentration of the dust, the presence of inert materials in the mixture, the presence of volatiles in the mixture, the composition of the atmosphere (oxygen concentration), the dimension of the particles (the violence of the explosion is inversely related to the size of the particles), the moisture content of the material, the intensity of the ignition source, and the turbulence of the mixture, among others.

To avoid explosions, certain security measures must be attended to. First, actions must be undertaken to reduce the risks and possibilities of occurrence. Second, a set of measures aimed at avoiding explosions and limiting their effects includes making the facility inert by cleaning and avoiding the accumulation of dust; adopting measures to contain explosions, in case they happen; using a compressor for dust or using continuous systems for the capture of dusts; decompressing the facility; thoroughly cleaning the mass of grain; periodically cleaning the system for dust capture by changing the filters within the periods defined by the manufacturers; providing information to the workers about the risks of explosion and the measures to adopt in case they occur; performing periodic maintenance on electrical and mechanical equipment and verifying the state of electrical cables; being careful with soldering equipment during maintenance; replacing metallic transporters with components made mostly of plastic; use of lighting appropriate for the environment without risk of explosion; and designing buildings and facilities that structurally include areas for easy rupture in case of an explosion.

2.7   QUALITY OF WHOLE GRAINS

The quality of the final product is influenced by the quality of the initial grain or seed, and later by the storage conditions. The care and attention must start during cultivation by adopting preventive measures against mechanical damage, attack from insects, and delays in harvest. If it is found necessary, a predrying of the product is performed and all transport equipment are cleaned so as to avoid contaminations.

To guarantee the quality of cereal grains, it is important that the storing place is thoroughly cleaned. The control of rodents must include a tight fencing of the area, as well as the use of poisonous baits. Insects can be fought off by physical (temperature and moisture of the product) as well as chemical means. Furthermore, the grain must be dried up to levels that are insufficient for the development of fungi. As a general rule, low moisture content and low temperature are ideal conditions.

Evans (2001) stated that rainfall before and during the harvest period can encourage ear diseases and premature sprouting, whereas high grain moisture levels will increase the associated costs. As an example, the delaying of the wheat harvest can result in grain with high α-amylase content, lower specific weight, and lower protein contents, factors that will seriously reduce the quality of the grain for bread making. Evans (2001) also stated that almost all grain in the United Kingdom is harvested by large, self-propelled combine harvesters, which are highly efficient and cause minimal physical damages to the grain when they are properly set and operated. In this way, quality parameters are guaranteed at harvest. The condition of the crop at harvest has a significant influence on combine performance; severely lodged crops not only produce inferior quality grain, but also reduce combined speed and efficiency. The presence of weeds also interferes with the harvesting process and often leads to higher grain moisture levels and contamination with weed seeds, which increase the costs of cleaning (Evans 2001).

Moreover, it is also important to consider the association between the physical and the chemical changes in the composition of food, which have made the need for biochemical and nutritional quality control tools for stored products increase greatly. In 2006, Rehman studied the effects of storage temperature and time on the nutritional quality of different cereal grains (wheat, maize, and rice). The author found a gradual decline in moisture, whereas the total available lysine contents decreased by 6.50% and 18.5% in wheat, 14.3% and 20.7% in maize, and 23.7% and 34.2% in rice during 6 months of storage at 25°C and 45°C, respectively (Rehman 2006). The protein and starch digestibility of cereal grains also decreased under the same storage conditions. Despite this, at low temperatures (10°C), no significant change in nutritional quality was observed during the storage of cereal grains. In view of these facts, the author concluded that the nutritional quality of cereal gains was adversely affected as a result of storage at elevated temperatures. Thus, it is suggested that cereal grains (wheat, maize, and rice) should not be stored at more than 25°C to minimize nutrient losses during storage.

Another important aspect is the increasing demand of the consumers for safe foods. Regarding this point, storage plays a major role to the final quality of the cereal. Therefore, grain storage needs modern technologies, clean facilities, nonaggressive and nontoxic environments, economic viability, and sustainability so as to reduce costs and provide healthy products, both for human and animal feed.

ACKNOWLEDGMENTS

The authors thank the reviewers for their valuable suggestions aimed at improving the present chapter: Blanka Vombergar, PhD (The Educational Centre Piramida Maribor, Higher Vocational College, Maribor, Slovenia); Isabel Neves Evaristo, PhD (Instituto Nacional de Investigação Agrária e Veterinária, Lisbon, Portugal); Tomris Altug Onogur, PhD (Food Engineering Department, Ege University, Izmir, Turkey); and Vesna Rafajlovska, PhD (Department of Food and Biotechnology, Ss. Cyril and Methodius University in Skopje, Republic of Macedonia).

REFERENCES

Abalone, R., Gastón, A., Cardoso, L., Bartosik, R.E., and Rodríguez, J.C. 2011a. Gas concentration in the interstitial atmosphere of a wheat silo-bag. Part I: Model development and validation. *Journal of Stored Products Research* 47: 268–275.
Abalone, R., Gastón, A., Cardoso, L., Bartosik, R.E., and Rodríguez, J.C. 2011b. Gas concentration in the interstitial atmosphere of a wheat silo-bag. Part II: Model sensitivity and effect of grain storage conditions. *Journal of Stored Products Research* 47: 276–283.
Abramson, D., Hulasare, R., White, N.D.G., Jayas, D.S., and Marquardt, R.R. 1999. Mycotoxin formation in hulless barley during granary storage at 15% and 19% moisture content. *Journal of Stored Products Research* 35: 297–305.

Banks, J. 1999. High moisture levels increase yield and quality. *Farming Ahead* 94: 39–40.

Bibby, I.P., and Conyers, S.T. 1998. Numerical simulations of gas exchange in leaky grain silos, using measured boundary conditions. *Journal of Stored Products Research* 34: 217–229.

Bragatto, S.A., and Barrella, W.D. 2001. Otimização do sistema de armazenagem de grãos: Um estudo de caso. *Produção Online* 1(1): 1–8.

Brod, F.P.R. 2005. Bem armazenados. *Cultivar: Caderno Técnico Máquinas* 43: 3–10.

Carrera-Rodríguez, M., Martínez-González, G.M., Navarrete-Bolaños, J.L., Botello-Álvarez, J.E., Rico-Martínez R., and Jiménez-Islas, H. 2011. Transient numerical study of the effect of ambient temperature on 2-D cereal grain storage in cylindrical silos. *Journal of Stored Products Research* 47: 106–122.

D'Arce, M.A.B.R. 2006. *Pós colheita e armazenamento de grãos*. São Paulo, Brazil: ESALQ/ USP.

De Martini, R.E., Prichoa, V.P., and Menegat, C.R. 2009. Vantagens e desvantagens da implantação de silo de armazenagem de grãos na granja De Martini. *Revista de Administração e Ciências Contábeis do IDEAU* 4(8): 1–17.

Dunkel, F.V. 1995. Applying current technologies to large-scale, underground grain storage. *Tunnelling and Underground Space Technology* 10: 477–496.

Evans, E.J. 2001. Cereal production methods. In *Cereal Processing Technology*, edited by G. Owens. Cambridge. UK: Woodhead Publishing Ltd.

Filho, A.F.L., Demito, A., Miranda, L., Costa, C.A., Heberle, E., and Volk, M.B.S. 2007. *Resultados comparativos do resfriamento artificial e aeração com ar ambiente durante a armazenagem de 16.000 t de milho a granel*. Viçoza, Minas Gerais, Brazil: Universidade Federal de Viçosa.

Graves, R.R., Rogers, R.F., Lyons, A.J., Jr., and Hesseltine, X.W. 1967. Bacterial and actinomycete flora of Kansas, Nebraska and Pacific Northwest wheat and wheat flour. *Cereal Chemistry* 44: 288–299.

Hoseney, C.R. 1990. *Principles of Cereal Science and Technology*. St. Paul, MN: American Association of Cereal Chemists.

Khatchatourian, O.A., and Binelo, O. 2008. Simulation of three-dimensional airflow in grain storage bins. *Biosystems Engineering* 101: 225–238.

Khatchatourian, O.A., and Oliveira, F.A. 2006. Mathematical modelling of airflow and thermal state in large aerated grain storage. *Biosystems Engineering* 95: 159–169.

Khatchatourian, O.A., Toniazzo, N.A., and Gortyshov, Y.F. 2009. Simulation of airflow in grain bulks under anisotropic conditions. *Biosystems Engineering* 104: 205–215.

Lopes, D.C., Martins, J.H., Melo, E.C., and Monteiro, P.M.B. 2006. Aeration simulation of stored grain under variable air ambient conditions. *Postharvest Biology and Technology* 42: 115–120.

Lopes, D.C., Martins, J.H., Lacerda Filho, A.F., Melo, E.C., Monteiro, P.M.B., and Queiroz, D.M. 2008. Aeration strategy for controlling grain storage based on simulation and on real data acquisition. *Computers and Electronics in Agriculture* 63: 140–146.

Meneguello, E.L. 2006. *Projeto e automação de fábricas de ração*. Erechim RS, Brazil: URIAUM.

Olstorpe, M., Schnürer, J., and Passoth, V. 2010. Microbial changes during storage of moist crimped cereal barley grain under Swedish farm conditions. *Animal Feed Science and Technology* 156: 37–46.

Pedersen, G.H., and Eckhoff, R.K. 1987. Initiation of grain dust explosions by heat generated during single impact between solid bodies. *Fire Safety Journal* 12: 153–164.

Rehman, Z.-U. 2006. Storage effects on nutritional quality of commonly consumed cereals. *Food Chemistry* 95: 53–57.

Ribereau-Gayon, R. 2000. Les explosions de poussières dans le stockage agroalimentaire: Les risques d'accident catastrophique et leurs conséquences médicales. *Médecine de Catastrophe—Urgences Collectives* 3: 13–20.

Silva, L.C. 2002. Stochastic Simulation of the Dynamic Behavior of Grain Storage Facilities. PhD Dissertation. Universidade Federal de Viçosa, Viçoza, Minas Gerais, Brazil.

Silva, L.C. 2005. *Secagem de grãos. Boletim Técnico: AG-04/05*. Campo Alegre, Brazil: Universidade Federal do Espírito Santo.

Silva, L.C. 2011. *Aeração de grãos armazenados. Boletim Técnico: AG-01/11*. Campo Alegre, Brazil: Universidade Federal do Espírito Santo.

Silva, L.C., Queiroz, D.M., Flores, R.A., and Melo, E.C. 2012. A simulation toolset for modeling grain storage facilities. *Journal of Stored Products Research* 48: 30–36.

Tefera, T., Kanampiu, F., De Groote, H., Hellin, J., Mugo, S., Kimenju, S., Beyene, Y., Boddupalli, P.M., Shiferaw, B., and Banziger, M. 2011. The metal silo: An effective grain storage technology for reducing post-harvest insect and pathogen losses in maize while improving smallholder farmers' food security in developing countries. *Crop Protection* 30: 240–245.

3 Malting

Luís F. Guido and Manuela M. Moreira

CONTENTS

3.1 INTRODUCTION

Approximately 96% of the malt produced in the world is used as the main raw material, along with water, to make beer. Likewise, malt is extremely important in Scotland for whisky production, but this only represents approximately 3% of the world's malt use. The rest is used as a very important ingredient in the food industry as natural flavoring, breakfast cereals, malted milk drinks, or malt vinegar. Although barley, wheat, and sorghum can be malted, barley malt is preferred for brewing and hence this chapter is focused on the principles of the production of barley malt.

Although the basic sequence of the malting process has remained unaltered for generations, the fundamental stages of the malting operations are briefly explained. The historical developments and the new requirements of the malting technology are then addressed. The biochemistry of germinating barley is rather complex and is beyond the scope of this chapter. Much of the information on the biochemical nature of the malting process can be found elsewhere (Briggs et al. 1981; Pollock 1962).

Market forces and consumers' demand for different beer types led the maltsters to produce colored and special malts; this is discussed in later sections. Malt is a natural product with indisputable health benefits. We are still learning what malt has to offer to increase confidence in food healthiness, but the recent findings are emphasized herein. Finally, the last paragraphs of the chapter are devoted to the current trends in the barley malting production, concerning the complex phenotype of the malting quality.

3.2 MALTING PROCESS

3.2.1 PRECLEANING, CLEANING, GRADING, AND STORAGE

After harvesting, barley must be prepared for storage during its period of dormancy, before conversion to malt, under conditions minimizing the risk of infestation from insects, fungal attack, etc. To achieve this, it is necessary to dry the wet or green grain from the fields for long-term storage, down to a moisture content of 12%. Grain with moisture levels between 10% and 12% can be stored for long periods, although initial drying down to 15% to 16% could be acceptable, providing that the grain is stored for short periods, and preferably cooled.

When barley arrives at a malting plant, it is rapidly precleaned to remove gross impurities before it is passed to the drier or to storage silos. Precleaning consists of a rough sieving process combined with, or followed by, aspiration of the grain stream with a flow of air. In this stage, items such as leaves, straw, stones, clods of earth, snail shells, barley awns, and other light material are removed (Gibson 2001).

Cleaning has to remove impurities of about the same size as barley, such as weed seeds of grains of other cereals, as well as broken grains, or undersized grains. The light material is removed with the help of cleaners and separators, whose sieves have a greater efficiency than those installed in the precleaning stage. Destoners, which separate the stones from the grain using their density difference, can be optionally installed. Finally, in the sieving equipment, barley for malting is graded by kernel size (width) into different categories, for example, grain sizes of more than 2.5 mm, 2.2 to 2.5 mm, and less than 2.2 mm. Broken kernels and grain of less than 2.2 mm width, removed as screenings, may be collected in a silo for eventual use as animal feed. It is important that all these technologies are fitted with a dust extraction system to minimize the risk of dust explosions.

Barley that has been processed ready for malting is stored in large-capacity circular steel bins, giving maximum versatility in the separation of varieties. An automatic weighing machine in this section controls the quantity of barley and assists with the stock records. If storing for longer periods, silos as large as 4000 m^3 may be used.

3.2.2 STEEPING

Steeping initiates malting and is a crucial step in producing quality malt. During steeping, barley takes up water and swells by one third. Respiration increases slowly at first and then more rapidly, causing the grain in steep to accumulate CO_2 and heat up; this is aggravated by the action of microorganisms associated with the barley.

The steeping stage consists of alternating periods when the grain is immersed in water, referred to as the "under water periods," and periods with the water drained from the grain, referred to as the "air rest periods." This combination is required to promote and maintain the efficiency of the germination (Bräuninger et al. 2000). To achieve satisfactory results during the subsequent malting, it is necessary to increase the water content of barley to approximately 42% to 46% during the steeping process, in a typical temperature range from 15°C to 20°C. When barley absorbs water, the embryo becomes active and uses the oxygen dissolved in the steeping water for respiratory purposes. During the first period under water, the dissolved oxygen in the water is rapidly adsorbed by the organic material and microflora in the grain. In case of prolonged failure of the oxygen supply, respiration with the release of CO_2 may induce fermentation with the production of ethanol. It was recently shown that steeping conditions affected the rate of recovery from oxygen deficiency, germination rate, and onset of α-amylase production (Wilhelmson et al. 2006). A lack of oxygen at any stage will stop the growth and enzyme production. Prolonged anaerobiosis is progressively damaging and finally lethal to the grain (Briggs et al. 1981).

Historically, barley was steeped for 3 days with a 44% increase in water content without the embryo showing any sign of growth. Today, pneumatic malting (see Section 3.2) is based on a steeping time of 2 days with aspiration of CO_2 and air-rest periods, which stimulates the grain to grow (Bräuninger et al. 2000). At the end of steeping, the chosen moisture content of the grain (within the range 42%–46% fresh weight basis) is sufficient to support modification, but not allow excessive growth. This is important because growth is associated with a decline in dry matter, lost as carbon dioxide, water, and rootlets (malting loss).

In addition to supplying water and oxygen to the barley, steeping is of considerable value in removing dirt from the grain. The film of dust and many of the microorganisms adhering to the grain are washed away by the steep water, and this cleaning action can be increased if special equipment is used to agitate the barley during steeping (Schuster 1962). Steep water leaches husk components that may impede germination and any musty smell that may be associated with the barley, and possibly some that prevent proper flocculation of yeast. Premature yeast flocculation is a sporadic problem for the malting and brewing industries, which can have significant economic and logistic implications. If available, the use of a washing screw or washing drum in the maltings before steeping, as well as the discarding of steep water, may alleviate the problem of premature yeast flocculation (Panteloglou et al. 2012).

Among the malting processes, the steeping step represents the most important water consumption and effluent generation stage. Water economy in malting is centered on reducing the volume of water used in steeping, notably by spray-steeping rather than successive immersions, as well as recirculating cooling and humidifying water. However, the reuse of steep-out waters to subsequent steeps resulted in a delayed germination and a deterioration of malt quality, mainly due to the presence of germination inhibitors. By using a membrane bioreactor for the efficient elimination of the germination inhibitors, treated waters were recycled into the steeping process and the resulting malts were satisfactory in terms of physicochemical parameters (Guiga et al. 2008). Nowadays, best practices in malting procedures have made single-step water steeping possible. This allows water consumption ratios as low as 3 m³/t of barley malt produced.

3.2.3 GERMINATION

The germination process is characterized by the growth of the embryo of the grain, manifested by the rootlets growth and increase in length of the shoot (acrospire), with the concomitant modification of the contents of the endosperm (Figure 3.1). To a certain extent, the germination and steeping steps overlap because many maltsters prefer the barley to be "chitted," that is to have the tip of the rootlet breaking through the husk, at the end of steeping. It is the objective of the maltster to obtain the necessary modification for a particular type of malt while reducing to a minimum the loss in weight consequent on the activity of the embryo. Subsequently, the growth of the acrospires should be limited between two-thirds and three-quarters of the length of the grain for pale malts, and between three-quarters and the full length of the grain for dark malts. However, malting loss resulting from respiration is inevitable and is generally in the order of 4% to 8%, depending on the type of barley and the malting procedures (Schuster 1962).

Germination rate and modification intensity are controlled by regulating the moisture content and temperature of the grain. Traditionally, the steeped grain spends between 4 and 6 days in humid and ventilated conditions. The temperature of germinating grain is controlled to between 14°C and 20°C by a flow of air through the bed; the germinating grain is humidified by atomizing water jets, which may be cooled by refrigeration. Also relevant, to ensure grain bed homogeneity throughout kernel germination, is a gentle revolving of barley along the germination period.

A large increase in the quantities of some of the hydrolytic enzymes present in the grain is one of the essential changes of modification. The alteration of the starchy endosperm begins due to the partial degradation of reserve substances (cell walls, gums, proteins, and starch), as shown in Figure 3.1. Consequently, a reduction in the

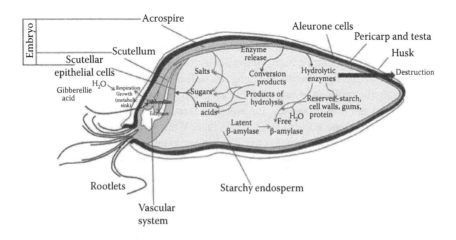

FIGURE 3.1 Some of the metabolic interrelationships occurring in malting barley.

strength of this tissue can be observed. Effectively, dry malt, in contrast with dry barley, is friable and readily crushed. Gums are degraded so that aqueous extracts of malt are less viscous than those of barley (Briggs et al. 1981).

The hydrolytic enzymes that catalyze modification are derived mainly from the scutellum and aleurone layer. In the initial stages of germination, the scutellum releases many hydrolytic enzymes that begin to degrade the cell walls of the crushed cell layer, and the walls, protein, and starch granules of the endosperm. In addition to enzymes, the scutellum releases gibberellins, which stimulate the production and release of numerous hydrolytic enzymes by the aleurone layer. α-Amylase is absent from the hydrated aleurone, and its *de novo* synthesis and release are independently triggered by the arrival of the hormone. It is estimated that in malt, approximately 85% of this enzyme originates from the aleurone layer and 15% from the scutellum layer (Briggs et al. 1981). β-Amylase, which exists in the original barley in an inactive form in the starchy endosperm, is liberated and the degradation of the components proceeds (Figure 3.1).

Proteases and carboxypeptidases accumulate in the starchy endosperm during germination. β-Glucan solubilase is a carboxypeptidase that converts the insoluble cell wall protein matrix to high molecular weight soluble but viscous β-glucans, which can create problems during wort separation and beer filtration. Those are broken down further to low molecular weight glucans and glucose by the endo-β-glucanases and β-glucosidases. These enzymes develop slowly in the malting process and are slightly sensitive to gibberellic acid. Gibberellic acid is an example of germination regulator which, like potassium bromate, can be added to the grain, to greatly accelerating the modification rate (Briggs et al. 1981). At this stage, the material is called green malt.

Inhomogeneity of malt modification is caused by uneven germination and uneven gibberellin production causing uneven distribution of endosperm-degrading enzymes and finally failure of functional and optimal quantities of endosperm-degrading enzymes to effectively break down resistant compact areas of the starchy endosperm (Palmer 1994). It is proposed that, for normal malting barley, variations in malt modification are related to the different percentages of grains containing high levels and different types of proteins, which resist enzymatic modification during malting. This kind of inhomogeneity of malt modification can cause brewhouse problems even if it cannot be detected with precision by standard malt analyses (Palmer 2000).

The metabolism of the growing grain is an exothermic process. Typically from 1 tonne of barley, 40 kg of starch will be converted to carbon dioxide and water with an associated energy production of 850 MJ of heat. This heat, which must be dissipated, may cause an increase in temperature of the air, and a temperature increase across the malt bed of 2°C is acceptable (Bräuninger et al. 2000).

3.2.4 KILNING

Kilning has the characteristics of a typical industrial drying process occurring in two distinct stages. First, moisture is removed from the green malt, reducing it from approximately 44% to 12%. With an upward air flow, this process takes about 12 h to pass through the bed for a single-deck kiln and 24 h for a double-deck kiln (see

Section 3.2.3). This phase of drying is rapid and is referred to as the "free-drying" or "withering" phase. The second phase of drying, in which the malt is dried from 12% to 4%, occurs in a much slower process, which is commonly referred to as the "falling rate" phase. At the end of the drying process, the kiln temperatures may be raised for 1 or 2 h in the "curing" stage, followed by a cooling period to achieve a temperature suitable for discharge and storage (Bräuninger et al. 2000).

Kilning not only dries the malt, preventing further growth and modification, but also removes undesirable flavors, partially or wholly destroys some hydrolytic and other enzymes, and develops flavor and color in the final product. Moreover, malt is kilned to produce a friable, readily milled stable product that may be stored for long periods, and from which roots may easily be removed (Briggs et al. 1981).

In double-deck kilns, the process is halted at the break point (when the air-off temperature closely approaches the air-on temperature) to permit discharge of finished malt from the lower bed and transfer of malt from the upper to the lower bed. Owing to longer exposure of the green malt to warm and moist conditions, there is more risk of microbiological problems in double-deck kilns. Significant differences were observed between the lower and upper malt bed in such kiln due to the moisture and temperature gradient, emphasizing the necessity to draw attention toward the heterogeneity of commercial malts (Guido et al. 2005).

The composition of malt changes during kilning, so that normally, the enzyme activity declines whereas its color, aroma, flavor, and extractable polyphenol content increases. Some enzymes, for example, α-glucosidase, are appreciably inactivated even at 45°C, which is well below kilning temperatures. At 80°C, the activities of enzymes, such as β-glucanase and β-amylase, are significantly reduced. α-Amylase is the most stable enzyme and its activity may even increase during kilning. The degree of enzyme destruction at any temperature is greater when the malt is wet. Highly enzymatic malts are kilned at lower temperatures and in a rapid air flow to ensure that the grain is cooled by evaporation. In roasted malts, on the contrary, enzyme destruction is complete. Two isoforms of lipoxygenase (*LOX-1* and *LOX-2*) are known to induce the oxidation of linoleic acid into the highly flavor-active aldehyde, *E*-2-nonenal, which is responsible for the cardboard flavor of aged beers. Both enzymes are extremely heat-sensitive and are extensively lost during most kilning regimes, with *LOX-1* being more resistant (Hugues et al. 1994).

Many changes that occur during kilning affect malt color and flavor. Melanoidin formation involves complex Maillard reaction and Amadori rearrangements involving the condensation of amino acids and reducing sugars, followed by polymerization and a series of reactions that give rise to colored polymeric substances, flavored compounds, aroma compounds, and reductones (Briggs et al. 1981). The presence of sulfur compounds, such as dimethyl sulfide (DMS), is responsible for an unpleasant "cooked vegetable" taste. During kilning, DMS precursor, which is formed during grain germination, is converted to free DMS which is volatilized and lost.

Thermal efficiency is important in kilning because this process consumes 85% to 90% of the energy used in a malting plant. Generally, it is economical to install heat recovery equipment to recover part of the heat being rejected by the kiln for preheating the incoming air. Therefore, it is currently best practice to install cogeneration plants together with malting plants.

3.2.5 SEPARATION OF CULMS

When kilning is complete, the heat is switched off and the grain is allowed to cool before it is stripped from the kiln in a stream of air at ambient temperatures. On its way to steel or concrete hopper-bottomed storage silos, the malt is "dressed" to remove dried rootlets, or culms, which are extremely hygroscopic, and must be stored out of contact with the air. They have a much lower economic value than the malt and are used for feeding cattle. Approximately 3% to 5% of the original barley is recovered as culms in the traditional malting processes (Briggs et al. 1981).

To break and remove rootlets, the cooled malt is agitated with beaters and screened by passing through a revolving perforated drum that retains the kernels and allows the culms to pass out through the perforations. The malt is, at the same time, aspirated with dry air to cool it and to remove the dust.

3.3 MALTING TECHNOLOGY

The technology of malting did not change for several centuries, in that barley that had recovered from dormancy was steeped in cold water for several days and then spread out to germinate on a barn floor. Although floor malting continued as the predominant on-site method of malt production until the late 1960s, many brewers and distillers experimented with novel types of mechanical pneumatic maltings and they have always been at the forefront of malting technology (Dolan 2003). Mechanical or pneumatic maltings were developed to replace floor maltings, in which steeped grain was simply spread out on a floor to grow; such practice was inefficient and labor intensive, and few floors remain today (Lewis and Young 2001).

3.3.1 STEEPING VESSELS

Originally, barley was steeped in rectangular cisterns or tanks throughout one end of the germinating floor. At intervals, the water covering the grain was drained away and was replaced with fresh well water. The steeped barley was then transferred to the floor manually.

Later, cylindroconical steeps (i.e., circular steeps with a vertical walled portion and a conical hopper below) became firmly established. These conical-bottomed steeps are economical in the use of water but are not ideal for grain aeration and consequent extraction of the carbon dioxide. Large simple conical steeps are disadvantageous because of the hydrostatic pressure occurring on the grain in the lower layers, and the difficulty in obtaining even treatment for all the grains in the bulk. The development of systems in which two or three cylindroconical vessels are arranged as a cascade improved the process due to the transfer between vessels increasing aeration and carbon dioxide removal.

The construction of the flat-bottomed steep vessels appeared in France in the 1960s, this being essentially a large, circular open-topped tank with a perforated floor onto which the grain was loaded (Figure 3.2). With capacities of 200 to 300 tonnes of barley, they had the indisputable advantage in the evenness of aeration, carbon dioxide extraction, and consequently, improved control of the temperature. The

FIGURE 3.2 Flat-bottomed steeping vessel with automatic loading and unloading by radial dischargers. (Photograph courtesy of Bühler GmbH.)

major disadvantages were the amounts of water required to fill the plenum before submerging the grain, and the difficulties in cleaning below the perforated floor. The most usual grain leveling and stripping device is called a giracleur, which consists of three arms designed to rotate at a very low speed around a central support and was controlled to rise or descend.

With increased demands on hygiene, there has been a tendency to revert to cylindroconical steeps of stainless steel construction. The current tendency, mainly since the 1990s, is to have a combination of a series of cylindroconical steep vessels with a flat-bottomed vessel with intermediate transfer. Such a compromise allows a substantial water economy in the conical steeps and a rapid and simple transfer to the flat-bottomed vessel, which is more often a pregermination vessel providing a high level of ventilation. Stainless steel is now used extensively for steep construction, together with the associated pipework, ducting, etc., allowing the highest possible standards of hygiene, and minimizing future maintenance costs. In addition, best practices in precleaning the barley before steeping, as well as water filtration and recirculation systems, allow substantial steep water saving and even a reduction in the number of steeps.

3.3.2 GERMINATION VESSELS

The original germination system was based on a thin layer of barley approximately 150 mm thick, spread on a flat-surfaced floor within a well-ventilated room. Temperature was controlled by opening or closing the windows at the sides of the floor, and by thickening or thinning the batch of germinating barley, referred to as the piece. As germination progressed on the floor for up to 21 days, successive pieces were moved in sequence. Such floor maltings had a low productive capacity, were costly in terms of manpower used, and only operated in the wintertime, when ambient temperatures were low. Thus, this method of germination has been replaced except for a small number of plants producing malt for specialty breweries.

Germination chambers nowadays are of two main types: drums or, much more commonly, boxes, which might be 3 to 10 times larger than drums. Drums are long horizontal metal cylinders, supported on rollers, in which the grain may be mixed, leveled, and turned by rotation about the long axis. The grain may be automatically sprayed while being turned. The majority of drums were restricted to approximately 30 tonnes capacity, whereas the capacity of rectangular boxes increased to approximately 150 tonnes or more.

The rectangular germination compartment, still often referred to as a Saladin box, is a rectangular, open-topped container with a false bottom or deck of slotted metal plates or wedge-wire (giving ~20% free space) on which the bed of grain rests. A fan-driven air flow is driven through a humidification and attemperation section and up through the perforated deck and bed of grain, offering excellent control of germination conditions.

Because the Saladin box operated as a batch system, requiring vital time to unload the vessel at the end of each germination cycle, another pneumatic system was invented, the Wanderhaufen or moving piece, which represented the first system of semicontinuous production. Unlike the Saladin boxes, each batch of germinating barley was moved at intervals along the germination box, thereby reducing the loading and unloading times. The Wanderhaufen offered several advantages, especially in the effectiveness of its turning, in its minimal labor equipment, and in maximizing the limited space at the dockside. The Lausmann transposal system is similar in concept, except that the grain is moved forward along a street divided into physically separated compartments. To move grain forward, the floor of a full compartment is gradually raised, while the adjacent floor of an empty compartment is slowly lowered.

Over the last 20 years, there has been a movement toward the introduction of much larger circular stainless steel vessels for germination. These are essentially large enclosed tanks, with the grain loaded onto an intermediate perforated floor (Gibson 2001). The circular germination vessel could be equipped with a fixed floor and an arrangement of turners mounted on a bridge traveling around the vessel, as in the system developed by Nordon (France). The rotating floor, developed by Bühler (Switzerland), in which the turners are arranged on a fixed bridge with the germinating barley moving relative to the turners offer an advantage from the cleaning aspect. The disadvantage of a such a system lies in its cost when compared with a fixed-floor vessel having manual cleaning (Bräuninger et al. 2000). It is now common practice to provide germinating vessels that are completely automated. Both loading and unloading can be achieved in this way, and during processing of the grain, the provision of appropriate sensors can control air volumes, humidity, temperatures, etc. (Gibson 2001).

3.3.3 KILNING VESSELS

A factor to be taken into account in kiln construction is the need for a higher standard of insulation to minimize heat loss. The kilning operation uses more energy, both in the form of heat and electrical power, than any other part of the malting process (perhaps up to 85%–90%). Thermal efficiency is thus very important in kilning. In some kilns, entering air is preheated by exchange with exit air, for useful savings (Lewis and Young 2001). Indirect heating of air for kilns is less efficient than direct firing in which the flue gases of the fuel pass through the malt. However, the combustion products of the fuel do not come in contact with the malt, such as arsenic and sulfur or any oxides of nitrogen, which can react with organic amines of malt to form the potentially carcinogenic nitramines. Therefore, indirect heating of kilns is used in most places to control nitrosamines.

As for the other parts of the process, the classical solution was a rectangular chamber composed of a perforated tipping floor. Rectangular kilns fitted with a venetian blind tipping floor were equipped for indirect firing using a heat transfer fluid or steam. This method of loading required considerable manual effort and this type of kiln disappeared in favor of circular kilns with a capacity of more than 100 tonnes.

In a single-deck kiln, the malt is turned regularly during drying and, therefore, the malt in the bed is relatively constant in moisture and other properties from top to bottom (Figure 3.3). The double-deck malt kiln with transfer of malt from the upper to the lower beds has been particularly appreciated by French and German maltsters up to the 1980s because it permitted batch sizes of up to 150 tonnes with a vessel diameter in the range of 18 to 20 m (Bräuninger et al. 2000). The air-off the

FIGURE 3.3   Circular single-floor kilning vessel. (Photograph courtesy of Bühler GmbH.)

lower kiln is always too hot and too low in volume to enter the upper kiln, which is loaded with more moist malt. The triple-deck kiln re-emerged toward the end of the 1980s because it represented an excellent compromise for tower maltings. It requires relatively simple equipment for loading and unloading compared with, for example, a giracleur. The significant feature of the triple-deck kiln is that the prebreak phase is divided between two kilns, thereby reducing the bed depth and, consequently, the fan power requirements. On reaching the break point, the two batches are transferred to a single kiln for the postbreak drying phase, where the air flow power is reduced due to lower resistance because of the shrinkage which has occurred (Bräuninger et al. 2000).

In some box maltings, the germination chamber may also serve as the kiln, after germination is complete, because the mechanical requirements for air flow and turning the grain are quite similar in both processes. These are called germination and kilning vessels, also known as "Flexies" in the United States and are popular in Great Britain and in Southern Africa (Lewis and Young 2001; Bräuninger et al. 2000). By reducing the three phases of the malting process into one automated steeping, germinating, and kilning vessel, the capital cost of these units should be competitive with improved hygiene and flexibility. Total quality, homogeneity, and traceability may be achievable, considering that the energy consumption is at least as good as current systems. Such an integrated and automated layout is particularly interesting in small and specialty products' malting plants.

An arrangement in a tower has the advantages of saving space, and allowing very efficient distribution of services and comparatively easy heat recovery. Such towers may have steeps at the top and the kilns are housed in the base of tower, whereas in others they are in a separate building (Briggs et al. 1981). The choice between tower and horizontal plant layout is mostly dependent of landfill prices or availability.

3.4 SPECIALTY AND COLORED MALTS

Colored and specialty malts are produced not for their enzyme content but rather for use by the brewer in relatively small quantities as a source of extra color, and distinct types of flavor are manufactured by using extreme variations of the basic procedures. There is much interest in these products for the opportunities they present for brewing different styles of beer (Bamforth 2003). Small percentages of special malts are used in grists for some beers, to which they impart color, special flavors and aromas, high antioxidant power, and palate fullness (Briggs et al. 1981).

Table 3.1 lists the products that result from special malting procedures applied to a range of raw materials, usually grouped into three types based on the material used, namely, raw cereals, pale malt, or green malt (germinating cereal). The product names and specifications shown in Table 3.1 may vary according to the manufacturer. It is possible to employ specialized roasting equipment for making colored products, however, the batch sizes of colored malts are often generally much smaller, typically 1 or 2 tonnes, than with pale malt products. In addition, colored malts often require conditions beyond the normal range of operation of the kiln with regard to the air-on temperature, the ability to operate with full air recirculation to produce a

TABLE 3.1
Specialty Malt Products

Product Type	Color Range (EBC units)	Moisture (%)	Descriptive
Roasted barley product	1440–1800	<3.5	Astringent, burnt, smoky
Pale malt products			
Amber malt	48–96	<3.5	Dry, baked
Chocolate malt	1200–1440	<3.0	Mocha, treacle, chocolate
Black malt	1440–1680	<3.0	Smoky, coffee
Green malt products			
Cara malt	25–40	<7.5	Sweet, caramel
Crystal malt	90–360	~3.5	Malty, caramel, toffee-like
Dark crystal malt	120–150	<3.5	Burnt toffee, caramel
Caramel malt	260.320	<3.5	Burnt toffee, caramel
Colored kilned malts			
Munich malt	10–15	~3.8	Grainy, malty (marked)
Vienna malt	7–10	~4.5	Grainy, malty (subtle)

stewing action, and to be able to offer a high degree of control over the process. An exception is the case of Munich and Vienna malts, which may be made using conventional kilns (Bräuninger et al. 2000).

To obtain the roasted barley, the dry grain is introduced into the drum roaster and is heated gradually over a period of approximately 90 min to a temperature of 215°C. At this temperature, heating is stopped and the temperature allowed rising to within the range 225°C to 230°C. As the maximum temperature is reached, water is injected into the roasting drum to lower the temperature and to prevent the combustion. The product, with a color of 1440 to 1800 EBC units (units for the color of beer defined by the European Brewery Convention), has a very astringent, almost burnt flavor, which is noticeable in traditional stouts.

Roasted malts are produced in a similar way to that of roast barley with dry pale malt introduced into the drum drier. Amber malts are prepared by kiln-drying well-modified malt to 3% to 4% moisture, and then by ambering in the kiln or a roasting drum by heating rapidly to approximately 93°C, in 15 to 20 min, and then gradually to 138°C to 149°C. The higher temperature is maintained until the correct color, 48 to 96 EBC units, is attained. This leaves an amber malt with a pleasant dry, baked flavor. Chocolate and black malt is prepared from plump uniform barleys having moderate nitrogen contents (1.5%–1.6%), germinated for 4 to 6 days, kiln-dried to approximately 5% moisture at which stage they may be stored. Later, they are roasted at a high temperature (221°C–233°C) for 2 to 2.5 h in a slowly rotating metal cylinder. The product should be uniform in appearance without charred grains with a color of 1200 to 1440 EBC units (chocolate malt) or 1440 to 1680 EBC units (black malt), and should contain 1.5% to 3.5% moisture. Chocolate malt shares many of the characteristics of black malt but, because it is roasted for a slightly shorter period and end temperatures are not so high, some of the harsher flavors of black malt are not so pronounced and color is 200 EBC lighter.

For producing crystal and caramel malts (green malt products), a fully modified malt, rich in sugars, is wetted or rewetted after light kilning, and the wet material (up to 50% moisture) is introduced in the roasting drum where it is warmed to approximately 65°C for 30 min. The combination of elevated temperatures and humidity encourages amylolysis and proteolysis. The drum is then opened and hot air is allowed through the cylinder to remove the free moisture from the grain for 50 min. The kiln temperature is then raised to between 120°C and 160°C, depending on the product type, with frequent color checks being made on the malt. The product has a color of 90 to 360 EBC units and contains about 3% moisture. The conversion of starch to sugar during processing produces a glassy crystallized endosperm that dominates its flavor profile, which has been described as malty, caramel, toffee-like, biscuity, and roasted. A pale crystal malt called cara malt (25–40 EBC units) has an almost completely glassy endosperm. It contains a greater degree of sweetness and stronger caramel flavor than crystal malt and the harsher nutty roasted flavors are not present (Bräuninger et al. 2000; Briggs et al. 1981).

Dark malts of the Munich type are often made from barley of somewhat higher protein content. The barley is fully steeped, the final moisture content being 45% or higher, and the germination is carried at higher temperatures and for a longer time than those used for pale malt. The green malt is subject to a stewing stage after loading onto the kiln where the recycling of saturated exhaust air raises the kiln temperature up to 100°C to 105°C (Schuster 1962). Vienna malt is made in a similar manner to Munich malt but with less intensive stewing and kilning. Consequently, the result is less highly colored malt (7–10 EBC units) with higher extract and soluble nitrogen (Bräuninger et al. 2000). The use of Munich and Vienna malt imparts body, sweetness, and a slightly reddish hue to the beer.

Recently, new special malts enriched in melanoidins could be found in certain malt suppliers. Melano malt is produced by a special germination process followed by a slow drying up to 130°C, allowing the melanoidins to form as part of the kilning process. The product, with a color of 75 to 85 EBC units, gives fullness and roundness to the beer color, improves flavor stability and promotes a red color in beer (Castle Malting 2011).

Malt whisky is made from malt having a high extract—and hence is well modified and has a moderate or low nitrogen content. Peat is burnt on the kiln so that the peat smoke condensate (reek) settles on the malt, increasing its phenol content and giving it a characteristic flavor. For reasons of tradition, gibberellic acid is not used in the preparation of this malt. Malt used for the production of grain whisky may have a high nitrogen content of 1.8% or more and may be made from six-rowed barley. For distilling purposes, as much carbohydrate as possible is fermented to alcohol. To achieve this, maximum enzyme development and survival is required during the preparation of the malt. Thus, the grain may be steeped and sprayed to a moisture content of 48% to 50% and is traditionally given a long period of low-temperature germination to yield the maximum quantity of enzymes. These are preserved by drying the malt at a low temperature (53°C–54.5°C; Briggs et al. 1981).

3.5 FUNCTIONAL PROPERTIES OF MALTED BARLEY

Interest in barley as a raw material for production of nutritiously valuable healthy foods was underlined in May 2006 when the U.S. Food and Drug Administration (FDA 2006) confirmed the qualified health claim linking barley to the prevention of cardiovascular diseases (as a result of high content of food fiber).

Barley is an excellent source of soluble and insoluble dietary fiber and other bioactive constituents, such as vitamin E (including tocotrienols), B complex vitamins, minerals, and phenolic compounds (Izydorczyk and Dexter 2008; Belcredi et al. 2009). Moreover, research on grain nutritious composition and clinical studies have demonstrated that consumption of barley grain is an effective tool for the control of type II diabetes, the digestive system diseases, disrupted immunity of the organism, and other civilization diseases (Manach et al. 2004; Liu 2007). Therefore, it is predicted that the health benefits of barley will stimulate interest among food producers and consumers in using barley for food purposes. Most of the more recent studies focus on the health benefits attributed to high amounts of dietary fibers (Li et al. 2003a,b; Izydorczyk and Dexter 2008). However, several reports have shown that antioxidant compounds are also one of the major health-promoting components in barley and malt (Dvorakova et al. 2008; Dabina-Bicka et al. 2011).

3.5.1 DIETARY FIBERS

In a barley grain, the content of dietary soluble fiber is mainly formed by β-glucans and arabinoxylans. The β-glucan contents range from 2.5% to 11.3%, which is generally higher than in other cereals, such as in oats (2.2%–7.8%), rye (1.2%–2.0%), and wheat (0.4%–1.4%) (Izydorczyk and Dexter 2008). The effectiveness of barley β-glucans in barley food products in lowering blood cholesterol (Behall et al. 2006; Choi et al. 2010) and glycemic index (Li et al. 2003b), and reducing the risk of heart disease (Liu et al. 2002) has been reported in several publications and is widely accepted. Nevertheless, despite the numerous health-promoting properties of barley, barley fiber, and particularly β-glucans, they are not desirable for the brewing and malting industries. They increase the density of malt or beer, retard wort and beer filtration, and cause nonbacterial colloidal hazes and sediments in the final product (Vis and Lorenz 1998; Zhang et al. 2001). Additionally, high β-glucan content is not favorable for animal feed either, especially the broiler industry (Newman and Newman 2008; Belcredi et al. 2009).

The nutritional values of other fiber components in barley, most notably arabinoxylans, have not been investigated to the same extent as those of β-glucans. However, recent studies revealed the positive effects of water-soluble maize, wheat and rye arabinoxylans on cecal fermentation, production of short-chain fatty acids, reduction of serum cholesterol, and improved adsorption of calcium and magnesium (Izydorczyk and Dexter 2008). Li et al. (2003a) have also reported that barley's insoluble fibers may be beneficial in helping the body maintain regular bowel function.

3.5.2 VITAMINS

Malt ingredients are all-natural and made only from grain and water, and so are a rich source of vitamins. They are mostly located in the embryo but also in the aleurone layer, and some partly move into the rootlets during malting and are lost to the malt (Briggs et al. 1981). Most of the vitamins found belong to the B complex group, such as thiamin (vitamin B_1), riboflavin (B_2), pyridoxine (B_6), and folate (B_9) (Hucker et al. 2012). B vitamins are wellknown to have multiple health benefits, and many nutritionists recommend taking a vitamin B complex supplement to ensure that one is receiving the amounts necessary for maintaining good physical health and mental well being (Fardet 2010). Thiamine (B_1) works to promote healthy nerves, improves mood, strengthens the heart, and may restore peripheral vascular resistance, whereas riboflavin (B_2) helps counteract migraines and participates in antioxidant defenses, and is possibly anticarcinogenic (for example, reduction in DNA damage). Pyridoxine (B_6) is needed for almost every function in the body, working as a coenzyme for numerous enzymes; plays a major role in heart, mental, and brain health (for example, depression, fatigue, insomnia, and epileptiform convulsions) (Fardet 2010). Other vitamins present include folates (B_9) and related substances, such as nicotinic acid. According to Jägerstad et al. (2005), the typical level of folate in unmalted cereals was between 0.5 and 1.0 mg/kg, but during the malting process, the folate levels in barley increased twofold to threefold, particularly during the first 2 days of germination. In the last decade, it has become clear that increasing folate intake may provide safer pregnancies and maintenance of normal plasma homocysteine levels and cognitive functions, as well as protection toward certain cancer forms, notably colon cancer (Feng Ma et al. 2003; Eitenmiller and Landen 1998). Moreover, a recent research from van der Gaag and collaborators (2000) reported that serum homocysteine increases after moderate consumption of red wine and spirits, but not after the moderate consumption of beer. Vitamin B_6 in beer seems to prevent the alcohol-induced increase in serum homocysteine. Furthermore, this increase in plasma vitamin B_6, as seen after beer and to a lesser extent after red wine and spirits consumption, may give beer drinkers additional protection against cardiovascular disease.

Vitamin E (α-tocopherol) is also a monophenolic compound present in barley and malt, which can quench free radicals. Vitamin E is the generic term used to describe a family of eight lipid-soluble antioxidants with two types of structures, the tocopherols (α-tocopherol, β-tocopherol, γ-tocopherol, δ-tocopherol) and tocotrienols (α-tocotrienol, β-tocotrienol, γ-tocotrienol, and δ-tocotrienol) (Liu 2007). Their antioxidant activity is based mainly on the tocopherol–tocopheril quinone redox system. The most important functions of vitamin E in the body are antioxidant activity and maintenance of membrane integrity (Randhir et al. 2008).

3.5.3 MINERALS

Minerals in barley and malt have nutritional importance and may also offer additional health benefits. They originate from the embryo and the aleurone layer in the endosperm, and constitute 3% of the grain's dry weight. About 75% of minerals are derived from the malt, whereas the remaining 25% originates from the water. The minerals include approximately 35% phosphates (expressed as phosphorous (5)-oxide P_2O_3), approximately 25% silicates (expressed as silicon dioxide, SiO_2), and approximately 20% potassium salts (expressed as potassium oxide, K_2O). The mineral composition of the malt depends on the variety, place where it was grown, atmospheric conditions, growing techniques and harvest, storage, and malting systems. Additionally, the malting technique is particularly important (Preedy 2009).

It was also reported that malt is low in calcium and is rich in magnesium, which may help protect against gallstones and kidney stone formation. This may be one reason why daily consumption of a glass of beer was reported to reduce the risk of kidney stones (Bamforth 2002).

Feng Ma et al. (2003) investigated the genotypic variation in silicon content of barley grain, which is thought to be essential in the diet for the proper development of collagen and bones. They also reported that silicon in the diet may prevent the absorption of aluminum in the gut, which is thought to be implicated in the etiology of neurodegenerative disorders such as Alzheimer's disease.

3.5.4 ANTIOXIDANTS

The antioxidant compounds present in barley extracts are complex, and their activities and mechanisms would largely depend on the composition and conditions of the test systems. Approximately 80% of the phenolic compounds present in beer are derived from barley malt, and the remaining comes from hops (Goupy et al. 1999). Polyphenols are concentrated in husks, pericarp, and testa with their contents ranging from 0.1% and 0.3% grain dry weight (Preedy 2009). Those phenolic compounds in malting barley include phenolic acids (benzoic and cinnamic acid derivatives), flavonoids, proanthocyanidins, tannins, and amino phenolic compounds (Zhao et al. 2008; Goupy et al. 1999). Flavan-3-ols constitute the major class of phenolics in barley and malt. They appear in monomer form, (+)-catechin and (−)-epicatechin, and polymer forms constituted mainly by units of (+)-catechin and (+)-gallocatechin. Monomeric, dimeric, and trimeric flavan-3-ols accounted for 58% to 68% of the total phenolic content, with a predominance of trimeric flavan-3-ols (Goupy et al. 1999; Dvorakova et al. 2008). Hydroxybenzoic and hydroxycinnamic acids, as free forms and as glycosidic esters, are also present in barley grains. These phenolic acids are known equally as primary antioxidants acting as free radical acceptors and chain breakers (Dvorakova et al. 2008). *trans*-Ferulic acid is the major compound besides *p*-coumaric and vanillic acids that have also been detected. They were mainly present in the aleurone layer and endosperm, and their glycosidic esters were detected in the husk, testa, and aleurone cells. Only trace amounts of insoluble bound phenolic acids were found in the endosperm of barley and malt. As previously mentioned, malt also contains several Maillard reaction products, which also possess antioxidant properties, playing a significant role in the malting and brewing process (Briggs et al. 1981; Preedy 2009; Shahidi and Naczk 2003). A recent study showed that finished malt from the hottest kilning regiment possessed the highest antioxidant activity, attributed to higher levels of Maillard reaction products. Modifying kilning conditions leads to changes in the release of bound ferulic acid and antioxidant activity with potential beneficial effects on flavor stability in malt and beer (Inns et al. 2011). However, the beneficial effects of melanoidins on the beer's flavor stability are far from being consensual, as dark sweet worts were less stable with high radical intensities, high Fe content, and a decreased ability to retain volatiles, suggesting that the Maillard reaction compounds formed during the roasting of malt are pro-oxidants (Hoff et al. 2012).

3.6 CURRENT TRENDS IN MALTING BARLEY PRODUCTION

In the future, it is imperative that the quality features of brewing cereals are defined in terms of genetic and environmental importance. The genetic, physiological, and structural mechanisms that control grain yield, germination, malt modification homogeneity, and product quality should be investigated thoroughly. Plant variety improvements in terms of productivity, pest resistance, extract content, and enzymatic activities are an important subject of crop innovation that will have a great relevance for maltsters.

Presently, there is a preference for cylindroconical steeps, with effective aeration and carbon dioxide extraction; this affords the best option for optimum hygienic conditions. Alkaline steeping seems to be a useful process for the development of gluten-free malt that could be employed for gluten-free beer production. Beers made from gluten-free raw materials (millet, sorghum, and buckwheat) have been in the market since 2006 (De Meo et al. 2011).

In the technology of kiln construction, one problem of great importance is how to obtain a uniform flow/temperature through the grain bed in the kiln. The bigger the kiln unit, the more difficult it appears to resolve the problems of malt heterogeneity and energy efficiency.

During the malting process, fungi naturally present in barley or in the malt house can lead to grain contamination by toxic metabolites, such as mycotoxins. Recent researches have shown that the addition of selected starter cultures in steeping water could prevent the synthesis of these toxins (Molimard et al. 2005; Mauch et al. 2011). Fungal infection of barley and malt is recognized to be a direct cause of spontaneous overfoaming of beer, a phenomenon known as gushing. Small fungal proteins, called hydrophobins, were found to act as gushing factors in beer (Sarlin et al. 2012). Yeasts isolated from industrial maltings can suppress *Fusarium* growth and the formation of these gushing factors (Laitila et al. 2007). The use of nitrogen or other appropriate gases to reduce the metabolic activity of microorganisms has been proposed as a means of stabilizing barley during storage. The possibility of introducing fragments of genetic material, which contain useful genes for improved disease resistance, has been extensively investigated.

Promising new high-quality barley breeding lines have been recently developed. As already discussed in Section 3.2.2, LOX catalyzes the hydroperoxidation of polyunsaturated fatty acids and serves as a precursor of E-2-nonenal that causes an off-flavor in beer. By performing mutation breeding, a *LOX-1*–null barley line was obtained, which can be effectively used to improve the flavor stability of beer without changing the other important beer qualities (Hirota et al. 2006). Diastatic power, a measure of the combined levels of amylolytic enzymes, is an important determinant of malt quality. Breeding lines with extremely high diastatic power, derived from *lys1* line *Yon Rkei 1363*, can now be easily selected (Nagamine and Kato 2008).

In recent years, a new technology, which allows massively parallel analysis on a single device, the so-called DNA chip, has become available. A barley gene chip array was used to conduct transcript profiling of differentially expressed genes during malting and the results provide candidate genes for malting quality phenotypes that need to be functionally validated (Lapitan et al. 2009). Approximately 50 of each gene conferring monogenic resistances and hundreds of quantitative trait loci for quantitative disease resistances have been recently reported in wheat and barley

(Miedaner and Korzun 2012). In the future, chip-based, high-throughput genotyping platforms and the introduction of genomic selection will open new avenues for molecular-based resistance breeding.

ACKNOWLEDGMENTS

The authors are grateful to Professor Paulo Almeida (FCUP) and Dr. Tiago Brandão (UNICER) for their precious help in reviewing this chapter. Manuela M. Moreira wishes to acknowledge the support of Fundação para a Ciência e a Tecnologia (FCT) for her PhD studentship (SFRH/BD/60577/2009).

REFERENCES

Bamforth, C.W. 2002. Nutritional aspects of beer—a review. *Nutrition Research* 22 (1–2):227–237.

Bamforth, C. 2003. *Beer: Tap Into the Art and Science of Brewing*. 2nd ed. New York: Oxford University Press, Inc.

Behall, K.M., D.J. Scholfield, and J.G. Hallfrisch. 2006. Barley β-glucan reduces plasma glucose and insulin responses compared with resistant starch in men. *Nutrition Research* 26 (12):644–650.

Belcredi, N., J. Ehrenbergerová, S. Běláková, and K. Vaculová. 2009. Barley grain as a source of health-beneficial substances. *Czech Journal of Food Sciences* 27:S242–S244.

Bräuninger, U., R. Brissart, S. Haydon, R. Morand, G. Palmer, R. Sauvage, and B. Seward. 2000. *Malting Technology: Manual of Good Practice*, edited by EBC Technology and Engineering Forum. Nürnberg: Fachverlag Hans Carl.

Briggs, D.E., J.S. Hough, R. Stevens, and T.W. Young. 1981. *Malting and Brewing Science: Malt and Sweet Wort*. 2nd ed. Vol. 1. London: Chapman & Hall.

Castle Malting. Specification Malt Château Melano Light (crop 2011). http://www.castle malting.com/Publications/SPECS_Malt_Chateau_Melano_Light_Crop2011_EN.pdf (accessed March 15, 2013).

Choi, J.S., H. Kim, M.H. Jung, S. Hong, and J. Song. 2010. Consumption of barley β-glucan ameliorates fatty liver and insulin resistance in mice fed a high-fat diet. *Molecular Nutrition and Food Research* 54 (7):1004–1013.

Dabina-Bicka, I., D. Karklina, and Z. Kruma. 2011. Polyphenols and vitamin E as potential antioxidants in barley and malt. Paper read at the 6th Baltic Conference on Food Science and Technology "FOODBALT-2011", at Latvia.

De Meo, B., G. Freeman, O. Marconi, C. Booer, G. Perretti, and P. Fantozzi. 2011. Behaviour of malted cereals and pseudo-cereals for gluten-free beer production. *Journal of the Institute of Brewing* 117 (4):541–546.

Dolan, T.C. 2003. Malt whiskies: Raw materials and processing. In *Whisky: Technology, Production and Marketing*, edited by I. Russell. London: Academic Press.

Dvorakova, M., M.M. Moreira, P. Dostalek, Z. Skulilova, L.F. Guido, and A.A. Barros. 2008. Characterization of monomeric and oligomeric flavan-3-ols from barley and malt by liquid chromatography–ultraviolet detection–electrospray ionization mass spectrometry. *Journal of Chromatography A* 1189 (1–2):398–405.

Eitenmiller, R.R., and W.O. Landen, Jr. 1998. Folate. In *Vitamin Analysis for the Health and Food Sciences*. Boca Raton, FL: CRC Press.

Fardet, A. 2010. New hypotheses for the health-protective mechanisms of whole-grain cereals: what is beyond fibre? *Nutrition Research Reviews* 23 (1):65–134.

FDA (U.S. Food and Drug Administration). 2006. Food labeling: health claims; soluble dietary fiber from certain foods and coronary heart disease. *Federal Register* 71 (98):29248–29250.

Feng Ma, J., A. Higashitani, K. Sato, and K. Takeda. 2003. Genotypic variation in silicon concentration of barley grain. *Plant and Soil* 249 (2):383–387.

Gibson, G. 2001. Malting. In *Cereals Processing Technology*, edited by G. Owens. Cambridge, MA: Woodhead Publishing Limited.

Goupy, P., M. Hugues, P. Boivin, and M.J. Amiot. 1999. Antioxidant composition and activity of barley (*Hordeum vulgare*) and malt extracts and of isolated phenolic compounds. *Journal of the Science of Food and Agriculture* 79 (12):1625–1634.

Guido, L.F., P. Boivin, N. Benismail, C.R. Goncalves, and A.A. Barros. 2005. An early development of the nonenal potential in the malting process. *European Food Research and Technology* 220 (2):200–206.

Guiga, W., P. Boivin, N. Ouarnier, F. Fournier, and M. Fick. 2008. Quantification of the inhibitory effect of steep effluents on barley germination. *Process Biochemistry* 43 (3):311–319.

Hirota, N., H. Kuroda, K. Takoi, T. Kaneko, H. Kaneda, I. Yoshida, M. Takashio, K. Ito, and K. Takeda. 2006. Brewing performance of malted lipoxygenase-1 null barley and effect on the flavor stability of beer. *Cereal Chemistry* 83 (3):250–254.

Hoff, S., M.N. Lund, M.A. Petersen, B.M. Jespersen, and M.L. Andersen. 2012. Influence of malt roasting on the oxidative stability of sweet wort. *Journal of Agricultural and Food Chemistry* 60 (22):5652–5659.

Hucker, B., L. Wakeling, and F. Vriesekoop. 2012. Investigations into the thiamine and riboflavin content of malt and the effects of malting and roasting on their final content. *Journal of Cereal Science* 56 (2):300–306.

Hugues, M., P. Boivin, F. Gauillard, J. Nicolas, J.M. Thiry, and F. Richardforget. 1994. Two lipoxygenases from germinated barley—heat and kilning stability. *Journal of Food Science* 59 (4):885–889.

Inns, E.L., L.A. Buggey, C. Booer, H.E. Nursten, and J.M. Ames. 2011. Effect of modification of the kilning regimen on levels of free ferulic acid and antioxidant activity in malt. *Journal of Agricultural and Food Chemistry* 59 (17):9335–9343.

Izydorczyk, M.S., and J.E. Dexter. 2008. Barley β-glucans and arabinoxylans: molecular structure, physicochemical properties, and uses in food products—a review. *Food Research International* 41 (9):850–868.

Jägerstad, M., V. Piironen, C. Walker et al. 2005. Increasing natural food folates through bioprocessing and biotechnology. *Trends in Food Science and Technology* 16 (6–7):298–306.

Laitila, A., T. Sarlin, E. Kotaviita, T. Huttunen, S. Home, and A. Wilhelmson. 2007. Yeasts isolated from industrial maltings can suppress *Fusarium* growth and formation of gushing factors. *Journal of Industrial Microbiology and Biotechnology* 34 (11):701–713.

Lapitan, N.L.V., A. Hess, B. Cooper, A.M. Botha, D. Badillo, H. Iyer, J. Menert, T. Close, L. Wright, G. Hanning, M. Tahir, and C. Lawrence. 2009. Differentially expressed genes during malting and correlation with malting quality phenotypes in barley (*Hordeum vulgare* L.). *Theoretical and Applied Genetics* 118 (5):937–952.

Lewis, M.J., and T.W. Young. 2001. Malting technology: malt, specialized malts, and non-malt adjuncts. In *Brewing*. New York: Kluwer Academic.

Li, J., T. Kaneko, L.-Q. Qin, J. Wang, and Y. Wang. 2003a. Effects of barley intake on glucose tolerance, lipid metabolism, and bowel function in women. *Nutrition* 19 (11–12):926–929.

Li, J., T. Kaneko, L.-Q. Qin, J. Wang, Y. Wang, and A. Sato. 2003b. Long-term effects of high dietary fiber intake on glucose tolerance and lipid metabolism in GK rats: comparison among barley, rice, and cornstarch. *Metabolism* 52 (9):1206–1210.

Liu, R.H. 2007. Whole grain phytochemicals and health. *Journal of Cereal Science* 46 (3):207–219.

Liu, S., J.E. Buring, H.D. Sesso, E.B. Rimm, W.C. Willett, and J.E. Manson. 2002. A prospective study of dietary fiber intake and risk of cardiovascular disease among women. *Journal of the American College of Cardiology* 39 (1):49–56.

Manach, C., A. Scalbert, C. Morand, C. Rémésy, and L. Jiménez. 2004. Polyphenols: food sources and bioavailability. *The American Journal of Clinical Nutrition* 79 (5):727–747.

Mauch, A., F. Jacob, A. Coffey, and E.K. Arendt. 2011. Part I. The use of *Lactobacillus plantarum* starter cultures to inhibit rootlet growth during germination of barley, reducing malting loss, and its influence on malt quality. *Journal of the American Society of Brewing Chemists* 69 (4):227–238.

Miedaner, T., and V. Korzun. 2012. Marker-assisted selection for disease resistance in wheat and barley breeding. *Phytopathology* 102 (6):560–566.

Molimard, P., C. Buchet, and P. Boivin. 2005. Use of *Geotrichum candidum* Starters during the Malting Process for Malt Bioprotection. Paper read at European Brewery Convention Congress at Prague.

Nagamine, T., and T. Kato. 2008. Recent advances and problems in malting barley breeding in Japan. *JARQ-Japan Agricultural Research Quarterly* 42 (4):237–243.

Newman, R.K., and C.W. Newman. 2008. *Barley for Food and Human Nutrition: Science, Technology, and Products.* 1st ed. New Jersey: Wiley.

Palmer, G.H. 1994. Standardisation of homogeneity. In *Monograph XXIII–European Brewery Convention.* Nürnberg: Verlag Hans Carl Getränke-Fachverlag.

Palmer, G.H. 2000. Malt performance is more related to inhomogeneity of protein and β-glucan breakdown than the standard malt analyses. *Journal of the Institute of Brewing* 106 (3):189–192.

Panteloglou, A.G., K.A. Smart, and D.J. Cook. 2012. Malt-induced premature yeast flocculation: current perspectives. *Journal of Industrial Microbiology and Biotechnology* 39 (6):813–822.

Pollock, J.R.A. 1962. The nature of the malting process. In *Barley and Malt–Biology, Biochemistry, Technology*, edited by A.H. Cook. London: Academic Press Inc.

Preedy, V.R. 2009. In *Beer in Health and Disease Prevention.* San Diego: Academic Press.

Randhir, R., Y. Kwon, and K. Shetty. 2008. Effect of thermal processing on phenolics, antioxidant activity and health-relevant functionality of select grain sprouts and seedlings. *Innovative Food Science and Emerging Technologies* 9 (3):355–364.

Sarlin, T., T. Kivioja, N. Kalkkinen, M.B. Linder, and T. Nakari-Setala. 2012. Identification and characterization of gushing-active hydrophobins from *Fusarium graminearum* and related species. *Journal of Basic Microbiology* 52 (2):184–194.

Schuster, K. 1962. Malting technology. In *Barley and Malt–Biology, Biochemistry, Technology*, edited by A.H. Cook. London: Academic Press Inc.

Shahidi, F., and M. Naczk. 2003. In *Phenolics in Food and Nutraceuticals.* Boca Raton, FL: CRC Press.

van der Gaag, M.S., J.B. Ubbink, P. Sillanaukee, S. Nikkari, and H.F.J. Hendriks. 2000. Effect of consumption of red wine, spirits, and beer on serum homocysteine. *The Lancet* 355 (9214):1522.

Vis, R.B., and K. Lorenz. 1998. Malting and brewing with a high β-glucan barley. *LWT—Food Science and Technology* 31 (1):20–26.

Wilhelmson, A., A. Laitila, A. Vilpola, J. Olkku, E. Kotavita, K. Fagerstedt, and S. Home. 2006. Oxygen deficiency in barley (*Hordeum vulgare*) grain during malting. *Journal of Agricultural and Food Chemistry* 54 (2):409–416.

Zhang, G., J. Chen, J. Wang, and S. Ding. 2001. Cultivar and environmental effects on $(1\rightarrow3,1\rightarrow4)$-β-D-glucan and protein content in malting barley. *Journal of Cereal Science* 34 (3):295–301.

Zhao, H., W. Fan, J. Dong, J. Lu, J. Chen, L. Shan, Y. Lin, and W. Kong. 2008. Evaluation of antioxidant activities and total phenolic contents of typical malting barley varieties. *Food Chemistry* 107 (1):296–304.

4 Rice Processing

*Anshu Singh, Mithu Das, Satish Bal,
and Rintu Banerjee*

CONTENTS

4.1 INTRODUCTION

With the increase in population, the prime objective of the agricultural sector is to attain food sufficiency to cope with the demand. Rice is the staple diet as it meets 21% of the calorific needs of the current world population (Fitzgerald et al. 2009). Therefore, sustainability in rice production is highly essential. Rice is cultivated in agro-ecological zones of many continents, but Asia has a major share in the world rice production, with China and India together covering a major geographical rice-producing region among Asian countries. The rice production for 2011 was 480.5 million t (FAO 2011) with an expected 3% increase, that is, by about 14 million t during 2011 to 2012 (http://www.fao.org/worldfoodsituation/wfs-home/csdb/en/).

The recent increase in rice production clearly signifies its importance in the agriculture and food sector, but still the demand is on the higher side and the factors responsible for the gap between production and demand could be listed as follows:

- Lack of irrigation facilities
- Preference of consumers for white head rice
- Increasing urbanization
- Climatic and weather variations
- Fungal and pest attacks

Rice crop belongs to semiaquatic grass cultivars, which can grow in deep water to dry land under diverse climatic and biotic environments. These grasses from the Graminaceae family are known for their single-seeded fruit biologically termed as *caryopsis*. Rice caryopsis consists of a distinct three-layered coat surrounding the endosperm and germ. It includes pericarp, nucellus, and seed coat, which are composed of crushed cells of varying thickness. In the case of colored rice, pigments reside in either the pericarp or seed coat. Next to the caryopsis coat lies the morphologically distinct aleurone layer of endospermic tissue. The thickness of the aleurone layer varies with the size and shape of the grain. Long grain rice has thin layers compared with the coarser, short grain rice. The aleuronic layer of rice stores phosphate and lipid. Starchy endospermic tissue consists of large polygonal starch granules and protein bodies in the subaleuronic layers. Glutein and prolamin are commonly occurring proteins in rice, in which glutein makes up 80% of the total protein. The amylose content in rice starchy endosperm affects the grain processing quality and governs its palatability. Apart from surrounding the starchy endosperm, the aleurone layer also covers the embryo, which is located on the ventral side of the caryopsis and consists of the scutelleum and embryonic axes.

Rice is always in demand both in urban and rural areas, but its palatability and acceptance depend on certain characteristics, which are either physical or biochemical in nature.

4.1.1 PHYSICAL CHARACTERISTICS OF RICE

Rice grains possess heterogeneous morphological profiles in terms of their size, shape, and color. All these characteristics are polygenic in nature and are easily

TABLE 4.1
Approximate Distribution of Rice Kernels on the Basis of its Grain Shape

Rice Grain		Length (Longitudinal Dimension) [mm]	Width (Dorsiventral Diameter) [mm]	Length to Width Ratio
Long	Rough rice	10–7.9	2.8–2.3	>3
	Brown rice	7.9–6.6	2.4–2.1	>3
Medium	Rough rice	8.4–7.9	3.2–2.3	2.1–2.9
	Brown rice	6.5–5.9	2.9–2.5	2.1–2.9
Short	Rough rice	7.5–7.4	3.6–3.1	2.1–2.4
	Brown rice	5.5–5.4	2.8–3.0	<2

Source: Bai, X. et al., *BMC Genetics*, 11:16, 2010; Slaton, N. et al., Grain characteristics of rice varieties. In *Cooperative Extension Service Rice Information*, pp. 1–8, 2000; Koutroubas, S.D. et al., *Field Crops Res.* 86:115–130, 2004.

influenced by the disparity in the environmental factors. Grain dimension is an important parameter that determines the commercial quality of rice (Khan et al. 2009). However, there is no unique and rapid method or equipment to measure the dimensions precisely. The most common and practical measure used by rice industries is manual and visual analysis. Although this approach is not accurate, it is a common choice because it provides quick results. Grain weight is generally estimated by 1000-grain weight, whereas grain shape includes the analysis of length, width, and length-to-width ratio. Accordingly, the rice grains are classified into long, medium, and short (Table 4.1).

Choice of rice varies with the country; the short and oblique or long and elongated intact rice grains with a smooth, translucent, white surface and with minimum fractures or cracks are considered to be the best (Wan et al. 2008). Pattern recognition concepts for morphological feature analysis are being used for rice characterization (Aggarwal and Mohan 2010). Wee et al. (2009) used another algorithm based on Zernike moment for rice sorting based on their color and shape. Evaluation of the physical characteristics of rice is the most important consideration for its storage, drying, and milling. Some of the parameters are analyzed, starting from rough rice to milled rice. The parameters presented in Table 4.2 need to be studied with a view toward standardizing the process.

4.1.2 BIOCHEMICAL CHARACTERISTICS OF RICE GRAIN

Like other cereal grains, the main constituent of the rice grain is starch, the properties of which directly affect the cooking qualities. This homopolysaccharide structure consists of a linear chain of amylose and a highly branched amylopectin chain. The ratio of amylose–amylopectin is mainly responsible for the physicochemical nature of starch and the texture of cooked rice. Starch granules rich in amylose content have a better ability to absorb water and expand in volume. The volume

TABLE 4.2
Physical Characteristics and their Role in Processing

Physical Properties	Effects
Principal axial dimensions	Facilitate selection of sieve and in calculating power during the rice milling process
Surface area and volume	Important for grain drying, aeration, heating, and cooling
Thousand grain weight	Calculation of the head rice yield
Bulk density and true density	Useful in sizing grain hoppers and storage facilities
Porosity	Affects the rate of heat and mass transfer of moisture during aeration and drying processes

Source: Zareiforoush, H. et al., *Res. J. Appl. Sci., Eng, Tech.* 1:132–139, 2009.

expansion is possible due to the greater capacity of amylose to form a hydrogen bond. Waxy or glutinous starch has higher amylopectin content whereas nonwaxy or non-glutinous starch bears high amylose content. Long chains of amylopectin in starch granules help in resisting the swelling and breakage of rice grains during cooking.

Apart from starch granules, rice proteins also play a significant role in quality characteristics. Based on their studies, Liang and King (2003) concluded that the amino acids interact with starch resulting in the modification of the latter.

The lipid or oil content of rice is in the form of lipid bodies or spherosomes (0.1–1 mm) concentrated in the bran fraction, where it can contribute up to 20% by mass. Unsaturated fatty acids, like oleic acid and linoleic acid in rice bran oil, are known to lower blood cholesterol levels. Bran is also a rich source of vitamins and minerals (phosphorus, potassium, magnesium, calcium, zinc, sodium, and iron). Phenolic acids are the most common type of antioxidant phenolics noticed in rice, which are derivatives of benzoic acid and cinnamic acid.

4.2 RICE PROCESSING

Rice processing could be defined as a set of technological activities carried out for maximizing the palatability and its functional properties to make it usable in various forms as food, feed, or industrial raw material. Rice processing is an age-old technology as, next to wheat, it has been a part of the staple diet of human beings for centuries. Some of the significant steps involved in rice processing are discussed in the following sections.

4.2.1 DRYING OF ROUGH RICE/PADDY

Harvested paddy has high moisture content (25%–40% dry basis; Atthajariyakul and Leephakpreeda 2005). A proper conditioning to reduce moisture up to the safe storage level of 12% to 14% dry basis is required to avoid spoilage, discoloration, and undesirable odor during storage (Araullo et al. 1987). Drying is the common method for conditioning of rough rice because the vapor pressure in a rice grain is greater than that of the surrounding environment (Lucia and Assennato 1994).

The moisture content in rice is not necessarily uniform from one season to the next; hence, the first basic step should be the analysis of moisture content to avoid loss in quality during processing because it directly affects the yield of head rice (Bonazzi et al. 1997). The moisture content in a rice grain sample is measured either on a wet or dry basis. Normally, the industries prefer to measure the moisture on a wet basis, that is, percentage weight of water in a sample divided by the total sample weight. Initial drying of rice starts in the field itself after the grain has reached maturity, when its dependency on the plant for water and nutrients is highly reduced. After harvesting, further drying involves the equipment and standard procedures, which vary from industry to industry, and involve common basic components such as hot air dryers, numerous storage bins, and a conveying system for moving the rice.

Thin layer and fixed bed drying are simple and affordable techniques for drying the paddy but are labor-intensive processes (Wongwises and Thongprasert 2000). In case of fixed bed drying, the flow rate of the air being too low creates a moisture gradient, whereas in thin-layer drying, all grains are fully exposed to a uniform drying condition. A continuous flow dryer is one of the common dryers where the grains flow down under gravity between two screens separated by some distance and, in the process, get exposed to the drying air. Such column dryers have air flow at an angle of 90° to the flow of rice. The important parameter is the flow rate of rice, air supply, temperature, rice layer thickness, and the residence period of grains in the dryer (Zhang et al. 2002). Tempering is done for uniform distribution of moisture throughout the bed. The other types of dryers are designed to mix the grain for affective drying as it flows through the large upright equipment. In the case of a baffle dryer, hot air flows through the openings between baffles. These baffles are 6 in. apart and rice flows in a zigzag pattern. A fluidized bed dryer for paddy drying was first studied by Sutherland and Ghaly (1990) who recognized it as a fast-drying piece of equipment, relative to a static bed. This dryer offers great advantage over a continuous flow dryer in terms of smaller drying chamber with high drying capacity due to the large air-to-product contact area (Soponronnarit 1999).

Equalization of moisture is highly essential during drying. Usually, rice is stored in temporary holding bins between passes, after drying every 20 to 30 min. The major goal of drying is that the rice should be dried with minimum quality loss and as quickly as possible at low drying temperatures.

4.2.2 PARBOILING

Parboiling of paddy and brown rice is a hydrothermal process by which the nutritional attributes are better retained and storage life extended. During the process, starch gelatinization leads to irreversible swelling and fusion of starch granules, changing the crystalline form of starch into the amorphous state (Bhattacharya 1985).

Parboiling of paddy takes places in three steps, which includes soaking, steaming, and drying. In the soaking process, void spaces in the hull and rice endosperm get hydrated, leading to volume expansion of rough rice kernel. Soaked paddy, when exposed to steam for a given period, results in starch gelatinization imparting hardness before milling. During parboiling, paddy moisture needs to be reduced to a level of 12% to 14% for safe storage and processing. Sometimes, tempering of parboiled rice is also done before drying to avoid the formation of cracks and fissures. Starch gelatinization directly affects the organoleptic characteristics of rice by influencing the A type amylopectin organization and transition form of amorphous crystalline and amylose–lipid complexes formation. It is also noticed that during the drying step, the rice protein undergoes disulfide bridge formation affecting the cooking qualities.

Although there are several types of parboiling processes, starting from the home scale to the soak-drain steam dry method, dry heat and pressure parboiling methods are generally adopted for the processing of rice. In the dry heat process, the soaked rice instead of steaming is subjected to intense heat through conduction heating at high temperature, while in pressure parboiling, the paddy is soaked for a short time or steamed at atmospheric pressure followed by steaming under high pressure to get the finished product of parboiled rice. Parboiling of rice is popular in the Asian region and is advantageous over ordinary direct milling (Roy et al. 2011) because it

- Provides easier dehulling
- Strengthens the rice and thus minimizes the yield of broken rice
- Retains the nutritional qualities better than the milled rice
- Prevents spoilage and extends the storage period

On the other hand, a frequent problem allied with the parboiling process is the discoloration of the finished rice grain and limited wettability of husk, which, in turn, affects the consumer acceptability (Luh and Mickus 1991; Kimura et al. 1993). To overcome the problems associated with husk, an alternative technology is used for the parboiling of brown rice. In this process, soaking of dehusked rice is carried out at initial temperatures of 70°C to 80°C, and cooling at room temperature for a few hours followed by open steaming for 10 to 20 min; drying and polishing are required to have better quality characteristics (Kar et al. 1999).

Some features of parboiling brown rice include the following:

- Color and odor development is minimized (Luh and Mikus 1980)
- Reduction in processing time and energy
- Economically viable process
- Drying is slow and difficult

4.2.2.1   Degree of Parboiling

The degree of parboiling is a measure of the extent of starch gelatinization as it influences many of the quality traits of parboiled rice (Islam et al. 2003). Degree of parboiling can be determined by measuring the soluble starch content because, during gelatinization, amylose molecules leach out from the micelle structure outside granules (Hermansson and Svegmark 1996). A method for the determination of soluble starch involves extraction from rice flour and its assessment by the iodine method. Apart from soluble starch estimation, other tests used for quality analysis after parboiling include water uptake, expanded volume, color of parboiled rice, solubility of proteins, and chalkiness to check the incomplete gelatinization of the starch.

4.2.2.2 Factors Affecting the Parboiling

The parboiling process easily gets affected by variations in the soaking temperature (Sareepuang et al. 2008). At room temperature, the soaking process is time-consuming. Hot soaking has an advantage over it, as it hastens the process, but it requires proper optimization because during steeping, temperature creates the moisture gradient leading to removal of husk even before the hydration of the kernel has occurred. Time of soaking is also an important parameter as high soaking times lead to foul odors from rice grains due to fermentation (Miah et al. 2002). Steaming time controls the hardness of grain and its color. Disparity in soaking and steaming causes a color change due to Maillard-type browning reaction between reducing sugar and amino acid. The method of drying employed regulates the parboiling rice milling quality (Elbert et al. 2000). Various biochemical characteristics, such as starch gelatinization and amylose content, influence the hardening of parboiled rice. Variations in these factors have detrimental effects, leading to grain deformation, and splitting, therefore, the optimization of process conditions is highly essential for parboiling.

4.2.2.3 Problems Associated with Parboiling

A major problem noticed during the parboiling of the paddy is the wettability of the husk, acting as a barrier for water movement into the kernel. The siliceous husk also bears poor thermal conductivity, slowing down the heat transfer process to the kernel. All these add up to increase considerable time and heat energy for proper parboiling of paddy rice. Besides this setback associated with the husk, it is also found to hinder the drying operations after parboiling. Longer parboiling times result in undesirable odor, and darker and harder grains that affect the parboiling quality (Bhattacharya 1996).

4.2.3 MILLING

Milling is applicable to cereal grains, covering a broad array of methods to make them suitable for consumption. Normally, milling primarily aims at the separation of the outer tissue to make the endosperm easily accessible for consumption as food. Traditional rice milling by hand-pounding the paddy was first noticed in the 1900s. Initial pounding of rice in a mortar with a wooden pestle removes the hull; further pounding takes away the bran, but the main demerit is the high yield of broken grains.

Initially, different activities such as cleaning, drying, grading, polishing, etc., were carried out manually. With time, mechanization resulted in the development of big rice mills consisting of multiple milling units to improve the quality of the finished product and recovery of the whole grain. These units consist of equipment for each separate process:

1. Dehulling or dehusking: removal of hulls from the paddy grains or rough rice, the resultant finished product being brown rice.
2. Whitening: exclusion of the bran layer from the brown rice; the end product is white rice.
3. Polishing: removing the loose bran that keeps adhering to the surface of white rice after whitening to get refined rice as the end product.
4. Separating the broken and brewer grains from milled grains.

The modern method of the milling process starts with cleaning up of the paddy to remove unwanted matters such as mud, stones, chaff, etc., by oscillating sieves, aspirators, or gravity separators. This cleaned batch is then fed to a dehusker where, with the help of rubber roller or steel huller, husk is separated. Hull aspirators trap the removed husk using an air trap system when coarse rice passes through the sieve. Separation of brown rice and rough rice is done using paddy separators on the basis of difference in density, surface smoothness, and buoyancy. The brown rice is then fed to whitening machines where polishing is done by a mild friction or abrasion created within it. Further refining to achieve better translucency and increasing the storability of milled rice is achieved by gentle brushing of grain through a polisher/refiner. The resulting polished rice and bran are separated and collected by second aspirators. Edible rice makes up to 65% to 72% with 28% to 35% as by-products and waste.

4.2.3.1 Milling Quality

Milling quality analysis is of prime importance in terms of its economic and marketing aspects (Yadav and Jindal 2008). Milling yield is one of the key factors affecting rice quality. Accurate measurement of head rice yield (HRY) and total milling rice yield (MRY), which includes head and broken rice, is highly required to evaluate the efficacy of equipment and process (Gravois 1998). The ratio of the weight of milled and head rice kernels to the total weight of the rough rice or paddy kernel gives the HRY (Yadav and Jindal 2001). Rice kernels less than three quarters of the original length of grain are termed as broken rice. Head rice is economically more valued than broken rice. Standardized procedures are used for official grading.

Apart from the milling yield, the extent of polishing to achieve whiteness of rice is crucial for its acceptance at the consumer level (Conway et al. 1991). The whiteness of milled rice is measured with a whiteness meter in terms of whiteness index in the linear range of 0 to 100, where 0 corresponds to perfect black surface and the 100 corresponds to the whiteness of the magnesium oxide fumes. The degree of milling is directly related to the whiteness of the grain and measures the extent to which the bran and germ layers have been removed from the rice endosperm. Milled rice is broadly classified into four sections by the Food and Agriculture Organization (FAO) revised model system for grading rice in International trade:

1. Undermilled rice (part of germ and most of pericarp removed during milling)
2. Reasonably milled rice (with germ and pericarp, most of the aleurone layers have been removed)
3. Well-milled rice (whole germ, pericarp, and all aleuronic layers have been removed)
4. Extra well-milled rice (only starchy endosperm is remaining)

The degree of milling can be determined by two means: (1) estimation of the quantity of removed bran or residual bran; and (2) estimation of other chemical constituents of rice, which progressively vary with milling.

1. *Estimation of the quantity of removed bran or residual bran.* It is based on an estimation of the amount of the bran removed from the milled rice grain (Pan 2004). It is necessary to weigh the rice before and after milling, when a desired whiteness level is achieved. Degree of milling can be calculated as

(%) Milling degree = (weight of milled rice/weight of rough rice) × 100

The use of dye to examine stained grains is also a common technique. Initially, Congo red with methylene blue, Sudan III, and alkaline alcohol were used for this purpose. The Japanese method of staining the rice grains by May-Grünwald's reagent (Chikubu and Shikano 1952), which consists of eosin and methylene blue dissolved in alcohol, was evaluated by Das et al. (2008) for analyzing the degree of polishing. The pericarp and testa of the kernel turned bluish green, the aleurone layer stained purple, and the endosperm turned purplish pink.

Another approach for estimating the degree of milling was the colored bran balance index, in which the magnified plane image is developed to calculate the index in terms of the percentage of milled rice area covered with bran. Evaluation of the chemical composition of bran by the solvent extraction method is also one of the quantitative tests (Bhattacharya and Sowbhagya 1972; Chen and Bergman 2005).

2. *Estimation of the other chemical constituents of rice, which progressively vary with milling.* Assessment of fat, ash, silica, and crude fiber contents is carried out to quantify the bran present on the surface of the milled grain. Petroleum ether extraction is employed to extract the surface fat from whole milled rice. Less surface fat concentration relates to more severe milling with high degree of milling (Siebenmorgen and Sun 1994). This is widely used as a reliable method, but it is time-consuming and requires suitable fat extraction apparatus. Another method using diode array near-infrared spectroscopy was reported (Saleh et al. 2008) for measuring bulk milled rice surface lipid content for measuring degree of milling.

4.2.3.2 Factors Affecting the Milling Process

Weak grains produced as a result of any activity will not be able to withstand the forces applied during milling and will end up with poor milling quality. Physical and mechanical properties of rice varieties affect the milling quality such as head rice yields and degree of milling (Shitanda et al. 2002). Grain hardness, size and shape, and surface ridge on the grain are some of the physical factors affecting the degree of milling (Liu et al. 2009). The whiteness of the rice grain varies with grain type, weather conditions during the crop year, and storage conditions. Broken rice during milling is the result of chalkiness and fissured grains. The portion of grain where chalk is visible is the feeble part that gets easily broken. Nonuniformity and chalkiness of grain affect overall processing with lower head rice yield. The moisture content of the harvested rice and the ambient temperature during kernel development play a critical role in head rice yields.

4.2.3.3   Problems Associated with the Milling Process

The following technical problems are associated with cereal polishing by mechanical milling:

- The abrasive hull of conventionally hulled cereal damages grain handling and grinding equipment, resulting in increased capital costs.
- Low nutritional profile content owing to the loss of phenolics and dietary fibers.

During the milling process, bran and most of the germ are removed, resulting in loss of fiber, minerals, vitamins, lignans, phytoestrogens, and phenolic compounds. The endospermic tissue aleurone layer, full of biofunctional molecules, also gets removed as part of the bran during roller milling. Another disadvantage of using the milling process lies in the fracture of endosperm resulting in broken grains.

4.2.4   Enzymatic Polishing of Rice

Milling of rice grains for human consumption requires the removal of the bran layer (Hoseney 1986) at the expense of nutrients. This problem of mechanical polishing can be overcome by exogenous enzyme treatment before polishing for improving the nutritional quality characteristics. With these objectives, new and novel polishing technologies have emerged. As these processing techniques are a bit different from the conventional rice-processing technologies, the authors feel that to appreciate the mechanism of enzymatic digestion, an understanding of the structural composition of the rice bran that plays a significant role is essential.

4.2.4.1   Structural Composition of Rice Bran

The major polysaccharides present in the bran are crude cellulose (9.6%–12.8% dry basis) and hemicellulose or pentosans (8.6%–10.9% dry basis), which constitute the major part of the insoluble dietary fiber.

Cellulose consists of linear β-1,4-linked D-glucopyranose chains that are condensed by hydrogen bonds into crystalline structures, called microfibrils (Figure 4.1). These microfibrils consist of large glucose chains and are linked by hemicelluloses. It forms crystals where intramolecular (O_3-H-O_5' and O_6-H-O_2') and intrastrand (O_6-H-O_3') hydrogen bonds hold the network flat, allowing the more hydrophobic ribbon faces to stack, thereby making it completely insoluble in normal aqueous solutions. In addition to this crystalline structure, cellulose contains noncrystalline

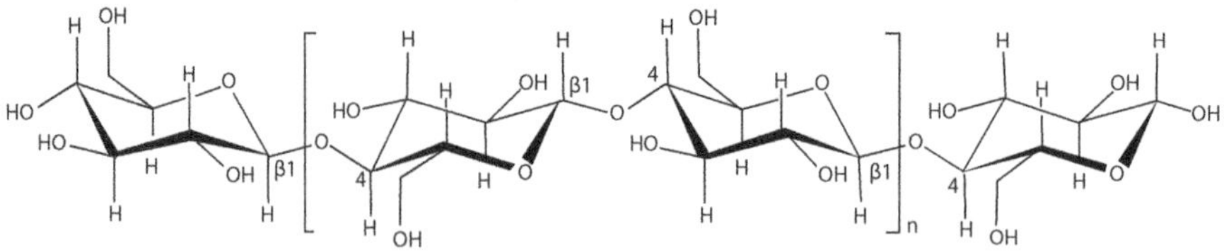

FIGURE 4.1   Structural unit of cellulose.

(amorphous) regions within the microfibrils. The major hemicellulose polymer in cereals is xylan. Xylan consists of a β-1,4–linked D-xylose backbone and can be substituted with different side groups such as α-L-arabinose, D-galactose, acetyl, feruloyl, p-coumaroyl, and D-glucuronic acid residues (Figure 4.2). Xyloglucans consist of β-1,4–linked D-glucose backbone substituted by D-xylose.

Rice bran pentosans are chiefly arabinoxylans composed of 1500 to 5000 sugar units, arabinose, and xylose with α-L-arabinose substituted either at C-2/C-3. They take up a twisted ribbon conformation with hydrophobic cavities. The complex polyphenolic polymer known as lignin (7.7%–12.0% dry basis) interacts with the cellulose fibrils, creating a rigid structure. The xylan layer, with its covalent linkage to lignin and its noncovalent interaction with cellulose, may be important in maintaining the integrity of the bran *in situ* and thereby imparts its resistant nature (Uffen 1997). Much of the nascent proteins of bran (16.36% dry basis) are bound in glycosidic linkages and held within the fibrous matrix of the bran.

Phenolic acids reported from rice endosperm cell walls are cinnamic acid, ferulic acid (FA), p-coumaric acid, and diferulic esters. A portion of the total cinnamate exists as ferulate dimers linked in various ways. Cinnamates and particularly FA are introduced into cell wall matrices attached to polysaccharides. Thus, FA becomes esterified to arabinose residues in primary cell wall arabinoxylan matrices. The attachment involves a covalent ester linkage between the carboxylate group of FA and the primary alcohol on the C-5 carbon of arabinosyl side chains attached to the xylan backbone. The formation of ferulate dimers facilitates covalent coupling of the polysaccharides by radically mediated dimerization. As the wall lignifies, ferulates and diferulates become involved in radical cross-linking reactions with lignin monomers to intimately incorporate the ferulates into lignin. The phenolic compounds exert a significant effect on the properties of the cell wall, which is mechanically strengthened by the cross-linking (Figure 4.3; Chrastil 1992).

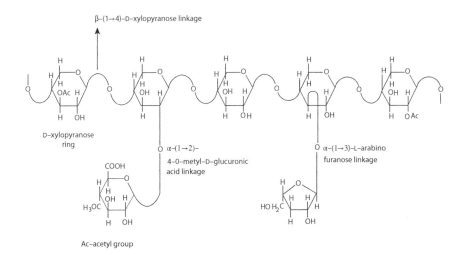

FIGURE 4.2 Xylan structure with side chains attached.

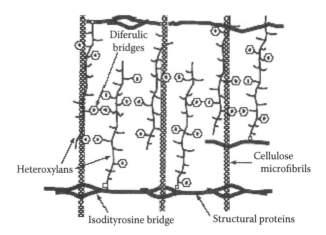

FIGURE 4.3   Schematic representation of cell wall structure of grains.

4.2.4.2   Enzymes Involved in Rice Processing

The enzymes that are mainly required for different types of processing of the food or related materials belong to food enzymes. The major application of enzymes such as cellulase, xylanase, proteases, amylases, pullulanase, etc., is found in different sectors starting from starch processing, sweetener production, baking, brewing, dairy products, distilling, juice and wine making, etc., and also as the feed enzymes in processing animal food.

4.2.4.2.1   Xylanase

The enzymes that break down hemicelluloses are referred to as hemicellulases. These are defined and classified according to the substrates on which they act and are collectively grouped as glycan hydrolases (EC 3.2.1). Three different types of xylanases are involved in xylan degradation.

(1) Endo-β-(1,4)-ᴅ-xylanase [β-(1,4)-ᴅ-xylan xylano hydrolase, EC 3.2.1.8]: these enzymes act randomly on xylan to produce large amounts of xylo-oligosaccharides of various chain lengths and belong to four different types:
   (a) Nonarabinose liberating endoxylanases: these cannot act on ʟ-arabinosyl–initiated branch points at β-(1,4) linkages and produce only xylobiose and xylose as the major end products. These enzymes can break down xylo-oligosaccharides as small as xylobiose.
   (b) Nonarabinose liberating endoxylanases: these cannot cleave branch points at α-(1,2) and α-(1,3), and produce mainly xylo-oligosaccharides larger than xylobiose. These endoxylanases have no action on xylotriose and xylobiose.
   (c) Arabinose-liberating endoxylanases: these can cleave the xylan chain at the branch points and produce mainly xylobiose, xylose, and arabinose.
   (d) Arabinose-liberating endoxylanases: these can hydrolyze the branch points and produce intermediate size xylo-oligosaccharides and arabinose.

(2) Exo-β-(1,4)-D-xylanase [β-(1,4)-D-xylan xylohydrolase]: these enzymes remove the single xylose units from the nonreducing end of the xylan chain.

(3) β-Xylosidase or xylobiase (EC 3.2.1.37): these enzymes hydrolyze disaccharides such as xylobiose and the higher xylo-oligosaccharides with decreasing specific affinity.

The xylanolytic enzyme system carrying out the xylan hydrolysis is usually composed of a repertoire of hydrolytic enzymes: β-1,4-endoxylanase, β-xylosidase, α-L-arabinofuranosidase, α-glucuronidase, acetyl xylan esterase, and phenolic acid (ferulic and p-coumaric acid) esterase. All these enzymes act cooperatively to convert xylan into its constituent sugars.

The xylanases have been reported mainly from bacteria, fungi, actinomycetes, and yeast. The presence of such a multifunctional xylanolytic enzyme system is quite widespread among fungi (Belancic et al. 1995). The members of the fungal genus *Aspergillus* are commonly used for the production of polysaccharide-degrading enzymes. This genus produces a wide spectrum of cell wall-degrading enzymes, allowing not only complete degradation of the polysaccharides but also tailored modifications by using specific enzymes purified from these fungi. Xylanases from *Trichoderma reesei*, *Trichoderma harzianum*, *Penicillium purpurogenum*, *Thermomyces lanuginosus*, *Fusarium oxysporum*, and *Cephalosporium* sp. are also well characterized (Beg et al. 2001). The major industrial applications of xylanases are in pulp and paper industries for biobleaching. Xylanases are also used in food and feed industries.

4.2.4.2.2 Cellulase

Cellulases have attracted much interest because of the diversity of their applications and also for facilitating the understanding of mechanisms of enzymatic hydrolysis of plant carbohydrate polymers. It is a hydrolytic enzyme that degrades cellulose to β-glucose. A cellulosic enzyme system consists of four major components: endo-β-glucanase (EC 3.2.1.4), exo-β-glucanase (cellobiohydrolase or glucanase; EC 3.2.1.91), exo-β-glucanase (exoglucanase or glucohydrolase; EC 3.2.1.74), and β-glucosidase (EC 3.2.1.21). The modes of action of each of these are described as follows:

1. Endo-β-glucanase, 1,4-β-D-glucan glucanohydrolase, CMCase, "random" scission of cellulose chains yielding glucose and cello-oligosaccharides. Generally, the endoglucanases randomly attack H3PO4-swollen cellulose, CM cellulose, and the amorphous regions of the cellulose and release cello-oligosaccharides.

2. Exo-β-glucanase (cellobiohydrolase, CBH or glucanase) releases cellobiose either from the reducing or nonreducing end. The CBHs hydrolyze H3PO4-swollen cellulose and Avicel sequentially by removing the cellobiose units from the nonreducing end of the cellulose chain.

3. Exo-β-glucanase (exoglucanase or glucohydrolase) releases glucose from the nonreducing end.

4. β-Glucosidase, cellobiase releases glucose from cellobiose and short-chain cello-oligosaccharides.

The endoglucanase and CBH act synergistically to affect the extensive hydrolysis of crystalline cellulose, whereas the β-glucosidase completes the hydrolysis by converting the resultant cellobiose to glucose.

Cellulase is chiefly produced by some fungi, bacteria (including symbiotic bacteria residing within termites and ruminant herbivores), protozoans, and ruminant herbivores. Fungal cellulases are produced in large amounts, which include all the components of a multienzyme system with different specificities and modes of action, acting in synergy for complete hydrolysis of cellulose (Wood and Garcia-Campayo 1990; Ogawa et al. 1991; Bhat and Bhat 1997). One of the most extensively studied cellulolytic microorganisms is the soft rot fungus, *T. reesei*. The cellulase systems of the aerobic fungi *Trichoderma viride*, *Penicillium pinophilum*, *Phanerochaete chrysosporium* (*Sporotrichum pulverulentum*), *Fusarium solani*, *Talaromyces emersonii*, *Trichoderma koningii*, and *Rhizopus oryzae* are also well studied (Murashima et al. 2002). Several cellulolytic anaerobic bacteria secrete multienzyme cellulase complexes called cellulosomes, which are difficult to disrupt, whereas others secrete very small amounts of cellulases and have been found to lack cellobiohydrolases (Mawadza et al. 2000).

Major industrial applications of cellulase are in the textile industry for biopolishing of fabrics; in animal feeds for improving nutritional quality and digestibility; for commercial food processing, for example, fruit juices, baking, etc.; and in paper and pulp industries for the de-inking of paper.

The application of specific cell wall-degrading enzymes for the hydrolysis of the bran layer in a selective manner is a new approach. The major constituents of rice bran are carbohydrates comprised mainly of cellulose and hemicelluloses (Choct 1997). With the application of the cellulase and xylanases, the hydrolysis of the bran layer is effective in minimizing the loss of bran and germ mass. These enzymes can be of any origin, microbial or plant, but microbial carbohydrate cleaving enzymes have always been the right choice due to low cost and better productivity. While choosing the microbial enzyme, the choice has to be made on the basis of a "generally regarded as safe" organism that is efficient in producing the respective enzyme.

Enzymatic hydrolysis parameters such as temperature, incubation time, pH, moisture content, and enzyme-to-grain ratio, are critical points that need to be examined for each particular rice variety to optimize the yield of biopolished rice. Das et al. (2008) considered all such parameters while treating the rice with cellulase and xylanase, and confirmed that all these factors act synergistically for the enzymatic degradation of bran layer. Variation in any of these parameters will have a direct effect on the nutritional content of rice and its cooking quality. Arora et al. (2007) noticed that an increase in enzyme concentration affects the cooking quality of Basmati rice. Apart from cellulase and xylanase, protease has also been explored in combination with other enzymes (Sarao et al. 2011). Biopolishing is a thrust area for research groups working on rice because it has the potential to replace mechanical polishing to an extent that rice with health-promoting components can reach the common man.

4.2.4.3 Advantages of Enzymatic Polishing of Rice

These techniques have an advantage over other processes in having a higher head yield of rice from the paddy, a better retention of important components of the bran layer (phenolics and dietary fiber), and a reduction in time of polishing (Morris and Bryce 2004).

4.2.5 New Approaches for Value-Added Food Grains

Dietary micronutrient deficiencies, for example, lack of vitamins A and B_1, iodine, iron, or zinc, are a major source of morbidity (increased susceptibility to disease) and mortality worldwide. These deficiencies affect children in particular, impairing their immune systems and normal development, causing disease and, ultimately, death. The best way to avoid micronutrient deficiencies is by intake of nutrient-dense diets. Due to growing health consciousness, some consumers have started using rice milled to lower degrees, brown rice, or even germinated brown rice.

4.2.5.1 Germinated Brown Rice

Nutrition of sprouted grains has been studied for decades. In Japan, eating soaked brown rice has long been a common practice (Kayahara 2002). However, germinated brown rice gained popularity because it is softer, nutritious, and quicker to cook than brown rice. During the process of sprouting, nutrients in the brown rice change drastically. The hydrolytic enzymes are activated and they decompose starch, non-starch polysaccharides, and proteins, which lead to an increase in oligosaccharides and amino acids, mainly γ-aminobutyric acid. The decomposition of high molecular weight polymers during germination leads to the generation of biofunctional substances and an improvement of the organoleptic qualities due to the softening of the texture. Because germinated brown rice is eaten as brown rice, the 10% loss in nutrients that usually occurs during the milling process in brown rice is eliminated. Germination makes the outer layers of the rice softer. Germinated brown rice's high nutrition content is quite critical when the food supply is short.

4.2.5.2 Enriched and Fortified Rice

The purpose of rice enrichment and fortification is to restore to milled rice the levels of B vitamins and minerals removed from the grain during milling. It is technically more difficult than enriching wheat flour because rice is consumed as a whole grain. Traditional methods include parboiling, acid parboiling with 1% acetic acid, thiamine enrichment, coating, and the production of artificial rice, dibenzoylthiamine enrichment, and multinutrient enrichment by adding a nutrient-enriched premix (Misaki and Yasumatsu 1985). The nutrient levels are the same as those of brown rice. This multinutrient-enriched rice is blended with milled rice at a ratio of 1:200. Only 10% of any nutrient is lost through ordinary washing before cooking and another 10% during cooking. Obstacles to the successful introduction of rice enrichment by the premix method include the following (FAO 1954):

- The cost of the imported premix.
- The difficulty in ensuring the correct proportion of the premix added to milled rice.
- The slightly greater cost of enriched rice as compared with that of ordinary rice, which affects its sale to lower income groups.
- The loss of added vitamins that may occur when enriched rice is cooked in excess water that is later discarded according to the current practice in some rice-eating countries.

TABLE 4.3

Preliminary Test for Assessing the Suitability of Processed Rice for its End Uses

Hulling percentage	Paddy is dehulled using a standard dehusker and the average whole-grain yield is calculated
Grain classification	Length, breadth of brown rice grains are measured using micrometer
	Based on the L/B ratio, grains are classified into long slender (LS), short slender (SS), medium slender (MS), long bold (LB), and short bold (SB)
Chalkiness of endosperm	Analyzed using a stereozoom microscope
	The rice grains are classified into white belly, white core, and white back, depending on its location on endosperm
Chalk index determination	Identification of the grain with more than 50% of chalkiness. Weighing and calculation of percentage of chalkiness
Color	Analyzed by optical transmission or reflectance measurement
	Color of rice lies in the range of white to dark gray to rosy
Test weight	Weight of a known volume of rice grain
	Measure of amount of foreign material with unfilled and immature grains
Moisture content	Analyzed by electric moisture meter, near infrared reflectance, etc.
Translucence index	Measured by colorimeter, indicates the endosperm translucency

Source: Lisle, A.J. et al., *Cereal Chem.* 77:627–632, 2000.

4.3 FUNCTIONALITY TEST FOR EVALUATION OF RICE QUALITY

The functionality test is considered an important quality analysis of rice because a good index of these characteristics will provide us a high-quality finished product. Table 4.3 shows the primary tests that are essential before starting up of any rice processing techniques.

4.3.1 COOKING AND EATING QUALITY

With rapid economic growth, the emphasis is on quality rather than quantity. The rice-processing industries are continuously establishing mills with modern equipment to maximize the marketing price of rice according to its quality. The quality of processed rice depends on its appearance and its cooking properties. The cooking and eating qualities of rice are assayed in several ways, as mentioned in Table 4.4 (Park et al. 2001), and its characteristics depend highly on the amylose content. Rice starch granules, rich in amylase, are nonsticky and become hard after cooling, whereas rice with a low amylose content is sticky in nature. Gel consistency, one of the factors measured to obtain eating quality, varies even at the same amylose content.

Textural analysis of cooked and raw rice provides us with an idea about the eating characteristics of the kernel. Previously taste panels were used to estimate the rice's texture by sensory evaluation. The requirement for a large number of samples and

TABLE 4.4

Parameters that Need to be Evaluated to Assess the Cooking and Eating Qualities of Rice

Sl. No.	Cooking Parameters	Index	Standard Method
1.	Amylose content	Amylose content is an index of resistance to disintegration during cooking and stickiness	Starch iodine blue test Based on amylose content, milled rice is classified as: waxy (1%–2% amylose) nonwaxy (>2% amylose) very low (2%–9% amylose) intermediate (20%–25% amylose) high (25%–33%) amylose
2.	Gelatinization temperature	The range of temperature within which the starch granules start to swell irreversibly in hot water	Alkali spreading value test. Alkali spreading value corresponds to gelatinization temperature as follows: 1–2 high (74.5°C–80°C) 3, high intermediate 4–5, intermediate (70°C–74°C), and 6–7, low (<70°C) Differential scanning calorimetery
3.	Gel consistency test/ gel texture	An index of peak viscosity, breakdown, hot paste viscosity, setback, cold paste viscosity of the rice (harder gel consistency is associated with harder cooked rice)	Amylography or viscography
4.	Water absorption	It is an index of amount of water absorbed by the grains when cooked for a length of time?	Water uptake ratio calculated as ratio of weight of cooked rice/ weight of raw kernel
5.	Swelling capacity	It gives an indication as to how well the grains will swell when cooked	The ratio of the final weight or volume of cooked rice to the initial weight or volume
6.	Cooking time	Indicates minimum cooking time required for particular rice	The time when 90% of the rice starch no longer shows opaque center while pressed between two glass plates

(continued)

TABLE 4.4 (Continued)

Parameters that Need to be Evaluated to Assess the Cooking and Eating Qualities of Rice

Sl. No.	Cooking Parameters	Index	Standard Method
7.	Hardness, stickiness, adhesiveness, chewiness, springiness gumminess, cohesiveness	Index for eating quality of rice	Texture analyzer (a texturometer double-bite technique simulating chewing action)
8.	Grain elongation	Indicator for volume expansion during cooking (length increase without increase in breadth is desirable for high-quality rice)	The ratio of the average length of cooked rice to the average length of raw kernel

Source: Hossain, M.S. et al., *Science Asia* 35:320–325, 2009; Danbaba, N. et al., *Int. Food Res. J.* 18:629–634, 2011.

the lack of a fixed point of reference due to variations in the sensory system, with daily physical and mental conditions, was a drawback for this evaluating system. Nowadays, using various instruments, simple and quick methods have been designed for assessing rice texture with overall consistency.

4.4 RICE UTILITY AND PRODUCT DEVELOPMENT

Rice has now entered into the miscellaneous sector of the food industry as a substitute to wheat flour to derive value-added products.

4.4.1 POPPED RICE

Popping up of rice is a traditional process, mostly common in Asian countries (Bhat Upadya et al. 2008). During the process, raw rice or rough rice is subjected to heat for 30 to 35 s at 240°C, or at 275°C for 40 to 45 s, creating pressure inside the grain; as a result, volume expansion of the kernels takes place (Murugesan and Bhattacharya 1989). The main factors such as chalkiness, white belly, hull, fissured grains, moisture content, and time of harvesting are influencing aspects for popped rice.

4.4.2 PUFFED RICE

The traditional process of puffing is done at normal atmospheric pressure. Initially, rough rice is steeped in cold or warm water followed by dry heating in an iron pan or clay pot with sand (240°C–270°C for 10–11 s) in which roasting occurs. This is then tempered for a few minutes and pounded to remove any portions of hull or bran, resulting in a yield of puffed rice of up to 70%. Another type of puffing is based on pressure differentials (Chandrasekhar and Chattopadhyay 1988). Process variables affecting the puffing process include the amylase–amylopectin ratio, addition of salts, and preconditioning (Mohapatra et al. 2012). Gun puffing (Juliano and Sakurai 1985) or oven puffing are some of the common techniques used to produce puffed rice. Gun-puffed rice has certain disadvantages such as blackening of the germ, lack of crispness (which is not acceptable to consumers), and high overall production cost. Oven-puffed rice is made with sugar, malt syrup, or salt and is highly accepted by consumers or bakeries due to its tenderness and crispness.

4.4.3 BEATEN RICE

Beaten rice is made from brown rice that is flattened into the form of flakes. This rice product is easily digestible and can be consumed after puffing, rehydration, or roasting in oil and it can also be eaten raw with milk or water. Processing of beaten rice includes cleaning of paddy, soaking in hot/warm water for about 45 min, followed by drying. After roasting in a continuous roaster, it is flattened to produce flakes in a roller flaker or edge runner flaker of the desired thickness. These flakes are passed through sieves to maintain a fairly even size. Beaten rice is relatively light in weight and is convenient for transporting and storage.

4.4.4 READY-TO-EAT RICE

Conventional rice preparation and cooking is a time-consuming process taking about 30 min. Currently, ready-to-eat rice is becoming popular because it is convenient for consumers (Table 4.5).

TABLE 4.5
Types of Quick-Cooking Rice

Retort rice	It is a type of cooked rice in which cooking is done in plastics, followed by pasteurization at 120°C to enhance its shelf life. It can be used both for brown rice and milled rice
Canned rice	Milled rice is placed in tin cans, steamed, sealed, and sterilized at 112°C for 80 min. Nowadays, seasoned rice is also becoming popular with added flavor and aroma
Cup rice	Rice is cooked, dried with an explosion puffing machine or fluidized dryer to obtain porous grains so that adding water to cup rice makes it ready to eat in 2–3 min
Gelatinized rice	Processing includes cooking and drying fast to minimize starch retrogradation
Frozen rice	Processing includes soaking, draining, steaming, boiling, and freezing. Cooked rice is then frozen by sublimation under high vacuum and instantly packed in plastic pouches in the freezer

Source: Ohtsubo, K. et al., Processed novel foodstuffs from prepigmented brown rice by a twin screw extruder. In *Rice Is Life: Scientific Perspectives for the 21st Century*, edited by K.L. Heong, 275–277, 2000.

4.4.5 Rice Flour

Milled head rice has always been the first preference of consumers. However, a number of rice products using rice flour have recently been developed with ethnic properties. Flour is produced from whole or broken grains by dry milling or wet milling process. The efficacy of using rice flour to replace wheat flour has attracted the processing industry to incorporate it into a number of formulations.

4.4.5.1 Infant Formulae

To meet the infant food requirements, the common method is mixing milk with cereal flour. Dehydrated cereal flour mixed with water forms a homogeneous smooth slurry for feeding infants who are intolerant or allergic to cow's milk-based and soymilk-based foods (Lasekan et al. 2006). The vitamin, mineral, and protein contents of milled brown rice make it a perfect ingredient for infant foods (Gastañaduy et al. 1990). Apart from rice flour, rice protein concentrates, isolates, or hydrolysates, fortified rice (lysine and threonine) have recently been used in infant foods (Koo and Lasekan 2007). Rice can be used in the formulation in precooked form after drum drying or extrusion. Particle size of rice flour and its moisture content play an important role in the formulation. The packaging material for infant formulae with rice flour requires special attention. The most likely material chosen for such a formulation should allow the transmission of water vapor and gas to avoid rancidity. Mainly, paperboard or preprinted cartons should be used for packaging.

4.4.5.2 Rice Flour for Baking

Rice flour is widely used in baked products because it does not contain gluten (Gallagher et al. 2004), and therefore its dough does not retain gases during baking. The addition of rice starch also improves the baking properties because of its amylose–amylopectin content, gelatinization temperature, and pasting behavior. Composite baking flour, made by adding rice flour to wheat flour, is used to make bread and flakes (Wanyo et al. 2009). Rice bran has also been exploited to make bread (Hu et al. 2009). The flour from medium-grained and short-grained rice varieties provides soft texture and stickiness, and is preferable to the long-grain type for making rice bread. Cakes and cookies are also produced with rice flour. Schober et al. (2003) produced cookies with better sensory qualities using rice flour, corn starch, soy flour, and potato starch. The addition of rice flour to bakery products reduces the fracture strength and increases crispness.

4.4.5.3 Processed Products

Processed foods made from rice are also in demand because of its attractive taste and aroma (Pszczola 2001). Rice noodles or vermicelli are commonly consumed in southern Asia, and is made by processing high-amylose rice flour by wet milling and water. Sometimes, corn flour is also added to enhance its functional attributes. Fine strands, whiteness, and translucency are required for assessing better quality noodles (Nura et al. 2011). Apart from flour particle size, gel hardness, swelling, and pasting properties are important physicochemical properties that affect the texture of rice noodles (Hormdok and Noomhorm 2007). Rice noodle preparation requires

TABLE 4.6
Fermented Products Derived from Rice

Types	Examples	Organism	Uses
Solid fermented foods	(Tseng et al. 2011) (bright reddish purple color)	Submerged and solid state fermentation using mold *Monascus purpureus*	Natural coloring agent in wine-making and exert besides several health-promoting features
Paste fermented products	Miso	Natural fermentation	Sauce for seasoning
Liquid fermented foods	Rice sake, vinegar and wine	*Aspergillus oryzae* (Coronel et al. 1981), *Zymomonas mobilis* (Abate et al. 1996), *Candida shehatae,* and *Saccharomyces cerevisiae* (Lebeau et al. 1997)	Sauce, wine, sake for cooking

boiling flour with water until dough formation, followed by kneading and extrusion. The extruded noodles are cooled, strained, and dried until the moisture content reaches 10% to 12%. Rice snacks are made by extrusion (Suksomboon et al. 2011) of both nonglutinous and glutinous rice, but glutinous rice is much preferred because of its sticky character and easy expansion into a porous texture. The processing of rice snacks involves grinding, steaming, kneading, cooling, pounding, drying, baking, seasoning (to add color and aroma), cutting, and packing. Rice fries—a type of snack—is liked by consumers because of its crisp exterior crust and fluffy interior (Shih and Daigle 1999). Pasta from rice flour has also been explored by Suteebut et al. (2009), who used jasmine rice.

4.4.6 FERMENTED PRODUCTS

Fermented rice foods are commonly preferred by consumers in Southeast Asia and can be classified into three forms (Table 4.6).

4.5 CONCLUSION

Brown rice is more preferred because of its high nutritional value but consumers' preference is for well-milled rice owing to its better appearance and palatability. Therefore, producing palatable rice, with the minimum amount of breakage while retaining the maximum possible amount of nutrients, has been the primary goal of the rice-processing industry. However, to do away with the losses incurred during mechanical milling and to produce a better quality, palatable brown rice retaining the maximum possible amount of nutrients, a novel attempt was made to develop an alternative polishing method called enzymatic polishing and use it on germinated brown rice. All the benefits of germinated rice can be obtained with the application of polysaccharide-degrading enzymes for the selective removal of the bran layer.

REFERENCES

Abate, C., D. Calloeri, E. Rodriguez, and O. Garro. 1996. Ethanol production by a mixed culture of flocculant strains of *Zymomonas mobilis* and *Saccharomyces* sp. *Applied Microbiology and Biotechnology* 45:580–83.

Aggarwal, A.K. and R. Mohan. 2010. Aspect ratio analysis using image processing for rice grain quality. *International Journal of Food Engineering* 6:1–14.

Araullo, E.V., D.B.D. Padua, and M. Graham. 1987. Rice postharvest technology. International Development Research Center, Ottawa. Report IDRC 053e, pp. 394.

Arora, G., V.K. Sehgal, and M. Arora. 2007. Optimization of process parameters for milling of enzymatically pretreated Basmati rice. *Journal of Food Engineering* 82:153–9.

Atthajariyakul, S. and T. Leephakpreeda. 2005. Fluidised bed paddy drying in optimal conditions via adaptive fuzzy logic control. *Journal of Food Engineering* 75:104–14.

Bai, X., L. Luo, W. Yan, M.R. Kovi, W. Zhan, and Y. Xing. 2010. Genetic dissection of rice grain shape using a recombinant inbred line population derived from two contrasting parents and fine mapping a pleiotropic quantitative trait locus *qGL7*. *BMC Genetics* 11:16.

Beg, Q.K., M. Kapoor, L. Mahajan, and G.S. Hoondal. 2001. Microbial xylanases and their industrial applications: A review. *Applied Microbiology Biotechnology* 56:326–38.

Belancic, A., J. Scarpa, A. Peirano, R. Diaz, J. Steiner, and J. Eyzayuirre. 1995. *Penicillium purpurogenum* produces several xylanases: purification and properties of two of the enzymes. *Journal of Biotechnology* 41:71–9.

Bhat Upadya, V.G., R.S. Bhat, V.V. Shenoy, and P.M. Salimath. 2008. Physico-Chemical Characterization of Popping—Special Rice Accessions* Karnataka. *Journal of Agricultural Science* 21:184–6.

Bhat, M.K. and S. Bhat. 1997. Cellulose degrading enzymes and their potential industrial applications. *Biotechnology Advances* 15:583–620.

Bhattacharya, K.R. and C.M. Sowbhagya. 1972. An improved alkali reaction test for rice quality. *International Journal of Food Science and Technology* 7:323–31.

Bhattacharya, K.R. 1985. Parboiling of rice. In *Rice: Chemistry and Technology*, ed. B.O. Juliano, 289–348. American Association of Cereals Inc., St. Paul, MN, USA.

Bhattacharya, S. 1996. Kinetics on color changes in rice due to parboiling. *Journal of Food Engineering* 29:99–106.

Bonazzi, C., M.A. Peuty, and A. Themelin. 1997. Influence of drying conditions on the processing quality of rough rice. *Drying Technology* 15:1141–51.

Chandrasekhar, P.R. and P.K. Chattopadhyay. 1988. Studies on microstructural changes of parboiled and puffed rice. *Journal of Food Processing and Preservation* 14:27–33.

Chen, M.H. and C.J. Bergman. 2005. A rapid procedure for analysing rice bran tocopherol, tocotrienol and γ-oryzanol contents. *Journal of Food Composition Analysis* 18:319–31.

Chikubu, T.S. and T. Shikano. 1952. *A New M.G. Method of Staining for Grain. Part 1. Application to Pressed Barley and Milled Rice*. Reports from Food Research Institute, Tokyo, pp. 75–8.

Choct, M. 1997. Feed non-starch polysaccharides: chemical structure and nutritional significance. *Feed Milling International* 13–26.

Chrastil, J. 1992. Correlations between the physicochemical and functional properties of rice. *Journal of Agricultural and Food Chemistry* 40:1683–6.

Conway, J.A., M. Sidik, and H. Halid. 1991. Quality/value relationships in milled rice stored in conventional warehouses. In *Proceedings of the Fourteenth ASEAN Seminar on Grain Postharvest Technology*, ed. O.J. Naewbanij and A.A. Manilay, 55–82. Manila, Philippines.

Coronel, L.M., A.O. Velasquez, and M.C. Castillo. 1981. Some factors affecting the production of rice wine using an isolate of *Aspergillus oryzae*. *Philippine Journal of Science* 110:1–9.

Danbaba, N., J.C. Anounye, A.S. Gana, M.E. Abo, and M.N. Ukwungwu. 2011. Grain quality characteristics of *Ofada* rice (*Oryza sativa* L.): cooking and eating quality. *International Food Research Journal* 18:629–34.

Das, M., S. Gupta, V. Kapoor, R. Banerjee, and S. Bal. 2008. Enzymatic polishing of rice—a new processing technology. *LWT—Food Science and Technology* 41:2079–84.

Elbert, G.M., P. Tolaba, and C. Suárez. 2000. Effects of drying conditions on head rice yield and browning index of parboiled rice. *Journal of Food Engineering* 47:37–1.

Fitzgerald, M.A., S.R. McCouch, and R.D. Hall. 2009. Not just a grain of rice: the quest for quality. *Trends in Plant Science* 14:133–9.

Food and Agriculture Organization (FAO). 1954. *Rice and Rice Diets—A nutritional survey*, published by FAO, 1954, Rome.

Food and Agriculture Organization (FAO). 2011. FAOSTAT. Statistical database. http://faostat.fao.org. Accessed date September 14, 2011.

Gallagher, E., T.R. Gormley, and E.K. Arendt. 2004. Recent advances in the formulations of gluten-free cereal based products. *Trends in Food Science and Technology* 15:143–52.

Gastañaduy, A., A. Cordano, and G.G. Graham. 1990. Acceptability, tolerance, and nutritional value of a rice-based infant formula. *Journal of Pediatric Gastroenterology and Nutrition* 11:240–6.

Gravois, K.A. 1998. Optimizing selection for rough rice yield, head rice, and total milled rice. *Euphytica* 101:151–56.

Hermansson, A.M. and K. Svegmark. 1996. Developments in the understanding of starch functionality. *Trends in Food Science and Technology* 7:345–53.

Hormdok, R. and A. Noomhorm. 2007. Hydrothermal treatments of rice starch for improvement of rice noodle quality. *LWT—Food Science and Technology* 40:1723–31.

Hoseney, R.C. 1986. *Principles of Cereal Science and Technology*. American Association of Cereal Chemicals, Inc., USA, pp. 19–20.

Hossain, M.S., A.K. Singh, and F. Zaman. 2009. Cooking and eating characteristics of some newly identified inter sub-specific (*indica/japonica*) rice hybrids. *Science Asia* 35:320–25.

Islam, M.R., N. Shimizu, and T.K. Energy. 2003. Requirement in parboiling and its relationship to some important quality indicators. *Journal of Food Engineering* 63:433–9.

Juliano, B.O. and J. Sakurai. 1985. Miscellaneous rice products. In *Rice: Chemistry and Technology*, ed. B.O. Juliano. American Association of Cereal Chemists, St Paul, MN.

Kar, N., R.K. Jain, and P.P. Srivastav. 1999. Parboiling of dehusked rice. *Journal of Food Engineering* 39:17–22.

Kayahara, H. 2002. Elucidation of functionality of GABA and probability for novel foodstuff. *Japan Food Science* 41:39–45.

Khan, A.R., S.S. Singh, M.A. Khan, O. Erenstein, R.G. Singh, and R.K. Gupta. 2009. Changing scenario of crop production through resource conservation technologies in Eastern Indo-Gangetic plains. In Proceedings of the 4th World Congress on Conservation Agriculture, World Congress on Conservation Agriculture, New Delhi, India.

Kimura, S., M. Yamada, I. Igaue, and T. Mitsui. 1993. Structure and function of the Golgi complex in rice cells: characterization of Golgi membrane glycoproteins. *Plant and Cell Physiology* 34:855–63.

Koo, W.W. and J.B. Lasekan. 2007. Rice protein-based infant formula: current status and future development. *Minerva Pediatrica* 59:35–41.

Koutroubas, S.D., F. Mazzini, B. Pons, and D.A. Ntanos. 2004. Grain quality variation and relationships with morpho-physiological traits in rice (*Oryza sativa* L.) genetic resources in Europe. *Field Crops Research* 86:115–30.

Lasekan, J.B., W.W. Koo, J. Walters, M. Neylan, and S. Luebbers. 2006. Growth, tolerance and biochemical measures in healthy infants fed a partially hydrolyzed rice protein-based formula: a randomized, blinded, prospective trial. *Journal of the American College of Nutrition* 2:12–19.

Lebeau, T., J. Jouenne, and G.A. Junter. 1997. Continuous alcoholic fermentation of glucose xylose mixtures by co-immobilized *Saccharomyces cerevisiae* and *Candida shehatae*. *Applied Microbiology and Biotechnology* 50:309–13.

Liang, X. and J.M. King. 2003. Pasting and crystalline property differences of commercial and isolated rice starch with added amino acids. *Journal of Food Science* 68:832–8.

Lisle, A.J., M. Martin, and M.A. Fitzgerald. 2000. Chalky and translucent rice grains differ in starch composition and structure and cooking properties. *Cereal Chemistry* 77:627–32.

Liu, T., D. Mao, S. Zhang, C. Xu, and Y. Xing. 2009. Fine mapping *SPP1*, a QTL controlling the number of spikelets per penicle, to a BAC clone in rice (*Oryza sativa*). *Theoretical and Applied Genetics* 118:1509–17.

Lucia, M. and D. Assennato. 1994. Agricultural engineering in development. FAO Agricultural Services Bulletin No. 93.

Luh, B.S. and R.R. Mikus. 1980. Parboiled rice in rice production and utilization. AVI: Westport, CT, pp. 501–42.

Luh, B.S. and R.R. Mickus. 1991. Parboiled rice. In *Rice volume II: Utilization*, ed. E.B.S. Luh, 51–88. New York, Van Nostrand Reinhold.

Mawadza C., R. Hatti-Kaul, R. Zvauya, and B. Mattiasson. 2000. Purification and characterization of cellulases produced by two *Bacillus* strains. *Journal of Biotechnology* 83:177–87.

Miah, M.A.K., A. Haque, M.P. Douglass, and B. Clarke. 2002. Parboiling of rice. Part I: Effect of hot soaking time on quality of milled rice. *International Journal of Food Science and Technology* 37:527–537. doi: 10.1046/j.1365-2621.2002.00610.x.

Misaki, M. and K. Yasumatsu. 1985. Rice enrichment and fortification. In *Rice Chemistry and Technology* (2nd ed.), ed. B.O. Juliano, 389–401. American Association of Cereal Chemistry, St Paul, Minnesota, USA.

Mohapatra, M., A. Kumar, and S.K. Das. 2012. Mixing characteristics of rice by tracer technique in an agitated cylindrical preconditioner developed for puffed rice production system. *Journal of Food Process Engineering* 35:784–791. doi: 10.1111/j.1745-4530.2010.00628.x.

Murashima, K., T. Nishimura, Y. Nakamura, J. Koga, T. Moriya, N. Sumida, T. Yaguchi, and T. Kono. 2002. Purification and characterization of new endo-1,4-β-D-glucanases from *Rhizopus oryzae*. *Enzyme and Microbial Technology* 30:319–26.

Murugesan, G. and K.R. Bhattacharya. 1989. The nature of starch in popped rice. *Carbohydrates Polymers* 10:215–25.

Nura, M., M. Kharidah, B. Jamilah, and K. Roselina. 2011. Textural properties of laksa noodle as affected by rice flour particle size. *International Food Research Journal* 18:1309–12.

Ogawa, K., D. Toyama, and N. Fujii. 1991. Microcrystalline cellulose-hydrolyzing cellulase (endo-cellulase) from *Trichoderma reesei* CDU-II. *Journal of General and Applied Microbiology* 37:249–59.

Ohtsubo, K., T. Okunishi, and K. Suzuki. 2000. Processed novel foodstuffs from prepigmented brown rice by a twin screw extruder. In *Rice Is Life: Scientific Perspectives for the 21st Century*, ed. K.L. Heong, 275–77. IRRI (International Rice Research Inst.), Manila, Philippines.

Pan, Z. 2004. *Annual Report, Comprehensive Research on Rice. Improvement of Consistency and Accuracy of Rice Simple Milling*. USDA ARS WRRC, Albany.

Park, J.K., S.S. Kim, and K.O. Kim. 2001. Effect of milling ratio on sensory properties of cooked and on physicochemical properties of milled and cooked rice. *Cereal Chemistry* 78:151–56.

Pszczola, D.E. 2001. Rice: not just for throwing. *Food Technology* 55:53–9.

Roy, P., T. Orikasa, H. Okadome, N. Nakamura, and T. Shiina. 2011. Processing conditions, rice properties, health and environment. *International Journal of Environmental Research and Public Health* 8:1957–76.

Saleh, M.I., J.F. Meullenet, and T.J. Siebenmorgen. 2008. Development and validation of prediction models for rice surface lipid content and color parameters using near-infrared spectroscopy: a basis for predicting rice degree of milling. *Cereal Chemistry* 8:787–91.

Sarao, L.K., M. Arora, V.K. Sehgal, and S. Bhatia. 2011. The use of fungal enzymes viz protease, cellulase, and xylanase for polishing rice. *International Journal of Food Safety* 13:26–37.

Sareepuang, K., S. Siriamornpun, L. Wiset, and N. Meeso. 2008. Effect of soaking temperature on physical, chemical and cooking properties of parboiled fragrant rice. *World Journal of Agricultural Sciences* 4:409–15.

Schober, T.J., C.M. Brien, D. McCarthy, A. Barnedde, and E.K. Arendt. 2003. Influence of gluten free flour mixes and fat powders in the quality of gluten free biscuits. *European Food Research Technology* 5:369–76.

Shih, F.F. and K.W. Daigle. 1999. Oil uptake properties of fried batters from rice flour. *Journal of Agricultural and Food Chemistry* 47:1611–15.

Shitanda, D., Y. Nishiyama, and S. Koide. 2002. Compressive strength of rough rice considering variation of contact area. *Journal of Food Engineering* 53:53–8.

Siebenmorgen, T.J. and H. Sun. 1994. Relationship between milled rice surface fat concentration and degree of milling as measured with a commercial milling meter. *Cereal Chemistry* 71:327–9.

Slaton, N., K. Moldenhauer, J. Gibbons, M. Blocker, J.C. Wilson, R. Dilday, J. Robinson, and B. Koen. 2000. Grain characteristics of rice varieties. University of Arkansas, Cooperative Extension Service, Rice Information Bulletin, No. 146.

Soponronnarit, S. 1999. Fluidised-bed paddy drying. *Science Asia* 25:51–6.

Suksomboon, A., K. Limroongreungrat, A. Sangnark, K. Thititumjariya, and A. Noomhorm. 2011. Effect of extrusion conditions on the physicochemical properties of a snack made from purple rice (Hom Nil) and soybean flour blend. *International Journal of Food Science and Technology* 46:201–8.

Suteebut, N., K. Petcharat, D. Tungsathitporn, and D. Sae-Tung. 2009. Pasta from organic Jasmine rice. *Asian Journal of Food and Agro-Industry* S349–S355.

Sutherland, J.W. and T.F. Ghaly. 1990. Rapid fluidized bed drying of paddy in the humid tropics. In: Naewbanij, J.O. (Ed.), Proceedings of the Thirteenth ASEAN Seminar on Grain Postharvest Tech., 4–7 September 1990, Brunei Darussalam, pp. 1–12.

Tseng, Y., J. Yang, C. Chen, and J. Mau. 2011. Quality and antioxidant properties of anka-enriched bread. *Journal of Food Processing and Preservation* 35:518–23.

Uffen, R.L. 1997. Xylan degradation: a glimpse at microbial diversity. *Journal of Industrial Microbiology and Biotechnology* 19:1–6.

Wan, X., J. Weng, H. Zhai, J. Wang, C. Lei, X. Liu, T. Guo, L. Jiang, N. Su, and J. Wan. 2008. Quantitative trait loci (QTL) analysis for rice grain width and fine mapping of an identified QTL allele gw-5 in a recombination hotspot region on chromosome 5. *Genetics* 179:2239–52.

Wanyo, P., C. Chomnawang, and S. Siriamornpun. 2009. Substitution of wheat flour with rice flour and rice bran in flake products: effects on chemical, physical and antioxidant properties. *World Applied Sciences Journal* 7:49–6.

Wee, C., P. Raveendran, and F. Takeda. 2009. Sorting of rice grains using Zernike moments. *Journal of Real-Time Image Processing* 4:353–63.

Wongwises, S. and M. Thongprasert. 2000. Thin layer and deep bed drying of long grain rough rice. *Drying Technology* 18:1583–99.

Wood, T.M. and V. Garcia-Campayo. 1990. Enzymology of cellulose degradation. *Biodegradation* 1:147–61.

Yadav, B.K. and V.K. Jindal. 2001. Monitoring milling quality of rice by image analysis. *Computers and Electronics in Agriculture* 33:19–33.

Yadav, B.K. and V.K. Jindal. 2008. Changes in head rice yield and whiteness during of rough rice (*Oryza sativa* L.). *Journal of Food Engineering* 86:113–21.

Zareiforoush, H., M.H. Komarizadeh, and M.R. Alizadeh. 2009. Effect of moisture content on some physical properties of paddy grains. *Research Journal of Applied Sciences, Engineering and Technology* 1:132–9.

Zhang, Q., S.X. Yang, G.S. Mittal, and S. Yi. 2002. YAE—automation and emerging technologies: prediction of performance indices and optimal parameters of rough rice drying using neural networks. *Biosystems Engineering* 83:281–90.

5 Dry Milling

Sara Beirão da Costa, Vítor D. Alves,
and Luísa Beirão da Costa

CONTENTS

5.1 INTRODUCTION

Cereals are probably the most ancient basic staple foods ever used by humans; dating back to even before agricultural practices had been established. However, in the raw state, their consumption is not attractive due to hard texture, lack of taste, and low digestibility. As soon as fire was discovered, primitive man learned how to improve those characteristics: at first, by toasting the whole grains, and thus trying to at least partially separate the harder outer layers; and later on (~10,000 BC), by baking bread and brewing as the first forms of food processing. Surely, these processes imply some kind of previous preparation such as the size reduction of grains. This process was traditionally performed in stone mills, first operated by hand, animal, water, or wind power and, more recently, using diesel or electrical energy.

Cereals' dry milling aims, most of the times, not only at size reduction but also at the mechanical separation of the starchy endosperm from kernels. Cereal grains can be milled into many different forms, from whole (dehusked) or cracked (grits) kernels to small particles (semolina and flour). The dehusking process is applicable to husky cereals like rice, barley, and oats. In the case of rice processing after husk removal, another operation is performed to remove bran layers and produce polished or white rice from brown rice. However, most cereals are not consumed in grain form but are further processed into different end products. Therefore, wheat, maize, or rye are milled to semolina or flour to be used mainly in baking bread, biscuits, or pasta products, whereas maize or oats are processed to coarsely broken grain to yield grits for flakes, puffed breakfast cereals, or animal feed.

The dissimilar physical properties of cereal grains justify tailor-made processes and thus dry milling may cover a wide range of procedures. Nevertheless, a cereal dry milling industrial unit basically covers some common types of operations: (i) reception and cleaning of grains, (ii) conditioning, (iii) size reduction, and (iv) classification and separation of fractions.

From these operations, size reduction is a crucial step because its operating parameters determine the resulting product characteristics.

This chapter intends to cover these topics based on important case studies, as justified by the economic importance of each one.

5.2 MILLING AS A UNIT OPERATION

The aim of the size reduction step is the production of smaller mass portions from larger mass pieces of the same material, and is associated with an increase in the overall surface area. The material is subjected to stress by the application of compression, impact, or shear forces, depending on the type of equipment used. In the dry milling process of cereals, not only is the size reduction of the grains performed but also the separation of the grain parts according to their composition. This is achieved by carefully selecting the type of forces to be applied. For example, compressive forces are commonly used for coarse crushing of hard materials and, if moderately applied, also enables control over the breakdown of the material (e.g., to crack open grains of wheat to facilitate the separation of the endosperm from the bran; Brennan 2012). Although the three types of forces may be present in every kind of mill, generally, only one of them is predominant. Whereas compression is dominant in roller mills, impact force is the most important in hammer mills, and the shear forces are the most intense in disk mills. The degree of the material size reduction is generally expressed as the size reduction ratio:

$$R = \frac{x_1}{x_2} \qquad (5.1)$$

where x_1 and x_2 are the average particle sizes of the feed and product, respectively. The size reduction ratio may vary from 8 in coarse crushing to more than 100 in fine grinding (Barbosa-Cánovas et al. 2005).

5.2.1 ENERGY CONSUMPTION

Even though it is impossible to estimate, with accuracy, the energy needed for particle size reduction of a given material, several empirical correlations have been proposed based on the following equation (Coulson and Richardson 1965):

$$\frac{dE}{dx} = -Cx^p \tag{5.2}$$

which indicates the energy (dE) required for a small size reduction (dx) per unit mass of material. When $p = -2$ and $C = K_R f_c$, from the integration of Equation 5.2, the Rittinger law is obtained:

$$E = K_R f_c \left(\frac{1}{x_2} - \frac{1}{x_1} \right) \tag{5.3}$$

where f_c represents the resistance of the material, K_R is the Rittinger constant, and x_1 and x_2 are the particle sizes of the feed and product, respectively. This equation is more appropriate for fine milling, in which there is a large increase in surface area per unit mass of material. On the other hand, Kick's law (Equation 5.4) is more suitable for describing energy consumption for coarse milling, as it is related to plastic deformation before particle rupture and the new surface area produced is significantly smaller. It is obtained from Equation 5.2, for $p = -1$ and $C = K_k f_c$:

$$E = K_k f_c \ln \frac{x_1}{x_2} \tag{5.4}$$

in which K_k is the Kick constant. Bond's law (Equation 5.5) is an intermediate approach between the other two and has been applied with success to a variety of materials undergoing coarse, intermediate, and fine grinding (Brennan 2012). It is obtained by the integration of Equation 5.2 for $p = -1.5$:

$$E = 2C \sqrt{\frac{1}{x_2}} \left[1 - \frac{1}{\left(x_1/x_2 \right)^{0.5}} \right] \tag{5.5}$$

It has been stated that only less than 2% of the energy supplied during the milling operation is effectively used for the size reduction of the material (Brennan 2012). This energy refers to the one applied in elastic deformation of the particles before fracture and to the plastic deformation that induces rupture. Most of the energy supplied is lost as heat in elastic distortion of the equipment, friction between particles and between particles and the equipment, friction within the parts of the equipment, as well as noise and vibration (Coulson and Richardson 1965).

5.2.2 Particle Size Distribution

The objective of most size reduction operations is to produce particles within a specified size range. Consequently, it is common practice to classify the particles coming from a mill into different size ranges. For that, several methods may be used including microscopy (e.g., optical, scanning electron and transmission electron microscopy), particle settling, flow classification (e.g., gravity and centrifugal elutriation, cyclonic separation), laser refraction (e.g., low angle laser light scattering), and sieving (Barbosa-Cánovas et al. 2005).

Sieving is the most widely used technique for this purpose because it is the simplest, least expensive, and most reproducible, and belongs to the techniques using the principle of geometry similarity. Moreover, it is considered the only method that allows us to obtain a particle size distribution based on the mass of particles in each size range (Barbosa-Cánovas et al. 2005).

The sieving operation is carried out using a set of sieves with different openings covering a wide range of sizes, from microns to centimeters. The size of the sieves is defined as the minimum square opening through which the particles can pass and is usually referred to as its mesh size (number of wires per linear inch). The size analysis is carried out by stacking the sieves in ascending order of opening size from the bottom to the top of the stack, feeding the powder to the upper sieve, followed by vibration during a fixed time and ending with the quantification of the weight fraction retained in each sieve (Barbosa-Cánovas et al. 2005). The results are usually expressed graphically in cumulative curves of undersized or oversized material as a function of particle size (sieve opening). Furthermore, frequency curves may be obtained by representing the derivative of the cumulative curve as a function of the particle size (Coulson and Richardson 1965).

The minimum particle size that may be processed by sieving is limited, either by the difficulty to produce sieve cloth fine enough to retain the powders or because of the very low gravity force over the particles, which does not keep them from adhering to one another and to the sieve.

5.2.3 Equipment Selection for Size Reduction Operation

The selection of the equipment to be used is closely related to the physical and chemical characteristics of the product to be processed. Some of the material characteristics that one should be aware of are described below.

5.2.3.1 Mechanical Properties of the Feed

The mechanical resistance of the solid material has a great effect, both on the degree of particle size reduction and on the energy consumption. When solid samples are subjected to tension, different correlations between the stress applied and the deformation experienced by the materials may be observed, depending on the mechanical properties of such materials (Figure 5.1).

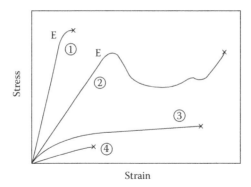

FIGURE 5.1 Typical stress–strain curves for various materials: 1, hard and brittle; 2, hard and ductile; 3, soft and ductile; 4 soft and brittle; E, end of the elastic region, X, breaking point.

Most materials show an elastic deformation within the first range of tension values applied, as demonstrated by a linear correlation between stress and strain. This deformation is reversible, meaning that the solid is able to recover its original size after removing the tension. There is no size reduction under these conditions. The energy supplied is used in the elastic deformation.

The higher the slope of the stress–strain curve in the elastic region becomes (elastic modulus), the harder and stronger the material is. After point E, the limit of the elastic region, the material experiences a plastic deformation, which is irreversible. At this stage, brittle materials will break rapidly (curves 1 and 4), but ductile materials (curves 2 and 3) will continue to deform, eventually reaching the breakup point (X). The overall mechanical resistance is proportional to the area under the stress–strain curve. As such, curve 2 refers to more resistant materials that will need more energy to break up than the materials of curves 1, 3, and 4.

Very hard materials generally need a higher residence time in the action zone, leading to a lower feed processing rate. To overcome this effect, mills with a higher processing capacity may be used. Hard materials are sometimes abrasive. As such, the working surfaces should be made of hard material, like manganese steel, and should be easy to remove and replace.

Fibrous materials have the ability to resist the propagation of cracks and are difficult to break, as fibers increase material toughness by relieving stress at the ends of the cracks. The most adequate mills for this type of materials are disk mills, pin-disk mills, or cutting devices (Brennan 2012).

5.2.3.2 Moisture Content of the Feed

Generally, for each solid material, there is an optimal moisture content that enables the desired processing rate and size reduction. In some cases, when feed moisture content is in excess, it may cause clogging, negatively affecting the throughput and efficiency. Moisture can also cause agglomeration, decreasing the free-flowing ability of the product. In addition, water may act as a plasticizer in hydrophilic materials, making them less brittle and more susceptible to elastic deformation. A common problem arising from the milling of very dry materials is the formation of dust, which can cause respiratory problems in operatives and is a fire and explosion hazard (Brennan 2012).

5.2.3.3 Temperature Sensitivity of the Feed

Heat is generated during the milling process due to friction within the equipment and to the deformation of the material in the elastic range. As a consequence, there is a temperature increase, which is particularly important when high operation speeds are applied. This temperature increase may affect the quality of the product being processed. Some mills are equipped with cooling jackets to maintain the temperature below the critical value identified for a specific product. Furthermore, a cryogenic compound (e.g., solid carbon dioxide, liquid nitrogen) may be mixed with the feed. Beyond decreasing the operating temperature, it may facilitate the milling process of fibrous materials by decreasing the temperature to below their glass transition temperature. In the case of cereal milling, the temperature control is not a critical point, as there are no compounds potentially degraded or lost by evaporation at the temperature values reached under ordinary milling processes.

5.2.4 Size Reduction Equipment

The type of mill most applied in cereal milling is the roller mill. These mills are quite versatile, being applied for either coarse or fine gridding, upon choosing the appropriate type of rolls and processing conditions. However, hammer and disk mills have also been used, namely, for wet milling of corn, for sorghum milling, for rice flour dehusking, as well as for processing by-products.

5.2.4.1 Roller Mills

A roller mill consists of two cylindrical steel rolls mounted horizontally, vertically or diagonally, and aligned in parallel to each other, which are rotating in opposite directions (Figure 5.2a). Each form of positioning has specific advantages. The clearance between the rolls is adjustable. An overload compression spring is usually fitted to protect the rolls' surfaces from damage in case a hard object is pulled between them. Rollers operate independently and rotate at different speeds (differential speed), driven by belts connected to electric engines. The solids composing the feed are introduced from above into the space between the rolls and grabbed and pulled through them where they are subjected to compressive and shearing forces promoting their breakup. The slower roll supports the solid, whereas the faster roll shears it.

The different actions performed by roller mills are attained by differences in the size of the rolls (according to the amount of solids, e.g., grain to be milled), the rolls' surface, and the differential speed. Large rolls, with diameters greater than 500 mm, normally rotate at speeds in the range of 50 to 300 rpm, whereas rolls with smaller diameter may be operated at higher speeds (Brennan 2012). The greater the differential speed is, the greater the shear force that is applied. The roll surfaces may be smooth or fluted, or they may have intermeshing teeth or lugs. The smooth and fluted surfaced rolls are the most used in cereal milling. On fluted rolls, the flute form is similar to an italic V (*V*), with one side larger than the other (Figure 5.2b).

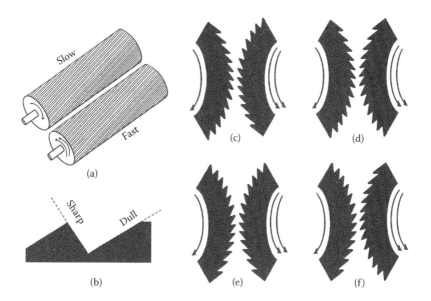

FIGURE 5.2 Schematic representation of roller mill parts: (a) fluted cylinders, (b) detail of typical British flutings, (c) sharp-to-dull, (d) sharp-to-sharp, (e) dull-to-dull, and (f) dull-to-sharp.

As the flutes are asymmetric, the rolls may be aligned with either the sharp or the dull side disposed toward the clearance. As such, the arrangement may be described as sharp-to-sharp, dull-to-dull, sharp-to-dull, and dull-to-sharp (Kent and Evers 1994; Figure 5.2c–f). The shearing operation is improved if flutes are aligned with different angles toward the roll axis. Besides those asymmetries, the flutes may also be designed with different depths. The different combinations induce different shearing forces during the milling process.

Large rolls with smooth surfaces are generally used for coarse crushing, usually achieving a reduction ratio of 4 or lower. However, smaller rolls, with different surface configurations and operating at higher speeds, can achieve higher ratios. To achieve high reduction ratios, the solid material may be processed by consecutive pairs of rolls, with decreasing clearance from one pair to the next. It is usual (e.g., in wheat milling) to have a separation stage between two sets of rolls, in which the particles are separated by size and composition.

The largest particle size that a pair of rolls can accommodate may be estimated by the following equation (Barbosa-Cánovas et al. 2005):

$$\cos \alpha = \frac{R_r + b}{R_r + R_f} \qquad (5.6)$$

where α is called the "nip" angle, that is, the angle formed by the two tangents common to the particle and to each one of the rolls, R_r (m) is the roll's radius, R_f (m) is the average feed particle radius, and b (m) is half of the clearance distance.

The "nip" angle may be predicted by Equation 5.7, which is valid when the particles are taken by the rolls by friction (Brennan 2012):

$$\mu = \tan\frac{\alpha}{2} \qquad (5.7)$$

in which μ is the friction coefficient between the particles and the roll's surface. From Equation 5.6, it is also possible to relate the roll diameter to the size reduction ratio that might be expected in the roller mill, considering that the clearance distance corresponds to the average particle diameter of the product. The theoretical mass flow rate (Q, kg/s) processed by a pair of rolls in a milling operation may be estimated by

$$Q = 4\pi R_r Nbl\rho \qquad (5.8)$$

where N is the roll rotation speed (rpm), l is the roll length (m), and ρ is the bulk density of the product (kg/m^3). It is generally accepted that the real mass flow rate is between 0.1 and 0.3 of the theoretical value (Barbosa-Cánovas et al. 2005; Brennan 2012).

5.2.4.2 Hammer Mills

These consist of ganged hammers (generally rectangular pieces of hardened steel), which are attached to a shaft that rotates at a high speed inside a grinding chamber. The solid material is typically fed by gravity into the mill's chamber and is crushed or shattered by repeated hammer impacts, collisions with the walls of the chamber, as well as particle–particle impacts. The chamber is fitted with a screen, which retains the coarse material for further grinding while allowing adequately sized material to pass as finished products. The final size of the product being ground may be designed by varying the screen size, the shaft speed, and the hammer configuration.

5.2.4.3 Disk Mills

Among the disk mills are the disk attrition and the pin-disk mills. The disk attrition mills may be composed either by one grooved disk rotating in proximity to a fixed grooved surface or by two counter-rotating disks with matching grooves, located close to each other in a casing. The feed is introduced through the center of the stationary surface or from the top into the clearance between the disks, depending on the mill configuration. In both cases, the feed material is broken down predominantly by shear forces. These mills are usually applied for milling fibrous materials such as corn and rice (Brennan 2012).

In the pin mills, two disks are positioned facing each other, separated by a small clearance. Both disks have concentric rows of pins, pegs, or teeth. The rows of one disk fit alternately into the rows of the other disk. There are two different operating configurations. In one of them, both disks may rotate either in the same or in opposite directions at different speeds. Alternatively, one of the disks is maintained static while the other is rotated. The feed is introduced through the central axis of the disks and the particles are accelerated radially outward through the mill by centrifugal forces and launched into the impact zone where they are subjected to impact and shear forces between the pins. Pin mills may be used for coarse grinding, fine grinding, and for deagglomeration. This flexibility is provided by varying the disk's rotation velocity. Whether this equipment is used, for example, for cracking wheat, creating powdered sugar, or as a cellulose fiber conditioner, large production throughput rates are usually attained.

5.3 DRY MILLING PROCESS

The aim of modern milling is to obtain, as much as possible, the grains' floury endosperm. The outer layers of the grain, identified as bran by millers, as well as the germ, should be separated to avoid compromising technological ability. Bran includes pericarp, seed coat, nucellar epidermis, and aleurone layer (Delcour and Hoseney 2010). However, these materials are interesting both from a nutritional as well as from an economical point of view, and so their recovery also increases added value to the industry. These two grain fractions present considerable amounts of protein, dietary fibers, B vitamins, tocopherol, and lipids (Delcour and Hoseney 2010).

The final product purity determines the extraction rate or yield, the last one increasing with bran contamination and resulting in higher ash content of the refined flour. The extraction yield, the amount of flour obtained from 100 kg of grain, may range from 100%, a theoretical value for whole meal, to different values depending on cereal type and flour ash content. Table 5.1 summarizes indicative figures for wheat and maize flours. Fraction separation was not completely attained in the traditional milling process and always yielded dark, fiber-rich flours.

5.3.1 RECEPTION AND CLEANING OF GRAINS

In an industrial unit, before the milling and fraction separations, cereals must be prepared in order to separate various naturally occurring materials mixed in the cereal grains, both to protect the processing equipment and to ensure the safety and quality of the final product.

Cleaning is the first stage of the cereals' milling process. Most of the processes involved in cleaning are of a physical nature, aiming to remove coarse and fine materials that may include metallic elements, straw, different kinds of other grains and seeds, stones, broken kernels, grains damaged by insects, molds or other types of

TABLE 5.1
Indicative Extraction Rate and Ash Content of Wheat and Maize Flours

Ash (%)	Extraction Rate (%)	
	Wheat	Maize
0.55	78	63
0.85	90	79
1.1	95	89
1.5	100	—

infestations, and dust. This step takes place in the screening room, where different equipment may be used based on different physical properties such as size, shape, specific gravity, and air resistance. The machines used include separator classifiers, destoners, indent separators, and scourers sometimes found in combined form.

5.3.1.1 Separation by Size/Shape

5.3.1.1.1 Separator Classifiers

Separator classifiers rely on a size separation principle by sifting, allowing the separation of materials much smaller or greater than the desired grain. An inclined double-sieved deck, with an adjustable slope, moves back and forth. At the first tray, coarse materials like odd large kernels, stones, straw or others are separated, whereas the bulk of the grain and smaller impurities pass through to the second sieve tray, which allows fine impurities (sand and broken kernels) to pass while the grain is collected. Aspiration systems may be combined to remove dust.

5.3.1.1.2 Indent Separators

Indent separators are used to isolate long grains from round ones. The separation is performed by a pocket system based on centrifugal force. The smaller grains/particles are thrown to the indented cylinder walls, lodged in the pockets, and raised until the point in which they fall due to gravity to a trough, to be removed by a worm screw while the bulk of the grains remain in the feed.

The pocked shape, rotating speed, and trough angle are important parameters to be controlled for high process efficiency. The values of the drum velocities should be adjusted to ensure that the retained particle will not fall down, either before the level of the trough is reached or from dropping out completely.

5.3.1.2 Separation by Specific Gravity and Air Resistance

5.3.1.2.1 Destoner

Despite the efficient cleaning performed in the previously described systems, many impurities, similar in size to grain, may pass along with it and must be removed. That is the case for small stones or even small pieces of glass or metal, which will be deleterious to the subsequent milling equipment.

Stones or other heavy materials are removed by taking advantage of their different specific gravities. In the past, water streams were used to dissipate and efficiently separate the bulk of materials—stones fell to the bottom of the tank while the grain stayed in suspension—allowing different collecting sites. This kind of technology was progressively abandoned, mainly due to cost and environmental constraints, moving on to air classification procedures. At the same time, some kind of motion is also applied, such as vibrations or oscillatory movements, to promote the materials' stratification. Inside a sieve box, an ascending air current stratifies materials according to their terminal velocities.

5.3.1.2.2 Spiral Seed Separator

The broken kernels, which can amount to as much as 3%, are difficult to recover on the separation equipment previously referred to, and flow together with vetch seed and other round particles. Only specific gravity separators may also remove (to some degree) these kinds of materials. However, other equipment taking advantage of gravity and centrifugal force may be very useful for this purpose. The design of the spiral separator presents inclined chutes with two spirals fed from a large hopper at the top, with an adjustable feed plate for each unit where stock is fed. The grain leaving the hopper runs over a cone divider, which spreads the feed evenly. As the stock moves down the spirals, the round seeds and broken round kernels travel at a much faster speed than elongated kernels. Their momentum increases until the round seeds run over the edge of the inner spirals, drop into the outer spiral, progressively outpacing the bulk of the grain, and shift toward the outer spirals. The inner spirals conduct the bulk of grain into one shoot while the outer ones separate the round seeds and broken kernels to another.

A range of equipment is also available to separate units on the basis of their optical, photometric, magnetic, and electrostatic properties. Taking advantage of new technologies, instead of the classic mechanical machinery, different materials can be sorted on the basis of differences in color, structure, shape, and size and even by contamination.

5.3.1.3 Magnetic Separators

Magnetic separators are fundamental equipment to be considered in a cereal milling unit to remove metallic elements (iron and steel) from grain, in this way preventing damages to subsequent machinery. The geometry of the equipment may assume different designs, such as magnets, drum magnets, or magnetized conveyor belts. In all, paramagnetic materials are retained while grains pass through.

5.3.1.4 Optical Sorting

Optical sorting technology led to a step change in the cleaning process. Different individual units may be separated automatically on the basis of color, structure, size, and shape differences.

Optical sorter machines operate in four main steps: the feed system, the inspection device, the image processor, and the ejection system (Hamid et al. 2004). The bulk material is introduced by a feeding system and analyzed on a unit-by-unit basis for different characteristics, allowing very high accuracy and efficiency. Modern equipment substituted the early low-resolution photodiode with high-resolution charged coupled devices that, by increasing optical resolution, can identify smaller defects (Hamid et al. 2004). Some equipment, such as Detox from BEST (Belgian Electronic Sorting Technology, Leuven, Belgium), allow improvements in safety by detecting and separating aflatoxin-contaminated grains. Fungus present in raw material reflects a unique low-intensity light when illuminated at an appropriate wavelength.

The use of an optical sorter may lead to considerable improvements in cleaning efficiency when compared with mechanical cleaning, with an added yield of 0.3% to 0.8% (Anon 2012). Some of these technologies allow substituting several of the previously mentioned machines with single systems.

5.3.1.5 Dedusting

Dedusting is an important operation to be performed after the separation of foreign materials mainly due to the morphological structure of cereals, such as wheat, oat,

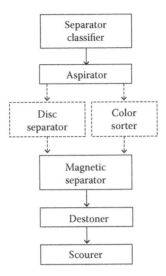

FIGURE 5.3 A sample diagram of a screening room (dashed boxes, alternative equipment).

and barley. In fact, besides dirt sticking to the surface, the existence of a deep groove increases the potential of impurities to be entrapped in that location and renders the cleaning procedures harder. Dedusting is usually accomplished in friction machines sometimes classified as scourers and brushes.

The kind and number of cleaning machines, as well as their positions in the screening room, are settled according to structure and capacity of the milling unit. Furthermore, the cleaning process also has to be fitted to the nature of the cereal crop and the kind and amount of expected impurities. An example diagram of a cleaning process is shown in Figure 5.3.

5.3.2 Conditioning

The purpose of the screening room is to supply grain that is free from impurities and in optimum condition for milling. The final stage of the screening room, after cleaning, is to "condition" the grains.

Conditioning is a preparatory operation intending to help the subsequent milling operations, allowing for the maximum fraction separation efficiency and extraction of flour, producing the desired uniform moisture content throughout the cereal mass, and ensuring that the quality parameters can be met. In this way, moisture is added in precise amounts to smooth the inner endosperm, toughen the outer layers, and make the germ more pliable, reducing powdering of the last two kinds of materials and subsequent darkening of the flour. The bran layers are also loosened to promote their separation from the endosperm. For example, if the moisture content of wheat grains is too high, it may deform rather than break down to release the endosperm. On the other hand, if it is too dry, the bran may be converted into fine particles, which may not be separated by the screens and may contaminate the white flour. In addition, conditioning also contributes to more efficient energy consumption. Water is added and has to penetrate into the grain. After damping, the grain is normally left to stand for 6 to 24 h to allow the moisture to penetrate evenly through the grain by osmosis. In the case of wheat, a usual resting time conditioning is approximately 9 to 10 h long. This type of damping and tempering gives an ideal distribution of water into wheat. At the end, moisture increases by up to 15% to 17.5%. The amount of water added and the tempering time depend on the variety of grain to be milled, initial moisture, prevailing climate, and specification of the finished product and endosperm structure–texture; for example, durum wheat needs a higher moistening conditioning than a soft one.

When conditioning other cereals, different conditions may be used. In the case of maize, and depending on the milling process, the final moisture content may range from 17% to 20% to 22%. Generally, resting time is approximately 1 to 2 h but may last for 24 h.

Rye conditioning, for flour production, generally aims at about 1% lower final moisture content than wheat because this cereal's endosperm is softer.

Conditioning, if not done correctly, may be a critical point in a milling unit. The period of operation, associated with high moisture levels and temperature, may allow microbiological development, compromising the milled product's safety and quality. To avoid this risk, water may be treated with ozone or chlorine to assure sanitation in this wet environment.

The addition of water is controlled by automatic systems and resting takes place when the wet grains are stored into specific bins.

5.3.3 MILLING

5.3.3.1 Wheat

After cleaning and conditioning, the grain is finally ready to begin the milling process. Nowadays, milling into flour or semolina is done with a roller mill gradual reduction process, which has two objectives: first, the separation of the endosperm from bran and germ; second, grinding the grain into fines. From each grinding stage, a blend of coarse, medium, and fine fractions including flour is obtained. Ideally, milling should separate as much endosperm as possible, minimizing the damage to the starch granules. The separation of different materials is achieved by sieving and purifying operations.

Grinding is a central point at a milling unit. This operation intends to break the grain into smaller portions, while achieving some separation of the individual parts of the grain. It is performed with roller mills, mostly with rolls diagonally oriented as this position favors efficient work.

Grinding is performed in a sequence of stages using grinding rolls of different types. The roller surfaces may be smooth or fluted according to the desired kind of action. Changes in roll grooving can make the roller mill useful for a wide range of actions. Furthermore, rolls may run sharp-to-sharp, dull-to-sharp, or dull-to-dull (Figure 5.2) according to the grain hardness characteristics or moisture conditions (or both). To increase the effectiveness of the desired shearing action, the flutes are carved into a slight spiral, inducing a scissor-like action during running. Thus, a cutting action is obtained opening the grain plainly and minimizing friction and damage.

Generally, the gradual reduction process, as applied in the current milling units, consists of two main systems:

- *The break system.* The objective of this system is to remove as much inner endosperm as possible from whole wheat grain or large grain fragments and its separation from the bran and germ, when passing by fluted pairs of rolls.

  The grooving size for each roll in the break system is governed by the size of feed particles. The differential speed is approximately 2.5:1. When conditioning is not efficiently achieved or the breaking rolls are not completely efficient (or both), the break system may also include scratching rolls, so-called because their grooving is much finer. Because they are like an extremely fine break roll, they can, if the pressure is very carefully and exactly adjusted, efficiently separate small portions of bran that remain attached to semolina, with a minimum of powdering. Most often, break systems consist of four or five roller mill stages. Each grinding stage produces a mix of coarse, medium, and fine fractions including flour.
- *The reduction system.* The objective of the reduction system is to grind the endosperm into finer particles producing flour.

  It consists of eight to 16 grinding stages, depending on the type of wheat, using smooth surface rolls. Those rolls run at a much lower differential speed than those of the break system (~1.25:1). The rolls may be classified into coarse and fine rolls referring not to the kind of surface that is always a smooth one, but to the distance between the rolls and whether they produce a coarse or fine grinding.

Proper adjustment and settings of the rolls ensure maximum output of high-quality flour. The miller should avoid overgrinding or using high grinding pressures because this results in broken bran particles mixed in the flour, damaged starch granules, and thus poorer flour quality. Overgrinding does not result in substitute by finer flour but is related to the amount of energy used. When grinding is too gentle, it will result in a waste of endosperm.

After size reduction steps, the small-sized materials should then be separated from the bran and returned to the appropriate rollers until the desired flour is obtained.

5.3.3.1.1 Classification by Size: Sieving

After each reduction operation in the rolls, the resulting product should be sent to classification according to size. Milled products are sent by pneumatic systems to multiple eccentric rotary box-type sifters – plansifters – consisting of a number of superimposed flat sieves with different mesh sizes. The purpose of the operation is to divide stock into coarse and fine middlings and flour. There may be as many as 20 sieves arranged in four different compartments, each one with a number of sieves showing the same range of openings. The mesh of individual sift surfaces depends on the roller mill to which the plansifter is coupled. The sifters separate the larger particles from the smaller ones that flow, as shown in Figure 5.4. The opened grain will pass through sieves in order to separate the fines. Larger particles are retained and separated in the upper sieves of each box while the finer particles sift to the next

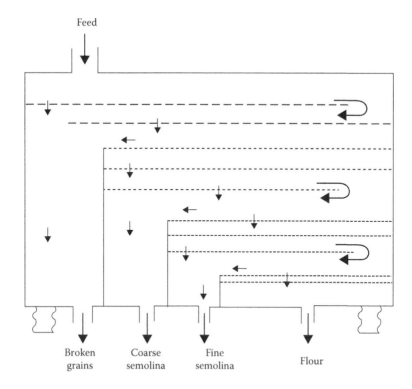

FIGURE 5.4 Example of a simplified scheme of materials flow along a plansifter.

box until they reach the bottom. The particles that pass through the finest sieve are considered flour. This process is repeated after each grinding/reduction operation, until the desired particle size is achieved. The particles larger than flour again proceed over to the rollers until it is broken down into flour.

Bran and germ are recovered together with finer semolina at the plansifters located after the fifth or sixth break roller mill. All the portions of the grain that do not break down into flour are still nutrient-rich products and are therefore used in feed production.

5.3.3.1.2 Classification by Composition: Purifying

After size classification, and for the same range of dimensions, it is necessary to separate materials from a composition point of view. Products resulting from the break system contain both bran and impure endosperm (with bran) particles. Purifying aims to remove pieces of bran and impure endosperm from completely pure ones. This operation must be performed before grinding on the reduction system to produce high-quality flour with low ash content. The main underlying physical principle involved in the process is the difference in terminal velocities among particles. Separation is achieved by blowing air currents, causing particle stratification according to size and composition.

Purifiers consist of long oscillatory sieves, divided into four sections, sloping down from the inlet (head) to the last collecting point (tail). The stock is uniformly fed at the head of sieve and should evenly cover it slowly, moving toward the tail by oscillatory motion and gradient. The different sections are subjected to increasing upward flows of air rising through the sieve valves, allowing the control of the air flow from the head to the tail. The motion is set to keep the stock permanently agitated. Therefore, the sieving period must be long enough to allow the complete stratification of the bulk material. The bolting cloth is progressively coarser from head to tail separating and grading fractions.

The separation of different materials is based on two main principles:

i. When particles, dissimilar in size but similar in density, are subjected to motion and shaken together, the smaller particles tend to stratify below the larger ones.
ii. Particles that are similar in size but different in density tend to stratify the heavier ones dropping below the lighter ones.

As a result, stratification in layers is produced in the purifiers by the following bottom-to-top order:

i. Smallest pure endosperm particles
ii. Larger pure endosperm particles
iii. Smaller particles containing bran
iv. Larger particles containing bran
v. Larger bran particles
vi. Smaller bran particles (raised by the air current)

Stratification in this way allows for separation along the purifier surface, although no sharp division may be expected among each section (Figure 5.5).

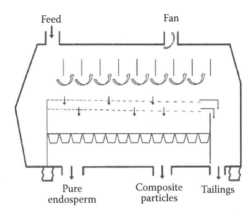

FIGURE 5.5   Example of a simplified scheme of the purifying process.

The outputs of purifiers go to reduction rolls unless further treatment is needed. Particles of endosperm still showing adhering bran are sent to scratching rolls. Bran particles are a by-product and are treated accordingly.

Purifiers are especially important in a mill for durum wheat semolina production to be used in pasta products because color defects or pigmentation should be avoided. The presence of bran must also be avoided as it may also induce the occurrence of fracture points in the pasta matrix.

Both plansifters' and purifiers' sieve mesh are set according to the range of dimensions to be treated. A series of gridding and reducing roll sifters and purifiers repeat the process until the desired extraction level is obtained.

Most often, cereal grains are infested with insects that compromise the end product quality. Flour or semolina may be subjected to a final treatment to guarantee their quality and stability during storage. Impact equipment, taking advantage of centrifugal force, provides a way to destroy insects from the egg stage to adult form. The flour is conducted by a screw and hurled between two steel rotating discs. The speed achieved is enough to kill all forms even those more resistant ones, such as eggs. An illustrative milling diagram is shown in Figure 5.6.

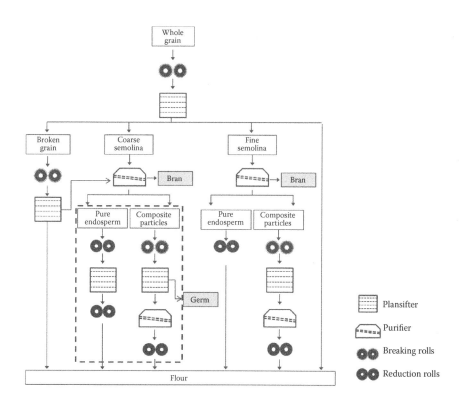

FIGURE 5.6　Illustrative diagram of wheat flour milling process (dashed line, repeat process).

5.3.3.2 Rye

Milling rye into flour is not considerably different from wheat milling; however, some adjustments in the process should be considered. The first one deals with the conditioning, as already mentioned.

The major difference is encountered in the reduction system that, unlike wheat milling, does not include smooth rolls. This fact is justified by the tendency of rye flour to flake. Instead of smooth rolls, very finely corrugated rolls are used. On the other hand, because rye products are more difficult to sift, a larger number of sifters should be considered.

5.3.3.3 Corn

The adjustments to be performed in a corn milling unit are mainly due to the considerable differences in grain morphology and composition when compared with wheat. Due to the high proportion of germ in the grain and because of its high fat content, an extra step in the process has to be considered. This step, degermination, aims to separate the germ (but some bran is also removed) and it is achieved by feeding the corn grain into an attrition device. Furthermore, the stock produced needs to be sorted after coming out from the degerminator.

Nowadays, degermination is performed in equipment working either horizontally or vertically. The Beall Degerminator (The Beall Degerminator Company, Decatur, IL), is essentially an attrition horizontal-working device with a cast iron cone section rotating at about 750 rpm inside a similar-shaped cone covering. The rotor has knob-like protuberances around the section and the shell has perforated screens. For efficient operation, corn generally needs a preconditioning step to approximately 20% to 22% moisture and a tempering time of about 2 h. A Buhler AG (Uzwil, Switzerland) Degerminator MHXM-M is also a horizontal device in which grain is processed between a roll rotor and a screen jacket. In this case, conditioning is not needed. The VBF series degerminator from Satake Corporation (Japan) differs from all previous machines by having a vertical configuration, allowing a more uniform action on the grain. It operates a vertically eccentric rotor and multiple screens exerting a rubbery action, resulting in germ and bran removal. Bran layers are removed by rubbing grain against grain resulting from its movement across specially designed textured surfaces. As the rotor turns, the clearance between the rotor and the screen is constantly changing, causing the grain to be subjected to alternate periods of compression and relaxation. For an efficient process, a short tempering time of 10 min is needed.

The dry degermination of corn in all the systems referred to ensures that after this step, the corn produced will have a low fat content. As previously noted, to obtain an efficient germ separation, conditioning at approximately 20% to 22% moisture content should be provided. At this level, besides germ separation, excessive fragmentation of endosperm is also avoided. However, at this moisture level, the grading and milling operations are difficult and therefore drying must be performed to approximately 15% to 16% humidity. Rolling and classification operations are set in a similar way as previously described for wheat milling.

5.3.3.4 Sorghum

Sorghum is a tropical crop and therefore the milling process is still performed by manual methods as industrial processes are still under development. However, most of the operations performed both by hand or industrially are similar. At the industrial level, the first attempt to produce sorghum flour included adapting wheat flour milling process and adding a decortication stage performed with attrition discs. However, this approach leads to low economic yields and was abandoned. Other approaches have been used during the last decades by introducing different kinds of decortication equipment, tempering conditions and mill types.

Nowadays, decortication of sorghum grain is generally conducted by dry abrasive actions and so does not imply a previous conditioning step. In fact, it has been demonstrated that tempering of the grain not only increases the amount of broken kernels and decreases the flour yield but also increases the fat and ash levels in the final products (Perten 1983, 1984). Currently, in Africa, the equipment most used for the decortication at the industrial level consists of a cylindrical box containing a number of evenly spaced carborundum stones rotating at 2000 rpm, the grain being decorticated by the abrasive action of the stones and also against the walls of the equipment. The extraction rate, in this case, defined as the amount of grain material produced as meal, is approximately 75% to 85%, with a relatively lower fat and ash content than those obtained using other methods (Eastman 1980). More recently, another kind of equipment, PeriTec, developed by Stake Corporation (Japan) for improving bran removal from wheat, was also applied to sorghum decortication. In this case, both abrasive and friction actions are applied, requiring a light conditioning step of increasing moisture content by 1% to 3% and a short resting period of 3 to 5 minutes (Kebakile 2008).

Reduction to flour is produced by roller or impact mills, which may or may not include a degermination step. When it is included in the process, the same kind of equipment used in maize degerming is used. Tempering conditions are quite dependent on sorghum variety as well as on the kind of mill to be used, but may range between 16% and 26% (Kebakile 2008).

REFERENCES

Anon. 2007. *Optical Sorter Z Series—A Revolution in Grain Sorting.* Buhler Sortex Lda. United Kingdom.

Anon. 2012. *Optical Sorter Sortex M MSOC.* Bühler AG. Switzerland.

Barbosa-Cánovas, G.V., E. Ortega-Rivas, P. Juliano, and H. Yan. 2005. *Food Powders: Physical Properties, Processing, and Functionality.* New York: Kluwer Academic/Plenum Publishers.

Brennan, J.G. 2012. Mixing, emulsification, and size reduction. In *Food Processing Handbook*, edited by Brennan, J.G., and A.S. Grandison, vol. 2, 363–406. Weinheim, Germany: Wiley-VCH Verlag & Co. KGaA.

Coulson, J.M., and F.F. Richardson. 1965. *Chemical Engineering, vol. II.* London: Pergamon Press Ltd.

Delcour, J.A., and R.C. Hoseney. 2010. *Principles of Cereal Science and Technology.* St. Paul, MN: AACC International, Inc.

Eastman, P. 1980. *An End to Pounding: A New Mechanical Flour Milling System in Use in Africa.* Ottawa, Canada: International Development Research Center.

Hamid, G., M.J. Honeywood, and S.C. Bee. 2004. Optical sorting for the cereal grain industry. *Food Science* 25 (10): 287–290.

Kebakile, M.M. 2008. Sorghum Dry-Milling Processes and their Influence on Meal and Porridge Quality. PhD diss., University of Pretoria.

Kent, N.L., and A.D. Evers. 1994. *Technology of Cereals: An Introduction for Students of Food Science and Agriculture*. New York: Elsevier Science Ltd.

Perten, H. 1983. Practical experience in processing and use of millet and sorghum in Senegal and Sudan. *Cereal Foods World* 28: 680–683.

Perten, H. 1984. Industrial Processing of Millet and Sorghum. Paper presented at Symposium ICC 11th Congress—The Processing of Sorghum and Millets: Criteria of Quality of Grains and Products for Human Foods, Vienna.

6 Flour Quality and Technological Abilities

Maria Papageorgiou and Adriana Skendi

CONTENTS

6.1   INTRODUCTION

Wheat flour, the most important ingredient in a recipe, is a key factor in the baking process determining the quality of the baker's products. When it comes to deciding about the quality of wheat, it should be taken into account that the term "wheat quality" has different meanings depending on its intended use. Therefore, the evaluation of the quality of flour must be made with the consideration of its end use. The suitability of wheat flour for one specific purpose does not mean that it fulfills the criteria for satisfactory results for another use (Halverson and Zeleny 1988). Wheat flour is used in baking, primarily to make white bread and numerous other bakery products. Generally, bread is made according to regional preferences, that is, French bread, Italian breads, Vienna bread, white and variety pan breads, Polish bread specialties, and steamed bread (in Far Eastern countries, especially in China). The ideal raw material for pasta is semolina from durum wheat. Although the hard-wheat farina produces good pasta except that it does not have the desired yellow color and is not as resistant to overcooking as that produced from semolina.

The technical information normally provided to the general user regarding wheat flours includes the following: flour type (e.g., patent, high-gluten, all-purpose, or wholemeal), whether or not the flour is made up of hard or soft wheat (or a blend), and the percentage protein content. The baking quality of flour, however, relates to both the amount and the quality of the gluten proteins and is also determined by the complex interactions of all the biochemical constituents present in flour. Generally, higher protein contents indicate a higher level of gluten, which results in a more elastic, better-textured bread.

In the following sections, we will try to elucidate the most important aspects related to flour quality and present the available tests to determine the quality.

6.2   INTERACTION OF MILLING AND FLOUR QUALITY

6.2.1   Tempering

It is possible, during the stages of milling, to affect the flour quality, for instance, by tempering. This is a routine procedure that enhances the efficiency of flour extraction. Tempering is the process of hydration of wheat in water for several hours. When the grain reaches the desired moisture content (15%–16%), then it is possible for the bran, endosperm, and germ to be separated effectively. After tempering, the bran becomes toughened—loses strength and stiffness but increases in elasticity and plasticity—so that it does not break during grinding. Also, the endosperm can be fractured but it does not become a very fine powder too fast. The quality of the flour can be changed when the conditioning (tempering) parameters (i.e., moisture, length of time, temperature, initial moisture) of wheat grain are changed. Kweon et al. (2009) showed that increasing the moisture content decreases the flour yield of soft red winter wheat grain and improves flour quality due to less bran contamination (Kweon et al. 2009). Furthermore, they observed that changes in wheat moisture content had the greater effect among the conditioning parameters.

6.2.2 Blending

Millers blend lots of grains of different varieties for uniformity of protein content and flour quality. The blending of grains with diverse quality attributes can be used to achieve improved quality and market value compared with the original lots of grain. Success in the blending process first depends on knowing the quality characteristics of the components, and then on various practical aspects, particularly the difficulties associated with sampling and mixing. A blend of hard red winter wheat flour and soft winter wheat is suggested to perform consistently well when making Italian style bread (Bilheux et al. 1989; Pyler 1988). However, the ability to blend flours is often limited by the storage capacity available for keeping the individual flour types separated (Carson and Edwards 2009).

6.2.3 Milling

After the stage of tempering, the wheat passes through five or six pairs of counter-rotating rolls, which comminute it. The goal of milling is to maximize the yield of flour with minimal contamination by bran or germ. Yield is affected by factors that influence easy separation of the endosperm from the bran (Steve et al. 1995).

The miller maximizes the flour quality and yield by using a combination of roll sets, that is, to reduce particle size, to separate bran from endosperm, and to classify the flour particles into various sizes. The total flour is called straight grade, but various streams are taken from it to make different types of flour.

Endosperm protein and mineral content are decreasing from the outside of the kernel to the center. At higher extraction rates, the amounts of aleurone and outer endosperm layer are larger than in flour from low extraction rates, that is, the ash and protein increase with the extraction rate. The proteins of cereal are unevenly distributed over the layers of the kernel. The water-insoluble protein fractions glutenin and gliadin, which are important for baking, are mainly found in the endosperm. The aleurone layer mainly contains water-soluble proteins that have no influence on the baking properties of wheat. On the other hand, the germ contains approximately 10% lipids. The milling process would destroy the structure in which the lipids are bound, leaving them exposed to oxygen, resulting in rapid oxidation (enzymatic and nonenzymatic), and, consequently, rancidity would occur. Wholemeal flour containing the wheat germ is therefore less stable than white flour.

However, to some extent, millers can handle the physical properties of the flour they produce through the grinding process and stream selection. By altering the grinding conditions, a reduction or an increase in the level of starch damage can be achieved, which in its turn affects flour's water-absorbing capacity. By further decreasing the extraction rate, the protein and the ash content decrease and the color grade improves in the resulting flour.

The yield of the fractions also varies, depending on the milling setup and the extraction rate used by the miller; however, typical distributions might be 75% as flour, 1% as purified germ, 12% as bran, and 12% as "shorts." Shorts is a stream derived mainly from the sifters during the break stages of milling. It includes bran flakes with endosperm that cannot be efficiently separated.

It is known that the milling process can affect the baking quality of the flour; excessive grinding is detrimental to flour quality. Shear forces developed during milling break the bran, for example, in hard wheat, whereas in soft wheat, the shear forces are redirected in a different way and the bran is less broken (Pomeranz and Williams 1990). Large shear forces and high temperatures at the roller surfaces are responsible for changes in flour properties.

Granulation or particle size may be considered as an important characteristic of flour. Fine average particle size accelerates the rate of flour hydration. Smaller particles usually mean more damage to the starch if the small size was achieved with ordinary cylinder mills. This results in higher water absorption (a characteristic that is valued in flour destined for plain bread manufacture) and better breakdown by the native and added enzymes, but also reduced stability of the dough.

Air classification is sometimes used as a means of producing special-purpose flours of desired protein–starch composition. This means the fractionation of flour, on the basis of particle size and density, by applying centrifugal force and air flow. The fractions differ in composition as well as in particle size and density, and are highly specialized for different uses. After air classification, soft wheat flour, which is initially low in protein, has a good protein shift (Peplinski et al. 1964), whereas normal hard wheat flour with protein content higher than that of soft wheat flour shows a lower protein shift (Peplinski et al. 1965; Stringfellow and Peplinski 1964). Air classification was used to obtain pure semolina (coarse particles) in durum wheat milling (Haddad et al. 1999).

6.3 PHYSICAL TESTS APPLIED TO DETERMINE FLOUR QUALITY

6.3.1 Color

Flour color often affects the color of the finished product and is therefore one of many flour specifications required by end users. In general, bright, white-colored flour is the most desirable for many products. Flour color evaluation is a complex process because there are different factors controlling it. It is affected not only by its grade or extraction rate but also by the color characteristics of the endosperm and the skin of the wheat kernel from which the flour is obtained (Yasunaga and Uemura 1962).

In particular, the presence of bran particles detracts from the whiteness of flour and of the products baked from it (Barnes 1986). It has been observed that the only significant color differences in the endosperm were in the yellowness due to carotenoids (Yasunaga and Uemura 1962). However, this contribution from the endosperm differs in different wheat varieties. Thus, flour color cannot be regarded as an accurate indicator of bran content between different samples (Barnes 1986). The apparent color of flours depends not only on genetic variety but also on the degree of weathering and vitreousness (Neel 1980).

Color testing of wheat flour is done for the purpose of evaluating either its whiteness, which determines the extent of oxidation of carotenoid pigments by bleaching agents, or the presence of bran particles. Quality control involving color evaluation is carried out in various ways. Because color perception differs from person to person, and depends on lighting and various other factors, many industries rely on human vision coupled with an instrumental system of color measurement. These instruments provide a quantitative measurement and are widely used to monitor the color of flour.

The Kent-Jones and Martin flour color grader (Kent-Jones and Martin 1950a, 1950b) and the Agtron Reflectance meter in the green mode (AACC Method 14-10) have been established as color grade quality factors by millers and bakers alike (Kent-Jones 1955). Both use the principle of incident light reflectance and measure the color of flour as a proportion of the light reflected from a standard surface. The Agtron instrument measures the reflectance of light from a flour–water slurry at 546 nm. The flour sample is calibrated against two disks of different reflectances to establish a range of brightness. In general, the purer the flour, the brighter the color and the higher the reflectance value. Straight-grade flour generally shows a color value of 80 to 85, patent flours yield higher values, whereas high-ash flours yield lower values.

More recently, dry flour measurements have been reported in terms of three-dimensional color values ($L^*a^*b^*$) in accordance with standards developed by the Commission Internationale de l'Éclairage (CIE Publication 1986). These three components of reflected light are measured with a tristimulus reflectometer Minolta Chroma Meter (Minolta 1994). Within this system, L^* is a measure of the lightness of the color of the sample (whiteness, 100 white–0 black), a^* measures the red (positive values +60) and green (negative values −60) characteristics, and b^* measures the yellow (positive values +60) and blue (negative values −60) characteristics. This method is widely used (Barros et al. 2010; McCaig 2002; Oliver et al. 1992). The perceived color of wheat flour was related to the ash and yellow pigment contents of the flour streams. L^* was correlated with flour ash content and b^* with flour yellow pigment content and flour color index (L^*-b^*) with both ash and yellow pigment (Oliver et al. 1993).

The operative miller compares the flours every hour against an accepted standard. This test is the "Pekar test," in which a flattened portion of flour is wet and dried to exaggerate the color differences (Posner 2003).

6.3.2 GRANULARITY (PARTICLE SIZE)

The milling industry produces different wheat flour types depending on the breadmaking process. Nowadays, millers have flour blenders for tailor-made flour because flour needs to be adapted to each bread or biscuit process. Large mills produce five or six principal flours and, depending on the type of bread or biscuits required, they could blend these flours to achieve the required flour quality. The granularity or particle size distribution of a flour is determined by the severity of the grinding and the mesh sizes of the sieves through which it is sifted during manufacture.

Excessive overgrinding can have detrimental effects because it is related to starch damage that, in turn, reduces the tolerance of dough to changes during fermentation. Fine flours have a whiter appearance compared with more granular ones, but this effect is not reflected in the color of the bread formed.

Granularity traditionally can be measured using sieves with a range of mesh sizes. Flour is submitted to a standardized and controlled sieving test. The flour passes through a range of flour sieves. Data are reported as the weight of material remaining on a specified sieve or sieves after sieving for a standard time, expressed as a percentage of the original weight of the sample (AACC Method 66-20.01).

Other methods such as microscopes (Wilson and Donelson 1970), sedimentation (Kent-Jones 1941; Kent-Jones et al. 1939), laser light scanning, or electrolyte displacement (Coulter counter; Williams 1970) can give more detailed information as well as a better description of the particle size distribution. The different techniques measure different parameters and each has its advantages and disadvantages.

Nowadays, digital image analysis coupled with light microscopy offers the ability to record physical parameters for each individual particle as well as their agglomerates. Although additional information can be obtained about the shape of the particles, that is, circularity, aspect ratio, and total perimeter, this method can be time-consuming.

A recent means of characterizing particle size is with a laser diffraction instrument. These instruments provide a complete size distribution without the physical separation required in sieve analysis. Flour particles are dispersed in a compatible liquid (a liquid that does not dissolve or agglomerate the particles) and analyzed. Laser diffraction sizing theory is based on the assumption that all the particles are spherically shaped. This can result in inaccurate diameter and volume estimates for flour particles.

6.4   CHEMICAL TESTS APPLIED TO DETERMINE FLOUR QUALITY

6.4.1   MOISTURE CONTENT

Moisture content is one of the most important factors for the determination of wheat flour quality; the lower the flour's moisture, the better its storage stability. It is closely related to the flour's shelf life and is important for determining its solid content (Gooding and Davies 1997). Moisture is also of great importance for the safe storage of cereals and their products regarding microorganisms, particularly certain species of fungi (Hoseney 1994). The deterioration in baking quality is less prevalent at lower moisture content, which can be attributed to retarded respiration and activity of microorganisms (Staudt and Ziegler 1973). When the moisture level exceeds 16%, the shelf life of the flour is greatly reduced. Generally, the moisture should be 14% to 15%, which when stored under appropriate conditions (relatively cool, dry, and aerated) will provide a long shelf life. Usually, the moisture content of flour could vary from 11% to 15% depending on the storage conditions and hygroscopic nature of the starch (Whiteley 1970).

Apart from the effect on wheat quality, moisture content can be an indicator of profitability in milling (Cornell and Hoveling 1998). The more water that is added, the more weight and profitability are gained from the wheat.

The moisture content can be determined by using the following different AACC and ICC standard methods: oven drying method (AACC Method 44-15A, 44-16, 44-18, 44-19, 44-20, ICC Standard 109/1, 110/1), dielectric meter (AACC Method 44-11), vacuum drying (AACC Method 44-40), and distillation with toluene (AACC Method 44-51).

Nowadays, a more common method is to use a calibrated near-infrared (NIR) spectrophotometer (Dowell et al. 2002; Gooding et al. 1997). By using NIR spectroscopy (ICC Recommendation No. 202), it is possible to obtain the result within seconds. This method uses light rays in the NIR range, which is then absorbed or reflected by the sample at different wavelengths. The calibration of the equipment enables the computer to determine the values for moisture and protein contents very quickly.

6.4.2 Mineral Content

Mineral content, more commonly known as ash content, changes with flour extraction rate. The mineral content is determined by ashing a flour sample in a muffle furnace at 900°C for up to 120 min. Only pure white ash has to be obtained in the porcelain crucible. After cooling, the crucible is weighed and the quantity of ash is stated as a percentage of dry matter (or on a basis of 14% flour humidity). The mineral content of flour may vary between a minimum and a maximum depending on the regulations in individual countries. In the production of bread with high volume yield, it makes a difference whether the flour used has a mineral content of 0.51% or 0.63%. The dough and bread volume varies with the mineral content. If the level of extraction includes little more than the endosperm of the grain, the protein content increases considerably in addition to the mineral content but, at the same time, it will likely be a sharp decrease in the volume yield. The ash content is determined according to ICC Standard 104/1. Other standardized procedures used are AACC Method 08-01, 08-02, and 08-03. A quick test is also possible using NIR spectroscopy based on ICC Recommendation No. 202 or AACC Method 08-21.

Where ash is considered as a grading factor, that is, German flour types 405, 550, 630, 812, and 1050, the ash range for each grade is quite wide. For example, flour could be considered as type 550 when the flour's ash content is between 0.490% and 0.580% (dry basis). Similarly, French flour type numbers are a factor of 10 smaller than those used in Germany.

6.4.3 Acidity Determination

The acidity of flour is an important indicator of its freshness. The lipids and phospholipids are broken down enzymatically during the storage of flour. This breakdown is favored by high moisture content, high temperatures, and a high degree of extraction. In some cases, this breakdown is desirable to improve the quality of flour; however, if the flour is stored for a long time, then it will not be suitable for baking. The lipases and phospholipases naturally present in flour cause an increase in the percentage of free fatty acids and phospholipids. Murray and Moss (1990) examined the fatty acid content in various milled wheat products stored for 12 weeks. They observed that fatty acid content increases during storage, but the increase depends on the milled wheat products. Moreover, they detected that increased storage temperature shows a higher rate of fatty acid release and moisture loss, but the rate of development of fat acidity lessened as sample moisture decreased.

By determining the acidity of the flour, taking the degree of extraction into account, it is possible to monitor the maturation and deterioration of flours. Because the changes in flour during storage consist mainly of an increase in the free fatty acids content, the sample is diluted in ethanol solution (67% v/v ethanol) before filtration. The filtrate is then titrated with sodium hydroxide solution to a pH of 8.5. The amount of 0.1 N of sodium hydroxide solution required (multiplied by two) is a measure of the degree of acidity. The official standardized methods to determine acidity in flours are ICC Standard 145 and AACC Method 02-31.

Higher alcoholic acidity is an indication of higher acidity of the germ oil in the flour, although there is no direct relationship or equivalence between the two.

6.4.4 Protein Determination

6.4.4.1 Quantitative Methods of Protein Determination

Protein is a fundamental quality test of wheat because it forms the basis for payment to farmers and is related to its end-product processing potential. The total amount of protein has a very important influence on the baking properties of wheat flour. The protein content increases with the degree of extraction. The increase results from the protein contained in the aleurone layer. In wheat flour, there are several protein fractions that differ in solubility. It is the gluten-forming protein fractions of the endosperm that determine the baking properties of wheat flours. The proteins from the outer layers and the germ have quite different properties. They are soluble in water or salt and therefore have no influence on the baking properties of flour.

The total protein content of the flour is determined by the Kjeldahl method (ICC Standard No. 105/2). The organic constituents are oxidized in the presence of a catalyst. The ammonia formed after another step is distilled and titrated. The calculated amount of nitrogen by titration is multiplied by 5.7.

Near-infrared reflectance and near-infrared transmission methods have also been developed to determine protein (AACC Method 39-10 and ICC Standard Method No. 159). NIR spectroscopy has long been a recognized method for prediction of the protein content of wheat. The NIR technique is empirical, which means that it requires a set of previously analyzed reference samples to calibrate the instrument before it can be used.

6.4.4.2 Wet Gluten Content

The wet gluten content is a measure of the amount of swollen gluten in the wheat flour, which can be determined by forming a paste from the flour sample and washing it out. Mechanical determination of the wet gluten content of wheat flour (ICC Standard 137/1) is carried out with Glutomatic equipment. According to this method, the paste is prepared and washed automatically inside the machine. The result has to be converted to correspond to a flour moisture content of 14%. It is possible to calculate the gluten index, which is determined by the proportion of gluten passing through the sieve. The higher the proportion of gluten that has not passed through the sieve, the higher the index and the better the quality of the measured gluten. Flours with a gluten index of more than 95 indicate strong gluten characteristics (too strong for bread making), whereas low values (up to 30) indicate weak gluten characteristics (Kulkarni et al. 1987). Flour with a wet gluten content of more than 27 in Europe is suitable for bread-making applications. Stojceska et al. (2007) considered the wet and dry gluten values (24.6–36.3 g/100 g and 8.8–12.5 g/100 g, respectively), which they found in four commercial brands of organic strong white flour, adequate for bread making. The gluten index for the abovementioned flours varied between 81 and 98.

6.4.5 STARCH: DAMAGED STARCH

Native starch is the main component of wheat grain (70%–75% dry weight) and its milling products. Starch shows little influence on the functional properties of wheat flours used in bread, cookie, and cake making, whereas damaged starch (mechanically damaged during the flour milling process) greatly influences the water absorption and fermentation time requirements of bread-making doughs, as well as the staling and crumb textural properties of bread. AACC recommends that Methods 76-11 or 76-13 should be employed to determine the starch content. Generally, in the quality testing of wheat flours, more attention is usually drawn to the physical condition of the starch granule rather than to the quantity of it. The determination of damaged starch in flour is performed by different techniques such as enzymatic (AACC Method 76-31.01 and ICC Method No. 164), iodometric, and NIR reflectance techniques. Damaged starch is also measured using an amperometric method (AACC Standard 76-33 and ICC 172) originally described by Medcalf and Gilles (1965), and incorporated by the Chopin Co. into their instrument. The Chopin SDmatic instrument employs an iodine reaction for the prediction of starch damage (Preston and Williams 2003).

It is recognized that mechanically damaged starch granules are much more susceptible to enzyme action compared with undamaged starch (Cornell and Hoveling 1998). An optimal level of damaged starch in flour is reflected in the effective water absorption (damaged starch absorbs approximately three times the amount of water compared with intact starch granules) and produces an increase in the maltose value that is responsible for the gas-evolving power during the baking process. To produce bread with satisfactory characteristics, an optimum relation should be maintained between the flour's protein content and the level of starch damage (Posner 2003). Optimum starch damage for bread making is approximately 24 g/100 g at 12 g/100 g protein content (Mailhot and Patton 1988). Rao et al. (1989) reported an optimal range of 14.1% to 16.5% for better pliability, texture, taste, and overall acceptability of chapatis. Hard wheat gives flour with much higher starch damage than soft wheat (Williams 1967).

6.4.6 FAT CONTENT

Determination of total fat content in cereal and cereal products follows the ICC Standard Method No. 136. Lipids occur in various endosperm membranes, aleurone layer, and starch granules or wheat. They affect the baking quality and storage stability of white flour. Factors such as the quantity of the lipids and proportions of nonpolar and polar lipids are of great importance for flour quality (Morrison 1978, 1979). Most of the polar lipids are found in bound form, complexed with protein and starch (Pomeranz 1988). Various processes such as milling, dough mixing, bread making, and staling of bread are affected by the lipids present in the flour (Turnbull and Rahman 2002). The quantity of the hexane-extractable free lipids in flour depends on the cultivar, and is higher in harder wheat cultivars than in soft wheat (Konopka et al. 2005; Panozzo et al. 1993). Greenblatt et al. (1995) identified galactolipids and phospholipids in greater amounts in soft wheat than in hard wheat (Greenblatt et al. 1995). Polar and nonpolar contents of free lipids or their fatty acid composition in commercially milled flour streams have also been assessed (Prabhasankar et al. 2000). This type of information can be used during the process of blending the streams to produce different traditional products.

Lipids are located inside as well as on the surface of the starch granules. The surface lipids are mostly free fatty acids, whereas those located inside the granules are lysophospholipids (Morrison and Gadan 1987). Higher levels of lipids have been generally associated with smaller starch granules (Gaines et al. 2000; Raeker et al. 1998). These high levels consist mostly of internal lipids (measured as total starch phosphorus; Konopka et al. 2005), which influence the swelling behavior of starch by inhibiting the movement of water into granules, resulting in reduced starch granule swelling, amylose leaching, and peak paste viscosity.

Lipids have been reported to markedly influence the baking quality of flours, particularly bread (Addo and Pomeranz 1992). Polar lipids have the greatest positive effect on loaf volume among the lipids investigated (Bekes et al. 1986; McCirnack et al. 1991) and show good correlation ($r = 0.877$; $p < 0.001$) with loaf volume (Chung et al. 1982). Fatty acids and particularly unsaturated fatty acids have the capacity to reinforce the gluten network by oxidation of S–H group through enzyme-coupled reactions (Addo and Pomeranz 1992; Morrison and Panpaprai 1975).

In their study, Sroan et al. (2009) observed that flour lipids significantly affect the volume of the loaves; incremental addition caused bread volume to decrease and, after reaching a minimum, to increase again. The amount of lipid (as low as 1%–1.5%) that was used in the study of MacRitchie and Gras (1973) showed significant effects on baking performance. It is suggested that probably due to their surface activity, lipids are adsorbed at the gas–liquid interface of the liquid lamellae and affect the stability of the gas cells (Mills et al. 2003; Paternotte et al. 1993). The expansion capacity of the gas cells influences loaf volume.

6.5 FUNCTIONAL TESTS APPLIED TO DETERMINE FLOUR'S QUALITY

6.5.1 Sedimentation Test (SDS Test)

The swelling properties of wheat flours are tested by determining the sedimentation value. Because gluten is insoluble in water and comes from the wheat flour, the method measures the volume of the swollen gluten proteins. On the one hand, this volume of sediment shows the amount of gluten, and on the other, it represents the swelling properties of the gluten. The sedimentation value combines the qualitative and quantitative elements of the measurement of wheat gluten.

The more sediment (in milliliters) that can be read off the measuring cylinder, the more suitable the flour is for making bakery products that require strong protein flours. Determination of the sedimentation value is carried out according to ICC Standard 116 and AACC Method 56-60. The result may lie between 8 mL for an extremely low protein content and 78 mL for flour with very strong gluten.

The higher the SDS value, the higher the potential baking capacity of wheat flour. The SDS sedimentation value possesses the greatest potential as a screening test because of its small sample size, high throughput, good correlation with loaf properties, growing location, and genetic differences in protein quality (Blackman and Gill 1980). Moreover, SDS value is dependent on the protein composition and is correlated with the protein content, hardness of the wheat, and loaf volume (Hrusková and Famera 2003).

This test is widely utilized to get information about differences in protein content as well as gluten quality because both these characteristics are of great importance regarding end-use quality, especially for flours with protein contents of less than 13% (Carter et al. 1999; De Villiers and Laubscher 1995). Osborne (1984) investigated NIR spectroscopy for predicting SDS sedimentation volume as a measurement of protein quality and concluded that there was no significant correlation with the NIR spectral data when the contribution of the protein content to the calibration was removed.

6.5.2   FALLING NUMBER TEST

The falling number value is a measure of the cereal α-amylase activity in wheat and flours. The enzyme breaks down large starch molecules and thereby reduces the starch's viscosity. The instrument measures the viscosity in a heated flour–water suspension by recording the time needed for a standardized object to move through the paste (AACC Method 56-81B). The α-amylase in the sample slowly breaks down the consistency of the suspension, converting the starch to dextrins, and the object gradually sinks under its own weight. The time needed is shorter if the starch is hydrolyzed by amylases, which indicate a high amylase activity (Delcour and Hoseney 2010). For the determination of the falling number, the ICC recommends Standard Method No. 107/1.

Excessive sugar production from high concentrations of α-amylase results in bread with sticky crumbs and poor texture. The sticky crumb is unwanted by the industry because it causes problems during mechanical cutting of the bread (Chamberlain et al. 1981; Posner and Hibbs 1997). It was observed that flours exhibiting low falling numbers also exhibited a decrease in their water-absorbing capacity, which affects loaf volume (Dowell et al. 2008).

If the falling number is too high, enzymes can be added to the flour in various ways to compensate. If the falling number is too low, enzyme activity can be controlled by adjusting the pH of the dough to avoid dry crumb and diminished loaf volumes (Perten 1964). An acceptable falling number for bread production is between 200 and 350 s.

6.5.3 AMYLOGRAPH TEST

An amylograph determines the gelatinization properties of the starch of a flour sample. An amylograph is a torsion viscometer that continuously records changes in viscosity of the flour slurry at a uniform rate of temperature increase or decrease of 1.5°C min⁻¹ under fixed shear conditions.

Based on AACC Method 22-10 (or the identical method of ICC Standard No. 126/1), a sample of flour is added to distilled water and mixed to make a slurry. The slurry is heated at a constant rate under continuous stirring until it reaches 95°C. The amylograph records the resistance of the slurry to the stirring action of pins or paddles as a viscosity curve on a graph paper. During heating, the starch granules swell and the slurry becomes a paste that has a higher peak viscosity. Peak viscosity is the maximum resistance of a heated flour and water slurry to mixing with pins. It is expressed in Brabender units. Apart from the maximum viscosity (peak viscosity), gelatinization temperature and temperature at peak viscosity are recorded as well.

The increase in peak viscosity is affected by the increase in temperature, by mechanical action of stirring, and by the action of the α-amylase present in flour. The maximum viscosity is a result of both gelatinization behavior of the flour and α-amylase activity. So, this method estimates α-amylase activity in wheat flour; high activity is related to low hot paste viscosity due to the liquefying effect of the enzyme.

The amylogram also reflects the effect of other amylolytic enzymes present in the wheat flour. Enzymes, which result from sprout damage, break down the starch into maltose, producing slurries with lower peak viscosities than sound flour.

Wheat varieties, climate conditions, and wheat milling processes influence the pasting properties of tested flours. Starch granule size and structure, amylose/amylopectin ratio, and molecular weight also influence the pasting properties of flours (Thomas and Atwell 1999). Mariotti et al. (2005) studied starch pasting properties excluding the α-enzymatic effect by modifying the amylograph procedure. They used a 1 mM AgNO₃ solution instead of distilled water to inhibit the activity of α-amylase.

6.5.4 RAPID VISCO ANALYZER TEST

The Rapid Visco Analyzer (RVA) apparatus is a cooking–stirring viscometer with ramped temperature and variable shear profiles optimized for testing the viscous properties of starch, grain, flour, and foods. The RVA was developed for measuring the pasting properties of starch and initially introduced as a rapid method to determine preharvest damage in cereals (Ross et al. 1987). Both the amylograph and the RVA operate on similar principles. The RVA has several advantages, including small sample size, the ability to set temperature profiles, and the ability to record the data electronically. The AACC has developed Method 76-21, whereas the ICC sets Standard No. 162 as the general pasting method for wheat or rye flour using RVA. Both AACC and ICC developed methods for determining "stirring number" using RVA (ICC Standard No. 161, AACC Method 22-08).

According to Morris (2004), the Visco Amylograph and RVA can assess existing starch damage and can also assess potential *in situ* starch damage, which might occur during processing. Similarly, they can assess the potential effects of added amylase.

6.5.5 Farinograph Test

Generally, farinograph measurements on flours are carried out using a Brabender farinograph (mixer bowl, 300 or 10 g) according to ICC Standard Method No. 115/1 (or AACC Method 54-21). The farinograph is an instrument that mixes dough, using two sigmoid blades, which turn at different speeds, folding the dough into itself and records this action on a graph called a farinogram (Walker and Hazelton 1996). The most widely used measurements obtained from the farinograph to assess flour properties are flour water absorption, dough development time, arrival time, dough stability, and degree of softening.

Water absorption relates to the volume of water required for the dough to reach the 500 Brabender unit line at the point of optimum development. It is expressed as a percentage of the flour (14% moisture base). This is generally considered as the amount of water needed for a flour to be optimally processed into its end products. Water absorption is an indicator of baking quality (MacRitchie 1984; Van Lill et al. 1995). Higher protein content flour results in higher water absorption (Van Lill and Smith 1997). Stability is an indication of the flour's tolerance to mixing. Stronger flours show higher stability values (D'Appolonia and Kunerth 1969; Eliasson and Larsson 1993; Miralbés 2004). Mailhot and Patton (1988) recommend a minimum dough stability of 7.5 min and a degree of softening of less than 75 (farinograph units) as appropriate for bread making (Mailhot and Patton 1988). Generally, flour with good bread-making characteristics has higher absorption, takes longer time to mix, and is more tolerant to overmixing than poor-quality bread flour.

6.5.6 Extensograph Test

An extensograph is used to measure the extensibility of the dough and its resistance to extension. Generally, the Brabender Extensograph measurements are performed according to ICC Standard Method No. 114/1 (or AACC Method 54-10). During this test, a cylindrical dough sample is clamped horizontally in a cradle and stretched by a hook, which is placed in the middle of the sample and moves downward until rupture after 45 min of resting (Kokelaar et al. 1996). Often, to record resistance and extensibility with time, the same dough will be re-formed, relaxed for another period of time, and stretched again. Because resistance to extension is a measure of dough strength and of the dough's capacity to retain gas, a higher resistance to extension means that more force is required to stretch the dough. The extensibility indicates the amount of elasticity in the dough and its ability to stretch without breaking. Different flour products have different requirements in terms of these two properties. For example, high strength (resistance) is required for flour doughs for pasta making, whereas low dough strength but high extensibility is associated with good biscuit quality. Intermediate properties such as moderate strength and high extensibility are appreciated to give optimum performance for bread making. A higher resistance to extension is desirable because a good bread-making dough should have an ability to retain gas during baking (Mailhot and Patton 1988). Values higher than 500 (extensograph units) are considered desirable.

Extensograms of the strong, intermediate, and weak flours show considerable morphological differences (Figure 6.1; Anderssen et al. 2004). Anderssen et al. (2004)

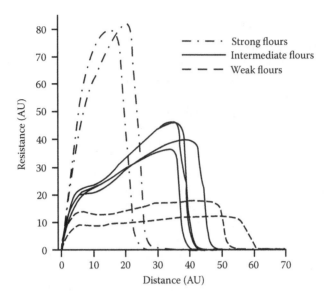

FIGURE 6.1 Comparison of representative extensograms for each of the eight flours. (From Anderssen, R. S. et al. *J. Cereal Sci.* 39: 195–203, 2004.)

observed that R_{max} represents an excellent basis for discriminating between wheat varieties on the basis of their glutenin composition. Their extensograms revealed that the amount of extension decreases as the flours (doughs) become stronger and the resistance to extension increases. At a molecular level, R_{max} occurs when the earlier phases of the extension have maximized the alignment of the macromolecules in the dough. Consequently, the alignment occurs more or less instantaneously in the stronger doughs and more rapidly in the medium strength doughs than the weak ones (Anderssen et al. 2004).

Another study reported that R_{max} correlated best with the glutenin fraction with molecular weight of more than 250,000 Da, whereas extensibility correlated with the flour protein content and was less affected by the molecular size distribution than R_{max} (Bangur et al. 1997). This means that only molecules that are larger than a critical molecular size contribute to dough strength. This critical value may correspond to the critical size for effective entanglements, a concept well established in polymer science for relating the physical properties of polymers (Ferry 1961).

6.5.7 ALVEOGRAPH TEST

The alveograph uses air pressure to inflate a thin sheet of dough, simulating the bubbles that are present in bread dough, causing it to stretch when rising. Alveograph measurements are usually performed using AACC Method 54-30A (or ICC Standard No. 121). Using the alveograph technique, different authors evaluated the bread-making (Chen and D'Appolonia; Janssen et al. 1996a; Khattak et al. 1974) and cookie-making (Bettge et al. 1989; Rasper et al. 1986) qualities of hard and soft wheat flour. The low and high P (maximum overpressure), L (average abscissa at bubble rupture), and W (deformation energy) values relate to weak and strong flour doughs, respectively (Bordes et al. 2008; Miralbés 2004; Walker and Hazelton 1996). These parameters are influenced by flour protein content (Van Lill and Smith 1997).

6.5.8 MIXOGRAPH TEST

The mixograph is an instrument that measures the rheological behavior of dough during mixing (Bordes et al. 2008; Walker and Hazelton 1996; Wikström and Bohlin 1996). It is used to classify wheat flours and predict water absorption in various dough processing systems as well as the quality of the finished product (Dobraszczyk and Schofield 2002; Khatkar et al. 1996; Lang et al. 1992). Mixograph analyses are carried out using AACC Method 54-40A. Finney and Shogren (1972) related peak times with loaf volume. They observed that flours showing peak times higher than 3 min, combined with protein contents of more than 13%, produced loaves with the same volume. Peak time is also influenced by protein content of the flour; flour having less than 12% protein takes longer to reach a peak (Hoseney 1994). As peak time increases, dough extensibility decreases, whereas dough stability, elasticity, and mixing tolerance increase. Long peak times are obtained for strong flours. Tolerance to overmixing is low for soft wheat flours, whereas stronger flours usually exhibit a tolerance to overmixing (Walker and Hazelton 1996). Parameters obtained from mixographs explained more than 90% of the variance observed in loaf volume (Neacşu et al. 2009; Wikström and Bohlin 1996). Peak times longer than 5 min usually exhibit too much tolerance, resulting in insufficient extensibility, which is undesirable for bread production. Medium to medium-long peak times usually exhibit acceptable tolerance and other dough-handling properties, making it desirable for bread production (Finney and Shogren 1972). Short peak times exhibit too much extensibility and too little elasticity for stable dough production (Finney et al. 1987).

Wikström and Bohlin (1996) and Ingelin (1997) observed that water absorption, determined on a mixograph, seems to be higher than water absorption determined on a farinograph. They postulated that this is due to the different mixing actions, the differences that occur in dough consistency, and the amount of water added to perform a mixogram, which all relate to the flour protein content of the sample being analyzed.

6.6 FUNDAMENTAL RHEOLOGICAL METHODS

Tests based on farinographs, mixographs, extensographs, etc., are useful for providing practical information for the baking industries, although they are not sufficient for interpreting the fundamental behavior of dough processing and baking quality. The availability of sophisticated rheometers in recent years has spawned a great deal of interest in characterizing wheat flour doughs and glutens of varying quality with fundamental rheological tests (Safari-Ardi and Phan-Thien 1998; Wikström and Eliasson 1998). These tests measure the well-defined properties of the material; the results are independent of the test equipment and can theoretically be used to model the flow conditions encountered by the dough during mixing, proofing, and baking.

The rheological properties of dough were measured using techniques such as dynamic oscillation (Autio et al. 2001; Safari-Ardi and Phan-Thien 1998), creep and stress relaxation (Bloksma 1990; Launey 1990; Wang and Sun 2002), and biaxial extension (van Vliet et al. 1992; Wikström and Bohlin 1999). Dynamic oscillatory measurements and stress–relaxation and creep–recovery tests, which measure small deformational rheological changes in polymers at low strains, not only require small amounts of sample but also provide information about the fundamental structural behavior of the material (Amemiya and Menjivar 1992; Campos et al. 1997; Khatkar and Schofield 2002a; Tronsmo et al. 2003). Rheological testing, especially in the linear viscoelastic region, has been used to follow the structure and properties of doughs and to study the functions of dough's ingredients (Janssen et al. 1996a; Miller and Hoseney 1999). These tests are performed using a rheometer equipped with a parallel plate configuration in which one plate is rotating in a sinusoidal motion and the other plate is stationary.

6.6.1 DYNAMIC OSCILLATION

In a dynamic oscillation shear measurement to the dough sample, a sinusoidal oscillating stress or strain with time is applied. When subjected to a sinusoidal strain, the viscoelastic material responds with a sinusoidal stress, which depends on the properties of the material. Dynamic oscillatory testing in the linear viscoelastic region at the same time measures the viscous and elastic characters of dough expressed as storage and loss moduli (G' and G'') and loss tangent (tan δ). G' values of gluten doughs can be a good indicator of loaf volume (Figure 6.2; Khatkar and Schofield 2002b). Doughs from poor quality bread-making flours had G' values of greater magnitude than those of the good quality bread-making flours (Figure 6.3; Khatkar

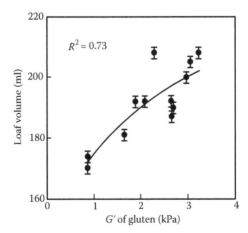

FIGURE 6.2   Relationship between G' of gluten (stress, 25 Pa; frequency, 1 Hz) and loaf volume for flours reconstituted using glutens of different wheat cultivars and a constant source of starch and water solubles. (From Khatkar, B. S. and Schofield, J. D., *J. Sci. Food and Agric.* 82: 823–826, 2002.)

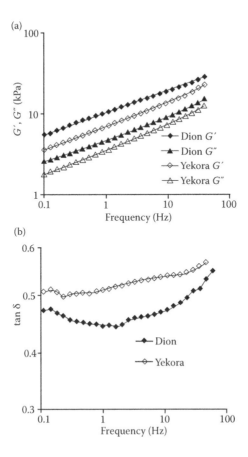

FIGURE 6.3 Variation in storage (G') and loss (G'') moduli (a) and tan δ (b) with frequency for control wheat Dion (poor bread-making quality flour) and Yekora (good bread-making quality flour) doughs measured at 25°C. (From Skendi, A. et al. *J. Food Eng.* 91: 594–601, 2009.)

and Schofield 2002b; Petrofsky and Hoseney 1995; Skendi et al. 2009). Dynamic rheological properties of gluten can describe the dough structure formed and the relationship to processing parameters of dough.

6.6.2 CREEP–RECOVERY

Creep–recovery is performed by subjecting the dough to a constant shear stress, with the shear strain being monitored as a function of time. The rheological testing of dough under small and large deformation with creep–recovery measurements is often used for dough characterization and to gain information on the structure of the composite material. The creep–recovery curves of doughs exhibit a typical viscoelastic behavior, combining both viscous fluid and elastic components (Steffe 1992). The creep and recovery responses of doughs from two wheat flours that differ in their bread-making quality are shown in Figure 6.4. According to Wang and

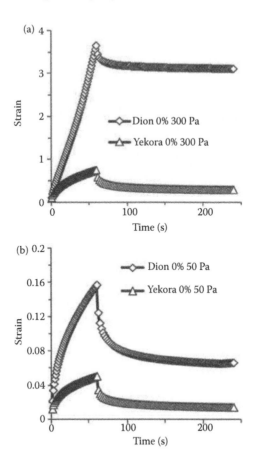

FIGURE 6.4 Typical creep–recovery curves of dough prepared from wheat flours Dion (poor bread-making quality flour) and Yekora (good bread-making quality flour) measured at 25°C under (a) high (300 Pa) and (b) low (50 Pa) load levels. (From Skendi, A. et al. *J. Food Eng.* 91: 594–601, 2009.)

Sun (2002), maximum creep strain could be used to describe dough rigidity (firmness). Generally, stronger doughs with greater resistance to deformation have smaller creep strain than softer doughs (Skendi et al. 2009; Wang and Sun 2002). There have been indications that the creep strain of isolated gluten could be used to discriminate the extensibility and baking potential of wheat flour (Janssen et al. 1996b). Results obtained from small deformation measurements performed on gluten are in agreement with those from traditional empirical large deformation rheological tests (Edwards et al. 1999; Khatkar et al. 1995).

6.6.3 STRESS RELAXATION

In a stress relaxation test, the dough is given an instantaneous constant strain, and the stress required to maintain the deformation is observed as a function of time.

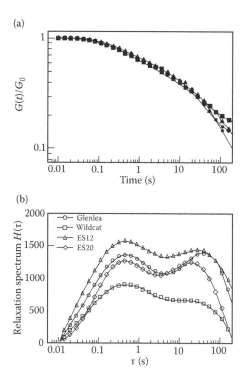

FIGURE 6.5 Normalized stress relaxation (a) and the corresponding relaxation spectra (b) of "extra-strong" flour–water doughs at 25°C. Applied shear strain and strain rise time were 0.05% and 0.2 s, respectively. Cultivars used included Glenlea at 12.8% protein, Wildcat at 14.0% protein, ES12 at 12.8% protein, and ES20 at 12.4% protein. (From Rao, V. K. et al. *J. Cereal Sci.* 31: 159–171, 2000.)

This test is convenient when the characterization of the linear viscoelastic region of the dough is needed, and it is also related to the molecular weight of the sample. Relaxation spectrum calculated for dough samples prepared from different cultivars is shown in Figure 6.5. Large molecules show long relaxation times. Strong flours that usually have high protein content show higher relaxation modulus [$G(t)$ and spectrum $H(t)$] over the whole relaxation time than those from weak flour that have low protein content (Li et al. 2003).

6.6.4 BIAXIAL EXTENSION

Biaxial extension test measures the large deformation of dough and gluten. During biaxial extension, the dough is stretched at equal rates in two perpendicular directions in one plane (Dobraszczyk and Morgenstern 2003). Results from this test are plotted as pressure versus drum distance trace of an inflating bubble from the dough sample. This test reproduces conditions experienced by the cell walls of the gas bubbles, that is, deformation that takes place around a growing cell, within the dough during proofing and oven rise (Dobraszczyk and Morgenstern 2003; Janssen et al.

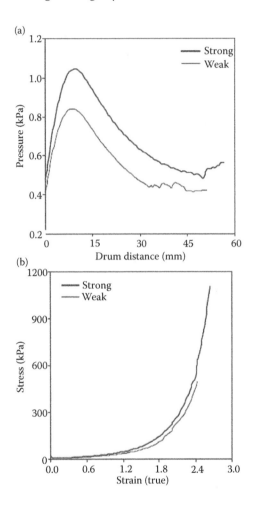

FIGURE 6.6 Pressure versus drum distance plots of inflating dough bubbles (a), and their corresponding stress versus strain plots (b) for doughs mixed at low speed to a work input of 10 kJ kg^{-1} (Strong, doughs from strong flour; Weak, doughs from weak flour). (From Chin, N. L. and Campbell, G. M., *J. Sci. Food and Agric.* 85: 2194–2202, 2005.)

1996a; Kokelaar et al. 1996; Rouille et al. 2005; Sliwinski et al. 2004). Doughs from strong flours show higher peak pressure and further drum distance before bubble rupture, suggesting the presence of a stronger gluten network that needs higher pressure to be broken down (Figure 6.6; Chin and Campbell 2005).

6.7 FACTORS INFLUENCING RHEOLOGICAL PROPERTIES

6.7.1 ADDED WATER

The water absorption capacity is the most important physical parameter affecting the farinogram and is a function of the wheat flour protein content and quality, the amount of starch damaged during milling, and the presence of nonstarch polysaccharides (Finney et al. 1987; Simmonds 1989). The absorption is the amount of water required to counter the farinograph curve on the 500–Brabender unit line for dough (Shuey 1984).

Water hydrates the proteins' fibrils and creates disulfide bonds between the proteins during mixing. Hydration gives rise to the formation of an apparent continuous water phase between gluten particles. Hydrated proteins form β-sheet structures that contribute to the network connectivity. The interactions between subunits in glutenins and aggregates involve intermolecular β-sheets located in their repetitive domains or in the region of chain entanglements. Entanglements, hydrogen bonding, hydrophobic interactions, and disulfide linkages are important for hydrated gluten (Khatkar 2004).

Generally, an optimum water level is required to develop dough with optimum gluten strength because the addition of too much water to the flour will result in a slurry, whereas too little water results in a slightly cohesive powder (Faubion and Hoseney 1989). Strong flours require higher water levels than weak flours due to the higher protein content. There is a general consensus that as the water content of the dough increases, within a given flour dough system, both the elastic (G') and the viscous moduli (G'') decrease (Amemiya and Menjivar 1992; Faubion and Hoseney 1989; Navickis et al. 1982; Skendi et al. 2010). Flour with higher water absorption produces more desired end products because it improves the texture of the bread (Simon 1987).

6.7.2 PROTEINS

Flour protein content (both soluble and insoluble) has an important influence on dough's rheological properties (Payne et al. 1987; Wall 1979). The level of protein is strongly influenced by environmental and cultivation conditions, whereas the quality of protein is linked to genetic composition and environmental interaction (Eagles et al. 2002). Although the protein content of flour is very important to bread quality, the quality of the gluten proteins is also important to bakers (Halverson and Zeleny 1988).

Wheat proteins have a molecular weight from 30,000 to more than 10 million Da (Hoseney 1994) and are primarily responsible for the viscoelastic behavior of dough (Spies 1989). Glutenins and gliadins are the two major storage proteins comprising gluten and are responsible for producing an appropriate balance of viscous and elastic properties in gluten and flour doughs (Shewry et al. 1986; Song and Zheng 2007). The ratio of gliadin/glutenin and high molecular weight/low molecular weight glutenin subunits have been proposed to explain gluten's viscoelastic properties (Popineau et al. 1994). It is widely accepted that gliadins confer viscous properties, whereas glutenins impart strength and elasticity (Wieser 2007).

Glutens from good bread-making wheat are cross-linked in a higher degree so the frequency dependence of G' is smaller than that of glutens from poor bread-making wheat (Mirsaeedghazi et al. 2008). The aggregation of glutenin within the gluten network brings an increase in the elastic modulus (Popineau et al. 1994). Glutenin has a major role in the difference between baking processes (Toufeili et al. 1999).

The amounts of soluble and insoluble glutenins mostly affect the dough strength parameters (Sapirstein and Fu 1998).

6.7.3 STARCH

The wheat starch contains approximately 25% amylose and 75% amylopectin (BeMiller and Huber 1996). Wheat starch granules exist in type B (spherical size, 1–3 μm diameter) and type A (larger granules, 20–45 μm diameter; Atwell 2001; Hoseney 1994).

During baking, water moves from hydrated gluten to starch granules causing gelatinization. Interactions between starch granules and gluten are possible, occurring via hydrogen bonding and thus preventing bread staling (Ottenhof and Farhat 2004).

Creep–recovery measurements of flour gels performed by Sasaki et al. (2008) showed that small differences in amylose content have a strong effect on the rheological parameters of flour gels. Furthermore, wheat flour gel with lower amylose content showed higher creep and recovery compliance.

6.7.4 NONSTARCH POLYSACCHARIDES

Arabinoxylans are the major nonstarch polysaccharides present in wheat, and their quantity and structure vary significantly among cultivars. Arabinoxylans consist of a xylan chain with β-(1 → 4)-linked D-xylopyranosyl residues (Xyl$p$) to which mostly single α-L-arabinofuranose units (Ara$f$) are linked, at the O-2 or O-3 positions of the xylose units as side residues (Cleemput et al. 1993; Izydorczyk and Biliaderis 1995). Moreover, some Ara$f$ units carry ferulic acid residues esterified to O-5 of Ara$f$ linked to O-3 of the xylose residues (Sosulski 1982). Arabinoxylans are classified on the basis of their extractability in water-extractable arabinoxylan and water-unextractable arabinoxylans. The arabinoxylan levels of flours have been related to the dough's properties (Jelaca and Hlynka 1971). Generally, arabinoxylans influence the functionality of gluten (Autio 2006), water balance (ability to absorb high amounts of water; Brennan and Cleary 2007; Prasad Rao et al. 2007), and rheological properties of dough, as well as retrogradation of starch.

One of the most important functional properties of arabinoxylans is their ability to absorb high quantities of water. In this context, arabinoxylans affect the farinograph water absorption of dough (Courtin and Delcour 1998; Cui et al. 1999; Jelaca and Hlynka 1971; Michniewicz et al. 1991) and are able to bind up to one-fourth of the water in a standard bread recipe (Bushuk 1966).

It was observed that the addition of arabinoxylans increased the dough stability time and affected dough development time (Delcour et al. 1999). It was also found to improve dough consistency (Jelaca and Hlynka 1972). Such effects depend on the quantity and molecular size of arabinoxylan (Biliaderis et al. 1995; Sasaki et al. 2004; Vanhamel et al. 1993); high average molecular weights exert greater effects on baking absorption and development time than their lower molecular weight counterparts (Biliaderis et al. 1995; Courtin and Delcour 1998). High molecular weight arabinoxylans enhance resistance to extension and decrease extensibility (Jelaca and Hlynka 1972). If present at concentrations higher than the optimum, arabinoxylans can result in a product of lower specific volume (Figure 6.7; Biliaderis et al. 1995). Upon the addition of nonstarch polysaccharides, a significant decrease in the crumb firmness, resulting in the improvement of bread texture (Wang et al. 2002) and reduced bread staling (Biliaderis et al. 1995), has been observed. It was suggested that the ability of nonstarch polysaccharides to form a network limits the movement of glutenin proteins and the formation of larger aggregates (Labat et al. 2002). This is based on the ability of water-extractable arabinoxylan to cross-link and form highly viscous aqueous solutions (Autio 2006; Figueroa-Espinoza and Rouau 1998; Labat et al. 2002; Skendi et al. 2011). In addition, ferulic acid linked with arabinoxylan is involved in the direct effect of arabinoxylans on gluten formation. This effect lowers the extensibility of the dough and gluten (Wang et al. 2002).

On the other hand, arabinoxylans stabilize gas cell walls by improving the mechanical properties of the gas bubble surface in the dough. Furthermore, the interactions of arabinoxylans with other carbohydrates or protein constituents may represent another possibility of their functional role in the dough system (Hoseney et al. 1969; Michniewicz et al. 1991; Roels et al. 1993; Saulnier et al. 1997). Other molecules present in the wheat flour, such as arabinogalactan peptides, also interact with gluten, resulting in a reduced water absorption and extensibility (Autio 2006).

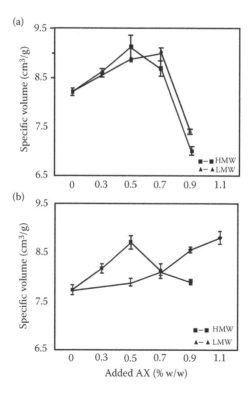

FIGURE 6.7 Effect of added arabinoxylans (HMW, high molecular weight; LMW, low molecular weight) on the specific volume of breads made from Canada western red spring (a) and Canada prairie spring (b) flours, cv. HY 368. (From Biliaderis, C. G. et al. *Food Chem.* 5: 165–171, 1995.)

REFERENCES

AACC International 2004. *Approved Methods of Analysis*. AACC Methods 14-10, 66-20.01, 44-15A, 44-16, 44-18, 44-19, 44-20, 44-11, 44-40, 44-51, 08-01, 08-02, 08-03, 08-21, 02-31, 39-10, 76-11, 76-13, 76-31.01, 76-33, 56-60, 56-81B, 22-10, 76-21, 22-08, 54-21, 54-10, 54-30A, 54-40A. St. Paul, MN: AACC International.

Addo, K., and Y. Pomeranz. 1992. Effects of lipids and emulsifiers on alveograph characteristics. *Cereal Chemistry* 69: 6–12.

Amemiya, J. I., and J. A. Menjivar. 1992. Comparison of small and large deformation measurements to characterize the rheology of wheat flour doughs. *Journal of Food Engineering* 16: 91–108.

Anderssen, R. S., F. Bekes, P. W. Gras, A. Nikolov, and J. T. Wood. 2004. Wheat-flour dough extensibility as a discriminator for wheat varieties. *Journal of Cereal Science* 39: 195–203.

Atwell, W. A. 2001. *Composition of Commercial Flour in Wheat Flour*. St. Paul, MN: American Association of Cereal Chemists.

Autio, K. 2006. Effects of cell wall components on the functionality of wheat gluten. *Biotechnology Advances* 24: 633–635.

Autio, K., L. Flander, A. Kinnunen, and R. Heinonen. 2001. Bread quality relationship with rheological measurements of wheat flour dough. *Cereal Chemistry* 78: 654–657.

Bangur, R., I. L. Batey, E. McKenzie, and F. MacRitchie. 1997. Dependence of extensograph parameters on wheat protein composition measured by SE-HPLC. *Journal of Cereal Science* 25: 237–241.

Barnes, P. J. 1986. The influence of wheat endosperm on flour colour grade. *Journal of Cereal Science* 4: 143–155.

Barros, F., J. N. Alviola, M. Tilley, Y. R. Chen, V. R. M. Pierucci, and L. W. Rooney. 2010. Predicting hot-press wheat tortilla quality using flour, dough and gluten properties. *Journal of Cereal Science* 52: 288–294.

Bekes, F., V. Zawistowska, R. R. Zillman, and W. Bushuk. 1986. Relationship between lipid content and composition and loaf volume of twenty-six common spring wheat. *Cereal Chemistry* 63: 327–331.

BeMiller, J. N., and K. C. Huber. 1996. Carbohydrates. In *Food Chemistry*, edited by Fennema, O. R., 83–154. New York: Marcel Dekker, Inc.

Bettge, A., G. Rubenthaler, and Y. Pomeranz. 1989. Alveograph algorithms to predict functional properties of wheat in bread and cookie baking. *Cereal Chemistry* 66: 81–86.

Bilheux, R., A. Escoffier, D. Herve, and J.-M. Pouradier. 1989. *Special and Decorative Breads*. New York: Van Nostrand Reinhold.

Biliaderis, C. G., M. S. Izydorczyk, and O. Rattan. 1995. Effect of arabinoxylans on breadmaking quality of wheat flours. *Food Chemistry* 5: 165–171.

Blackman, J. A., and A. A. Gill. 1980. A comparison of some small-scale tests for breadmaking quality used in wheat breeding. *The Journal of Agricultural Science* 95: 29–34.

Bloksma, A. H. 1990. Rheology of the breadmaking process. *Cereal Foods World* 35: 228–236.

Bordes, J., G. Branlard, F. X. Oury, G. Charmet, and F. Balfourier. 2008. Agronomic characteristics, grain quality and flour rheology of 372 bread wheats in a worldwide core collection. *Journal of Cereal Science* 48: 569–579.

Brennan, C. S., and L. J. Cleary. 2007. Utilisation Glucagel® in the β-glucan enrichment of breads: a physicochemical and nutritional evaluation. *Food Research International* 40: 291–296.

Bushuk, W. 1966. Distribution of water in dough and bread. *The Bakers Digest* 40: 38–40.

Campos, D. T., J. F. Steffe, and P. K. W. Ng. 1997. Rheological behavior of undeveloped and developed wheat dough. *Cereal Chemistry* 74: 489–494.

Carson, G. R., and N. M. Edwards. 2009. Criteria of wheat and flour quality. In *Wheat Chemistry and Technology*, edited by Khan, K., and P. R. Shewry, 100–110. St. Paul, MN: AACC International, Inc.

Carter, B. P., C. F. Morris, and J. A. Anderson. 1999. Optimising the SDS sedimentation test for end-use quality selection in a soft white and club wheat breeding program. *Cereal Chemistry* 76: 907–911.

Chamberlain, N., T. H. Collins, and E. E. McDermott. 1981. Alpha-amylase and bread properties. *International Journal of Food Science and Technology* 16: 127–152.

Chen, J., and B. L. D'Appolonia. 1985. Alveograph studies on hard red spring wheat flour. *Cereal Foods World* 30: 862–870.

Chin, N. L., and G. M. Campbell. 2005. Dough aeration and rheology: Part 2. Effects of flour type, mixing speed and total work input on aeration and rheology of bread dough. *Journal of the Science of Food and Agriculture* 85: 2194–2202.

Chung, O. K., Y. Pomeranz, and K. F. Finney. 1982. Relation of polar lipid content to mixing requirement and loaf volume potential of hard red wheat flour. *Cereal Chemistry* 59: 14–20.

CIE Publication. 1986. *CIE Publication No. 15.2, Colorimetry*. Austria, Wien: CIE Central Bureau Kegelgasse 27 A-1030.

Cleemput, G., S. P. Roels, M. Vanoort, P. J. Grobet, and J. A. Delcour. 1993. Heterogeneity in the structure of water-soluble arabinoxylans in European wheat flours of variable bread-making quality. *Cereal Chemistry* 70: 324–329.

Cornell, H. J., and A. W. Hoveling. 1998. *Wheat: Chemistry and Utilization*. Cleveland: CRC Press.

Courtin, C. M., and J. A. Delcour. 1998. Physicochemical and bread-making properties of low molecular weight wheat-derived arabinoxylans. *Journal of Agricultural and Food Chemistry* 46: 4066–4073.

Cui, W. W., P. J. Wood, J. Weisz, and M. U. Beer. 1999. Nonstarch polysaccharides from pre-processed wheat bran: Carbohydrate analysis and novel rheological properties. *Cereal Chemistry* 76: 129–133.

D'Appolonia, B. L., and W. H. Kunerth. 1969. *The Farinograph Handbook*. St. Paul, MN: American Association of Cereal Chemists, Inc.

De Villiers, O. T., and E. W. Laubscher. 1995. Use of the SDS-test to predict the protein content and bread volume of wheat cultivars. *South African Journal of Plant and Soil* 12: 140–142.

Delcour, J. A., and R. C. Hoseney. 2010. *Principles of Cereal Science and Technology*. St. Paul, MN: AACC International.

Delcour, J. A., H. Van Win, and P. J. Grobet. 1999. Distribution and structural variation of arabinoxylans in common wheat mill streams. *Journal of Agricultural and Food Chemistry* 47: 271–275.

Dobraszczyk, B. J., and M. P. Morgenstern. 2003. Rheology and the breadmaking process. *Journal of Cereal Science* 38: 229–245.

Dobraszczyk, B. J., and J. D. Schofield. 2002. Rapid assessment and prediction of wheat and gluten baking quality with the 2-G direct drive mixograph using multivariate statistical analysis. *Cereal Chemistry* 79: 607–612.

Dowell, F. E., E. B. Maghirang, R. O. Pierce, G. L. Lookhart, S. R. Bean, F. Xie, M. S. Caley et al. Relationship of bread quality to kernel, flour and dough properties. *Cereal Chemistry* 85: 82–91.

Dowell, F. E., T. C. Pearson, E. B. Maghirang, F. Xie, and D. T. Wicklow. 2002. Reflectance and transmittance spectroscopy applied to detecting fumonisin in single corn kernels infected with *Fusarium verticillioides*. *Cereal Chemistry* 79: 222–226.

Eagles, H. A., G. J. Hollamby, and R. F. Eastwood. 2002. Genetic and environmental variation for grain quality traits routinely evaluated in southern Australian wheat breeding programs. *Australian Journal of Agricultural Research* 53: 1047–1057.

Edwards, N. M., J. E. Dexter, M. G. Scanlon, and S. Cenkowski. 1999. Relationship of creep-recovery and dynamic oscillatory measurements to durum wheat physical dough properties. *Cereal Chemistry* 76: 638–645.

Eliasson, A., and K. Larsson. 1993. *Cereals in Bread Making, a Molecular Colloidal Approach*. New York: Marcel Dekker, Inc.

Faubion, J. M., and R. C. Hoseney. 1989. The viscoelastic properties of wheat flour doughs. In *Dough Rheology and Baked Product Texture*, edited by Faridi, H. A., and J. M. Faubion, 29–66. New York: Van Nostrand Reinhold.

Ferry, J. D. 1961. *Viscoelastic Properties of Polymers*. New York: Wiley.

Figueroa-Espinoza, M. C., and X. Rouau. 1998. Oxidative cross-linking of pentosans by a fungal laccase and horseradish peroxidase: Mechanism of linkage between feruloylated arabinoxylans. *Cereal Chemistry* 75: 259–265.

Finney, K. F., and M. D. Shogren. 1972. A 10 g mixograph for determining and predicting functional properties of wheat flours. *The Bakers Digest* 46: 32–35, 38–42.

Finney, K. F., W. T. Yamazaki, V. L. Youngs, and G. L. Rubenthaler. 1987. Quality of hard, soft, and durum wheat. In *Wheat and Wheat Improvement*, edited by Heyne, E. G., 677–748. Madison, WI: ASA/CSSA/SSSA.

Gaines, C. S., M. O. Raeker, M. Tilley, P. L. Finney, J. D. Wilson, D. B. Bechtel, R. J. Martin et al. 2000. Associations of starch gel hardness, granule size, waxy allelic expression, thermal pasting, milling quality, and kernel texture of 12 soft wheat cultivars. *Cereal Chemistry* 77: 163–168.

Gooding, M. J., and W. P. Davies. 1997. *Wheat Production and Utilization: Systems, Quality and the Environment*. Wallingford, UK: CAB International.

Gooding, M. J., G. Smith, W. P. Davies, and P. S. Kettlewell. 1997. The use of residual maximum likelihood to model grain quality characters of wheat with variety, climatic and nitrogen fertilizer effects. *Journal of Agricultural Science* 128: 135–142.

Greenblatt, G. A., A. D. Bettge, and C. F. Morris. 1995. Relationship between endosperm texture and the occurrence of friabilin and bound polar lipids on wheat-starch. *Cereal Chemistry* 72: 172–176.

Haddad, Y., F. Mabille, A. Mermet, J. Abecassis, and J. C. Benet. 1999. Rheological properties of wheat endosperm with a view on grinding behaviour. *Powder Technology* 105: 89–94.

Halverson, J., and L. Zeleny. 1988. Criteria for wheat quality. In *Wheat: Chemistry and Technology*, edited by Pomeranz, Y., 25. St. Paul, MN: American Association of Cereal Chemists.

Hoseney, R. C. 1994. Proteins of cereals. *Principles of Cereal Science and Technology*. 64–101. St. Paul, MN: American Association of Cereal Chemists, Inc.

Hoseney, R. C., K. F. Finney, M. D. Shogren, and Y. Pomeranz. 1969. Functional (bread-making) and biochemical properties of wheat flour components. 2. Role of water-solubles. *Cereal Chemistry* 46: 117–125.

Hrusková, M., and O. Famera. 2003. Prediction of wheat and flour Zeleny sedimentation value using NIR technique. *Czech Journal of Food Science* 21: 91–96.

ICC–International Association for Cereal Science and Technology 2001. *Standard Methods of the International Association for Cereal Science and Technology*. ICC Standards 109/1, 110/1, 202, 104/1, 145, 105/2, 159, 137/1, 164, 172, 136, 116, 107/1, 126/1, 162, 161, 115/1, 114/1, 121. Vienna, Austria: ICC–International Association for Cereal Science and Technology.

Ingelin, M. E. 1997. Comparison of two recording dough mixers. In *The Mixograph Handbook*, edited by Walker, C. E., J. L. Hazelton, and M. D. Shogren, 5–11. Lincoln, NE: National Manufacturing Division, TMCO.

Izydorczyk, M. S., and C. G. Biliaderis. 1995. Cereal arabinoxylans: advances in structure and physicochemical properties. *Carbohydrate Polymers* 28: 33–48.

Janssen, A. M., T. van Vliet, and J. M. Vereijken. 1996a. Fundamental and empirical rheological behaviour of wheat flour doughs and comparison with bread making performance. *Journal of Cereal Science* 23: 43–54.

Janssen, A. M., T. van Vliet, and J. M. Vereijken. 1996b. Rheological behaviour of wheat glutens at small and large deformations. Comparison of two glutens differing in bread making potential. *Journal of Cereal Science* 23: 19–31.

Jelaca, S. L., and I. Hlynka. 1971. Water-binding capacity of wheat flour crude pentosans and their relation to mixing characteristics of dough. *Cereal Chemistry* 48: 211–222.

Jelaca, S. L., and I. Hlynka. 1972. Effect of wheat-flour pentosans in dough, gluten and bread. *Cereal Chemistry* 49: 489–495.

Kent-Jones, D. W. 1941. Studies in flour granularity. *Cereal Chemistry* 18: 358–369.

Kent-Jones, D. W. 1955. Flour color grader. The advantages of direct measurement of flour color over the ash test. *The Baker's Digest* 29: 178–180.

Kent-Jones, D. W., and W. Martin. 1950a. Experiments in the photoelectric recording of flour grade by measurements of reflecting power. *Analyst* 75: 133–142.

Kent-Jones, D. W., and W. Martin. 1950b. A photoelectric method of determining the color of flour as affected by grade, by measurements of reflecting power. *Analyst* 75: 127–133.

Kent-Jones, D. W., E. G. Richardson, and R. C. Spalding. 1939. Flour granularity. *Journal of the Society of Chemical Industry, London* 58: 261–267.

Khatkar, B. S. 2004. Dynamic rheological properties and bread-making qualities of wheat gluten: Effects of urea and dithiothreitol. *Journal of the Science of Food and Agriculture* 85: 337–341.

Khatkar, B. S., A. E. Bell, and J. D. Schofield. 1995. The dynamic rheological properties of glutens and gluten sub-fractions from wheats of good and poor bread making quality. *Journal of Cereal Science* 22: 29–44.

Khatkar, B. S., A. E. Bell, and J. D. Schofield. 1996. A comparative study of the inter-relationships between mixograph parameters and bread making qualities of wheat flours and glutens. *Journal of the Science of Food and Agriculture* 72: 71–85.

Khatkar, B. S., and J. D. Schofield. 2002a. Dynamic rheology of wheat flour dough. I. Non-linear viscoelastic behaviour. *Journal of the Science of Food and Agriculture* 82: 827–829.

Khatkar, B. S., and J. D. Schofield. 2002b. Dynamic rheology of wheat flour dough. II. Assessment of dough strength and bread-making quality. *Journal of the Science of Food and Agriculture* 82: 823–826.

Khattak, S., B. L. D'Appolonia, and O. J. Banasik. 1974. Use of the alveograph for quality evaluation of hard red spring wheat. *Cereal Chemistry* 51: 355–360.

Kokelaar, J. J., T. van Vliet, and A. Prins. 1996. Strain hardening properties and extensibility of flour and gluten doughs in relation to breadmaking performance. *Journal of Cereal Science* 24: 199–214.

Konopka, W., D. Rotkiewicz, and M. Tanska. 2005. Wheat endosperm hardness. Part II. Relationships to content and composition of flour lipids. *European Food Research and Technology* 220: 20–24.

Kulkarni, R. G., J. G. Ponte, and K. Kulp. 1987. Significance of gluten content as an index of flour quality. *Cereal Chemistry* 64: 1–3.

Kweon, M., R. Martin, and E. Souza. 2009. Effect of tempering conditions on milling performance and flour functionality. *Cereal Chemistry* 86: 12–17.

Labat, E., X. Rouau, and M. H. Morel. 2002. Effect of flour water-extractable pentosans on molecular associations in gluten during mixing. *LWT—Food Science and Technology* 35: 185–189.

Lang, C. E., E. K. Neises, and C. E. Walker. 1992. Effects of additives on flour water dough mixograms. *Cereal Chemistry* 69: 587–591.

Launey, B. 1990. A simplified nonlinear model for describing the viscoelastic properties of wheat flour doughs at high shear strain. *Cereal Chemistry* 67: 25–31.

Li, W., B. J. Dobraszczyk, and J. D. Schofield. 2003. Stress relaxation behaviour of wheat dough, gluten and gluten protein fractions. *Cereal Chemistry* 80: 333–338.

MacRitchie, F. 1984. Baking quality of wheat flours. *Advances in Food Research* 29: 201–277.

MacRitchie, F., and P. W. Gras. 1973. The role of flour lipids in baking. *Cereal Chemistry* 50: 292–302.

Mailhot, W. C., and J. C. Patton. 1988. Criteria of flour quality. In *Wheat: Chemistry and Technology*, edited by Pomeranz, Y. vol. II, 69–90. St. Paul, MN: American Association of Cereal Chemists.

Mariotti, M., M. Zardi, M. Lucisano, and M. A. Pagani. 2005. Influence of the heating rate on the pasting properties of various flours. *Starch—Stärke* 57: 564–572.

McCaig, T. N. 2002. Extending the use of visible/near-infrared reflectance spectrophotometers to measure colour of food and agricultural products. *Food Research International* 35: 731–736.

McCirnack, G., J. Panozzo, and F. MacRitchie. 1991. Contributions to bread making of inherent variations in lipid content and composition of wheat cultivators. II. Fractionation and reconstitution studies. *Journal of Cereal Science* 13: 263–274.

Medcalf, D. G., and K. A. Gilles. 1965. Determination of starch damage by rate of iodine absorption. *Cereal Chemistry* 42: 546–557.

Michniewicz, J., C. G. Biliaderis, and W. Bushuk. 1991. Effect of added pentosans on some physical and technological characteristics of dough and gluten. *Cereal Chemistry* 68: 252–258.

Miller, K. A., and R. C. Hoseney. 1999. Dynamic rheological properties of wheat starch-gluten doughs. *Cereal Chemistry* 76: 105–109.

Mills, E. N. C., P. J. Wilde, L. J. Salt, and P. Skeggs. 2003. Bubble formation and stabilization in bread dough. *Food and Bioproducts Processing* 81: 189–193.

Minolta. 1994. *Precise Color Communication: Color Control From Feeling to Instrumentation*. Osaka, Japan: Minolta Co. Ltd.

Miralbés, C. 2004. Quality control in the milling industry using near-infrared transmittance spectroscopy. *Food Chemistry* 88: 621–628.

Mirsaeedghazi, H., Z. Emam-Djomeh, and S. M. A. Mousavi. 2008. Rheometric measurement of dough rheological characteristics and factors affecting it. *International Journal of Agriculture and Biology* 10: 112–119.

Morris, C. F. 2004. Cereals/grain—quality attributes. In *Encyclopedia of Grain Science*, edited by Colin, W., 238–254. Oxford: Elsevier.

Morrison, W. R. 1978. Cereal lipids. In *Advances in Cereal Science and Technology*, edited by Pomeranz, Y., vol. 2, 221–348. St. Paul, MN: American Association of Cereal Chemists.

Morrison, W. R. 1979. Lipids in wheat and their importance in wheat products. In *Recent Advances in Biotechnology of Cereals. Annual of Proceedings of Physiochemical Society of Europe,* edited by Laidman, D. C., R. G. Wyn Jones Laidman, and R. G. Wyn Jones, 313–335. London: Academic Press.

Morrison, W. R., and H. J. Gadan. 1987. The amylose and lipid contents of starch granules in developing wheat endosperm. *Journal of Cereal Science* 5: 263–275.

Morrison, W. R., and R. Panpaprai. 1975. Oxidation of free and esterified linoleic and linolenic acids in bread doughs by wheat and soya lipoxygenases. *Journal of Science Food and Agriculture* 26: 1225–1236.

Murray, L. F., and R. Moss. 1990. Estimation of fat acidity in milled wheat products. Part II. Using colorimetry to ascertain the effect of storage time and conditions. *Journal of Cereal Science* 11: 179–184.

Navickis, L. L., R. A. Anderson, E. B. Bagley, and B. K. Jasberg. 1982. Viscoelastic properties of wheat-flour doughs—variation of dynamic moduli with water and protein content. *Journal of Texture Studies* 13: 249–264.

Neacşu, A., G. Stanciu, and N. N. Săulescu. 2009. Most suitable mixing parameters for use in breeding bread wheat for processing quality. *Cereal Research Communications* 37: 83–92.

Neel, D. V. 1980. A Milling and Baking Comparison of Hard White Wheat versus Hard Red Wheat with Flour Color and Extraction Level as Primary Quality Factors. Deparment of Grain Science and Industry, Msc. Manhattan, KS: Kansas State University.

Oliver, J. R., A. B. Blakeney, and H. M. Allen. 1992. Measurement of flour color in color space parameters. *Cereal Chemistry* 69: 546–551.

Oliver, J. R., A. B. Blakeney, and H. M. Allen. 1993. The colour of flour streams as related to ash and pigment contents. *Journal of Cereal Science* 17: 169–182.

Osborne, B. G. 1984. Investigations into the use of near infrared reflectance spectroscopy for the quality assessment of wheat with respect to its potential for bread baking. *Journal of the Science of Food and Agriculture* 35: 106–110.

Ottenhof, M. A., and I. A. Farhat. 2004. The effect of gluten on the retrogradation of wheat starch. *Journal of Cereal Science*: 269–274.

Panozzo, J. F., M. C. Hannah, L. O'Brien, and F. Bekes. 1993. The relationship of free lipids and flour protein to breadmaking quality. *Journal of Cereal Science* 17: 47–62.

Paternotte, T. A., R. Orsel, and R. J. Hamer. 1993. Interactions between flour proteins and flour lipids at the liquid/air interface. *Gluten Proteins*, 207–217. Detmold, Germany: Association of Cereal Research.

Payne, P. I., M. A. Nightingale, A. F. Krttiger, and L. M. Holt. 1987. The relationship between HMW glutenin subunit composition and the bread-making quality of British grown wheat varieties. *Journal of the Science of Food and Agriculture* 40: 51–56.

Peplinski, A. J., A. C. Stringfellow, and E. L. J. Griffin. 1964. Air classification of Indiana, Ohio, and Michigan sodt wheat flours. *Northwest Miller* 27: 10.

Peplinski, A. J., A. C. Stringfellow, and E. L. J. Griffin. 1965. Air classification response of durum and hard red spring wheat flours. *Northwest Miller* 272: 34.

Perten, H. 1964. Application of the falling number method for evaluating alpha-amylase activity. *Cereal Chemistry* 41: 127–140.

Petrofsky, K. E., and R. C. Hoseney. 1995. Rheological properties of dough made with starch and gluten from several cereal sources. *Cereal Chemistry* 72: 53–58.

Pomeranz, Y. 1988. *Wheat Chemistry and Technology*. St. Paul, MN: Amercian Association of Cereal Chemists.

Pomeranz, Y., and P. C. Williams. 1990. Wheat hardness: its genetic, structural, and biochemical background, measurement, and significance. In *Advances in Cereal Science and Technology*, edited by Pomeranz, Y., 471–544. St Paul, MN: American Association of Cereal Chemists.

Popineau, Y., M. Cornec, J. Lefebvre, and B. Marchylo. 1994. Influence of high Mr glutenin subunits on glutenin polymers and rheological properties of glutens and gluten subfractions of near-isogenic lines of wheat Sicco. *Journal of Cereal Science* 19: 231–241.

Posner, E. S. 2003. Milling: characteristics of milled products. In *Encyclopedia of Food Sciences and Nutrition*, edited by Benjamin, C., 2nd ed., 3997–4005. Oxford: Academic Press.

Posner, E. S., and A. N. Hibbs. 1997. *Wheat Flour Milling*. St. Paul, MN: American Association of Cereal Chemists.

Prabhasankar, P., M. V. Kumar, B. R. Lokesh, and H. P. Rao. 2000. Distribution of free lipids and their fractions in wheat flour milled streams. *Food Chemistry* 71: 97–103.

Preston, K. R., and P. C. Williams. 2003. Flour: analysis of wheat flours. In *Encyclopedia of Food Sciences and Nutrition*, edited by Benjamin, C., 2543–2550. Oxford: Academic Press.

Pyler, E. J. 1988. *Baking Science and Technology*. Kansas City: Sosland Publishing Co.

Raeker, M. O., C. S. Gaines, P. L. Finney, and T. Donelson. 1998. Granule size distribution and chemical composition of starches from 12 soft wheat cultivars. *Cereal Chemistry* 75: 721–728.

Rao, H. P., K. Leelavathi, and S. R. Shurpalekar. 1989. Effect of damaged starch on the chapatti-making quality of whole wheat flour. *Cereal Chemistry* 66: 329–333.

Rao, V. K., S. J. Mulvaney, and J. E. Dexter. 2000. Rheological characterisation of long- and short-mixing flours based on stress-relaxation. *Journal of Cereal Science* 31: 159–171.

Rao, R. S. P., R. S. Manohar, and G. Muralikrishna. 2007. Functional properties of water soluble non-starch polysaccharides from rice and ragi: effect on dough characteristics and baking quality. *LWT—Food Science and Technology* 40: 1678–1686.

Rasper, V. F., M. L. Pico, and R. G. Fulcher. 1986. Alveography in quality assessment of soft white winter wheat cultivars. *Cereal Chemistry* 63: 395–400.

Roels, S. P., G. Cleemput, X. Vandewalle, M. Nys, and J. A. Delcour. 1993. Bread volume potential of variable-quality flours with constant protein level as determined by factors governing mixing time and baking absorption levels. *Cereal Chemistry* 70: 318–323.

Ross, A. S., C. E. Walker, R. I. Booth, R. A. Orth, and C. W. Wrigley. 1987. The Rapid ViscoAnalyser: a new technique for the evaluation of sprout damage. *Cereal Foods World* 32: 827–829.

Rouille, J., G. Valle, J. Lefebvre, E. Sliwinski, and T. van Vliet. 2005. Shear and extensional properties of bread doughs affected by their minor components. *Journal of Cereal Science* 42: 45–57.

Safari-Ardi, M., and N. Phan-Thien. 1998. Stress relaxation and oscillatory tests to distinguish between doughs prepared from wheat flours of different varietal origin. *Cereal Chemistry* 75: 80–84.

Sapirstein, H. D., and B. X. Fu. 1998. Intercultivar variation in the quantity of monomeric proteins, soluble and insoluble glutenin, and residue protein in wheat flour and relationships to breadmaking quality. *Cereal Chemistry* 75: 500–507.

Sasaki, T., K. Kaoru, and Y. Takeshi. 2004. Effect of water-soluble and insoluble non-starch polysaccharides isolated from wheat flour on the rheological properties of wheat starch gel. *Carbohydrate Polymers* 57: 451–458.

Sasaki, T., T. Yasui, and K. Kohyama. 2008. Influence of starch and gluten characteristics on rheological properties of wheat flour gel at small and large deformation. *Cereal Chemistry* 85: 329–334.

Saulnier, L., R. Andersson, and P. Aman. 1997. A study of the polysaccharide components in gluten. *Journal of Cereal Science* 25: 121–127.

Shewry, P. R., A. S. Tatham, J. Forde, M. Kreis, and B. J. Miflin. 1986. The classification and nomenclature of wheat gluten proteins: a reassessment. *Journal of Cereal Science* 4: 97–106.

Shuey, W. C. 1984. The Farinograph. In *The Farinograph Handbook*, edited by D'Appolonia, B. L. and W. H. Kunerth, 1–6. St. Paul, MN: American Association of Cereal Chemists, Inc.

Simmonds, D. H. 1989. *Inherent Quality Factors in Wheat. Wheat and Wheat Quality in Australia*. Melbourne: Australia Wheat Board.

Simon, S. J. 1987. More wheat with superior baking quality is needed. *Cereal Foods World* 32: 323–326.

Skendi, A., C. G. Biliaderis, M. S. Izydorczyk, M. Zervou, and P. Zoumpoulakis. 2011. Structural variation and rheological properties of water-extractable arabinoxylans from six Greek wheat cultivars. *Food Chemistry* 126: 526–536.

Skendi, A., M. Papageorgiou, and C. G. Biliaderis. 2009. Effect of barley β-glucan molecular size and level on wheat dough rheological properties. *Journal of Food Engineering* 91: 594–601.

Skendi, A., M. Papageorgiou, and C. G. Biliaderis. 2010. Influence of water and barley β-glucan addition on wheat dough viscoelasticity. *Food Research International* 43: 57–65.

Sliwinski, E. L., P. Kolster, and T. van Vliet. 2004. Large-deformation properties of wheat dough in uni- and biaxial extension. Part I. Flour dough. *Rheologica Acta* 43: 306–320.

Song, Y., and Q. Zheng. 2007. Dynamic rheological properties of wheat flour dough and proteins. *Trends in Food Science and Technology* 18: 132–138.

Sosulski, F., K. Krygier, and L. Hogge. 1982. Free, esterified, and insoluble-bound phenolic-acids. III. Composition of phenolic-acids in cereal and potato flours. *Journal of Agricultural and Food Chemistry* 30: 337–340.

Spies, R. 1989. Application of rheology in the bread industry. In *Dough Rheology and Baked Product Texture*, edited by Faridi, H. and J. M. Faubion, 343–359. New York: Springer.

Sroan, B. S., S. R. Bean, and F. MacRitchie. 2009. Mechanism of gas cell stabilization in bread making. I. The primary gluten–starch matrix. *Journal of Cereal Science* 49: 32–40.

Staudt, E., and E. Ziegler. 1973. *Flour Chemistry*. Engineering Works 9240. Uzwil, Switzerland: Bhuler Brothers Ltd.

Steffe, J. F. 1992. *Rheological Methods in Food Process Engineering*. Michigan: Freeman Press.

Steve, F. S., K. B. Robert, L. F. Patrick, and E. G. Edward. 1995. Relationship of test weight and kernel properties to milling and baking quality in soft winter wheat. *Crop Science* 35: 949–953.

Stojceska, V., F. Butler, E. Gallagher, and D. Keehan. 2007. A comparison of the ability of several small and large deformation rheological measurements of wheat dough to predict baking behaviour. *Journal of Food Engineering* 83: 475–482.

Stringfellow, A. C., and A. J. Peplinski. 1964. Air classification of Kansas hard red winter wheat flours. *Northwest Miller* 270: 19.

Thomas, D. J., and W. A. Atwell, 1999. *Starches*. St. Paul, MN: Eagan Press.

Toufeili, I., B. Smail, S. Shadarevian, R. Baalbaki, B. S. Khatkar, A. E. Bell, and J. D. Schofield. 1999. The role of gluten proteins in the baking of Arabic bread. *Journal of Cereal Science* 30: 255–265.

Tronsmo, K. M., E. M. Magnus, P. Baardseth, J. D. Schofield, A. Aamodt, and E. M. Færgestad. 2003. Comparison of small and large deformation rheological properties of wheat dough and gluten. *Cereal Chemistry* 80: 587–595.

Turnbull, K. M., and S. Rahman. 2002. Endosperm texture in wheat. *Journal of Cereal Science* 36: 327–337.

Van Lill, D., and M. F. Smith. 1997. A quality assurance strategy for wheat (*Triticum aestivum* L.) where growth environment predominates. *South African Journal of Plant and Soil* 14: 183–191.

Van Lill, D., J. L. Purchase, M. F. Smith, G. A. Agenbag, and O. T. De Viliers. 1995. Multivariate assessment of environmental effects on hard red winter wheat. I. Principle components analysis on yield and bread making characteristics. *South African Journal of Plant and Soil* 12: 158–163.

van Vliet, T., A. M. Janssen, A. H. Bloksma, and P. Walstra. 1992. Strain hardening of dough as a requirement for gas retention. *Journal of Texture studies* 23: 439–460.

Vanhamel, S., G. Cleemput, J. A. Delcour, M. Nys, and P. L. Darius. 1993. Physicochemical and functional properties of rye non-starch polysaccharides. IV. The effect of high molecular weight water-soluble pentosans on wheatbread quality in a straight-dough procedure. *Cereal Chemistry* 70: 306–311.

Walker, C. E., and J. L. Hazelton. 1996. Dough rheological tests. *Cereal Foods World* 41: 23–28.

Wall, J. S. 1979. The role of wheat proteins in determining baking quality. In *Recent Advances in the Biochemistry of Cereals*, edited by Laidman, D. L. and R. G. Wyn Jones, 275–311. London: Academic Press.

Wang, F. C., and X. S. Sun. 2002. Creep-recovery of wheat flour doughs and relationship to other physical dough tests and breadmaking performance. *Cereal Chemistry* 79: 567–571.

Wang, M., R. J. Hamer, T. van Vliet, and G. Oudgenoeg. 2002. Interaction of water extractable pentosans with gluten protein: effect on dough properties and gluten quality. *Journal of Cereal Science* 36: 25–37.

Whiteley, R. 1970. *Biscuit Manufacture*. London: Applied Science Publishers, Ltd.

Wieser, H. 2007. Chemistry of gluten proteins. *Food Microbiology* 24: 115–119.

Wikström, K., and L. Bohlin. 1996. Multivariate analysis as a tool to predict bread volume from mixogram parameters. *Cereal Chemistry* 73: 686–690.

Wikström, K., and L. Bohlin, 1999. Extensional flow studies of wheat flour dough. I. Experimental method for measurements in contraction flow geometry and application to flours varying in breadmaking performance. *Journal of Cereal Science* 29: 217–226.

Wikström, K., and A. C. Eliasson. 1998. Effects of enzymes and oxidizing agents on shear stress relaxation of wheat flour dough: Additions of protease, glucose oxidase, ascorbic acid, and potassium bromate. *Cereal Chemistry* 75: 331–337.

Williams, P. C. 1967. Relation of starch damage and related characteristics to kernel hardness in Australian wheat varieties. *Cereal Chemistry* 44: 383–391.

Williams, P. C. 1970. Particle size analysis of flour with the Coulter counter. *Cereal Science Today* 15: 102–107.

Wilson, J. T., and D. H. Donelson. 1970. Comparison of flour particle size distribution measured by electrical resistivity and microscopy. *Cereal Chemistry* 47: 126–134.

Yasunaga, T., and M. Uemura. 1962. Evaluation of color characteristics of flours obtained from various types and varieties of wheat. *Cereal Chemistry* 39: 171–183.

7 Bread Making

Duška Ćurić, Dubravka Novotni,
and Bojana Smerdel

CONTENTS

7.1   INTRODUCTION

Bread making is an ancient art that has been known to humankind for thousands of years. Today, the bakery market is one of the largest in the food industry. In the European Union, the market for bread amounts to 32.1 million tons, 73% of which are fresh products, 25% are prepacked products of long shelf life, and 2% are prepacked part-baked or frozen products (Gira 2007). By volume, 61% of the breads come from industrial production, 34% from artisan bakers, and 5% from in-store retailers and others. Industrial bread is sold mostly in Northern European countries such as Estonia, Ireland, and Sweden, whereas artisanal breads prevail in the Mediterranean

(Italy, Greece, and France). Bread is a popular staple food throughout the world. Around the globe, there are many varieties of bread that differ according to the ingredients and preparation methods used. The basic ingredients for bread making are flour, yeast, salt, and water. White wheat flour is the most often used flour for making bread. Due to the increased awareness of the importance of healthy eating, along with environmental issues nowadays, breads containing whole grain, multigrain, or other functional ingredients are becoming increasingly available in the market. Besides being a healthy food, bread has to be a tasty and convenient product.

7.2　BREAD-MAKING PROCESSES

Traditionally, bread making was a long process, but it has been drastically shortened with the mechanization of the bakery. Continuous systems of high-intensity mixers, proof boxes, tunnel ovens, coolers, and packaging machines have been developed. Today, the plant bread production process takes approximately 4 h from beginning to end. Most of the small artisan bakeries have been replaced by large bakery companies. The industrialization of bakeries is aimed at centralized production, diversified offerings, and consistent quality of bread. Bread is produced mainly according to two methods: the sponge and dough method and the straight dough method. With technological innovations and design of improvers, several modifications of these methods have arisen.

7.2.1　Sponge and Dough Process

The sponge and dough process is a traditional method that was originally developed in the United States and is widely accepted in Asia. In this method, one third or a half of the total amount of flour is mixed with part of water and total amount of yeast. It is allowed to ferment for 3 to 20 h at 21°C to 27°C, depending on the amount of yeast, consistency of the dough, temperature, and bread type. The preferment is called sponge or biga. After the first fermentation is ended, the remaining ingredients are added and mixed to form the dough. The dough is allowed to rest/proof for 15 to 30 min. After dividing and shaping, the dough is proofed for another hour. Baked bread has a very soft and porous crumb, with many bubbles of irregular size and distribution. Its flavor is distinguished and rich. It can be baked from weaker, and thus cheaper, flour and less yeast is required. The disadvantage of this process is that more bowls for the sponge and more space are needed.

7.2.2　Sourdough Bread-Making Process

Spontaneous sourdough fermentation is one of the oldest biotechnological processes. In the middle of the nineteenth century, after the introduction of baker's yeast as a leavening agent for bread production, the use of sourdough was almost completely abolished. Nowadays, the interest in sourdough has been renewed because of its positive effects on nutritive value, texture, flavor, and the shelf life of bread (Katina et al. 2006). Traditionally, a mixture of flour and water was fermented in several refreshing stages with microbes originating from the flour ("type I" sourdough). The dominating microbes of sourdough are homofermentative or heterofermentative lactic acid bacteria from the genus *Lactobacillus*, *Leuconostoc*, and *Pediococcus*, which produce organic acids. Lactic acid bacteria are often accompanied by a smaller population of yeast from the genera *Candida* or *Saccharomyces*. Nowadays, defined single-strain and multistrain culture preparations are commercially available. Sourdough is fermented for up to 24 or even 48 h at 22°C to 30°C until reaching pH ≤ 4, or until a total titratable acidity of 10 to 13 mL for 0.1 M NaOH is reached. Dough becomes acidic due to the production of organic acids by lactic acid bacteria. The major product of homofermentative bacteria is lactic acid, whereas heterofermentative bacteria produce a mixture of lactic and acetic acid. Sourdough is mostly produced in specialized plants from different cereals and marketed as a stabilized liquid or pasty ("type II") and dried ("type III") sourdough. The application of type II and type III sourdough in bread making requires the addition of bakery yeast for leavening. Traditional acidic sourdoughs are used in Mediterranean and Northern European countries, and in the San Francisco area in the United States, but yeasted preferment similar to the sponge and dough method is more commonly used than acidic sourdoughs. Nevertheless, sourdough is highly recommended in rye, whole grain, and gluten-free baking.

7.2.3 STRAIGHT DOUGH

Straight or direct dough baking is especially popular in small bakeries. All ingredients are mixed together at the same time. The bulk fermentation lasts from 0.5 to 3 h, depending on the quality of the flour, yeast level, dough temperature, and variety of bread produced. The dough is usually remixed or punched once or several times during fermentation. After dividing and shaping, dough is proofed once more for approximately half an hour and then baked. The resulting breads have fine and uniform crumb structures, but lower volume and harder crumb than breads produced by indirect methods.

A variation of the straight dough system—the no-time method—was developed in the 1960s. It refers to having no resting time or a short resting time of up to 30 min before dividing the bulk of the dough. It requires high-intensity mixers and a larger quantity of yeast. After mixing all the ingredients, the dough is divided and shaped, proofed, and baked. The disadvantage is that this kind of bread has limited flavor.

In the 1960s, innovative modification of a straight no-time process called the Chorleywood bread-making process was created by the British Flour and Baking Research Association at Chorleywood, UK. Dough is prepared by high-speed mixing of all ingredients for 3 to 5 min. After a short resting time, the dough is conventionally shaped, proofed, and baked. The process is suitable for weak flours and produces bread of high quality with respect to volume, color, and shelf life. It is usually a batch process but can be a continuous process. Large bakeries commonly use this method because it considerably reduces the cost and time of production as it requires only 1 to 2 h for completion.

7.2.4 CONTINUOUS SYSTEM

The continuous mixing method was introduced in the United States in the 1950s. The most common are the Do-Maker and Amflow systems. A liquid preferment, brew, or liquid sponge is prepared and fermented in stainless steel tanks under controlled conditions for several hours. Optionally, the fermented mixture is cooled by refrigerated coils between the walls of the tanks until needed. Preferment is mixed with other ingredients and then pumped to a developer that kneads the dough at high speed under pressure for 1 to 5 min. Air can be injected into the developer. There is no bulk fermentation, or if there is, it is achieved in the machine. The dough is pumped from the developer into an integrated divider, and then proofed and baked. In some bakeries, the dough is divided and shaped by conventional equipment, rather than extruding the dough directly into the pan. The bread produced by this method has a finer, more uniform structure and more consistent quality compared with other systems. However, it was not well accepted by consumers and now its use is limited to pizza dough.

7.2.5 COOLED OR FROZEN DOUGH

Methods of dough cooling and freezing were commercially introduced in the 1950s (Giannou and Tzia 2007). By the end of the 1990s, specific improvers for preproofed frozen dough had been developed. Such processes are gaining popularity in the baking industry because dough can be prepared in advance and finished when needed within 2 h, even on the selling spot. This reduces the required nighttime work but increases the energy demand. Although any of the bread-making processes can be applied, dough for interrupted proofing is mostly prepared using a no-time procedure in which yeast activity is minimized before freezing to enhance its cryoresistance (Lallemand Baking Update 1996).

Yeast fermentation of dough can be significantly slowed down by refrigeration. The following day, dough is warmed up, fully proofed, and then baked. This is usually done in a specialized, automated retarded proofer, which is a combination of a refrigerator and proofer with a high-humidity atmosphere. With this technology, at least 3 h is needed to prepare bread that has the same properties as conventionally baked breads. The interruption of fermentation, by lowering the temperature, is known to have a positive effect on dough structure and also on baked bread aroma and texture.

For prolonging shelf life, dough can be frozen in an unfermented or prefermented state. The freezing process demands more energy and skill than the chilling process. The unfermented frozen dough process requires higher amounts of freeze-tolerant yeast, low preparation temperatures, and tunnel freezing. After shaping, dough is frozen at a slow rate to preserve viable yeast cells that will be needed for postponed fermentation (Neyreneuf and Delpuech 1993). In the prefermented frozen dough method, the dough is partially fermented and then quickly frozen. After short thawing, it is baked. Temperature stability (±3°C) is imperative and a storage temperature of around −20°C is recommended to ensure a shelf life of 16 weeks (Phimolsiripol et al. 2008).

7.3 BREAD-MAKING STEPS

There are a few basic steps that are common for all bread-making processes: mixing of wheat flour and water, together with yeast, salt, and other ingredients; development of a gluten network in the dough through the application of energy during mixing; incorporation of air bubbles within the dough during mixing; continued development of the gluten structure created to modify the rheological properties of the dough and to improve its ability to expand when gas pressures increase into the dough; creation and modification of particular flavor compounds in the dough; dividing the dough mass into unit pieces; preliminary modification of the shape of the unit pieces; a short rest to further modify the physical and rheological properties of the dough pieces; shaping of the dough pieces; fermentation and expansion of the dough during proofing; further expansion of the dough pieces and fixation of the final bread structure during baking; and cooling and storage of the final product before consumption (Cauvain 2003).

7.3.1 Dough Mixing

In essence, mixing is the homogenization of the ingredients, whereas kneading is the development of the gluten structure after the initial mixing. Mixing contributes three main functions in the formation of dough structure: homogenization, air inclusion, and gluten development via mechanical energy input (Bloksma and Bushuk 1988). In the mixing of flour and water, water is absorbed by flour's components—the native and amorphous gluten proteins and the amorphous, noncrystalline parts of starch's components (Levine and Slade 1990). The water content of the hydrated flour is sufficient to allow substantial mobility to hydrated and solubilized molecules in the system (Roos 2003).

When dough is subjected to mechanical perturbation, it shows viscoelastic behavior: the mechanical force applied to the dough results in dimensional changes that are partially but not fully reversed when the force is removed. Nonetheless, dough stores some of the mechanical energy expended as elastic potential energy (Belton 2003). The molecular size and structure of the gluten polymers that make up the major structural components of wheat are related to their rheological properties. Interactions between polymer chain entanglements and branching are the mechanisms that determine the rheology of high molecular weight polymers. Dynamic shear plateau modulus is essentially independent of variations in gluten molecular weight among wheat varieties. The secondary structural and rheological properties of the insoluble polymer fractions are accountable for varying baking performances (Dobraszczyk 2004). The mixing speed and energy (work input) must be above a certain value to develop the gluten network and to produce bread of satisfactory quality (Kilborn and Tipples 1972). An optimum work input or mixing time varies depending on the mixer type, flour composition, and ingredients (Mani et al. 1992). Lower energy input is needed for flour of weak gluten content, whereas higher energy input is recommended for strong flour. If the mixing energy is larger than the optimum, the dough becomes wetter; it starts to stick to trough walls and its gas-holding ability drops. Dough mixing affects dough rheology and, conversely, the rheology of the dough defines the time and energy input required to achieve optimal development (Campbell and Shah 1999).

In most bread-making processes, dough development continues during resting after mixing, except in no-time doughs where the mixing is completely responsible for delivering a dough having the appropriate structure and rheological properties for further processing (Millar et al. 2005).

The production of a defined cellular structure in the baked bread depends entirely on the creation of gas bubbles in the dough during mixing and their retention during subsequent processing. Gas is occluded and concentrated in the liquid phase of dough only during the mixing stage of bread making (Baker and Mize 1946). However, the subsequent stages (dividing, sheeting, and molding) cause the subdivision of already existing gas cells, thus improving their number and size distribution. Two gases present in the dough after mixing are oxygen and nitrogen, but the oxygen is relatively quickly used up by the yeast cells within the dough (Chamberlain 1979). With the removal of oxygen, the only gas that remains entrapped is nitrogen, and this plays a major role by providing bubble nuclei into which the CO_2 produced later by yeast fermentation can diffuse and cause dough expansion. Bubble size is generally important as it can influence bubble growth during proofing, and hence the texture of finished bread (Shimiya and Nakamura 1997; Shah et al. 1998). Mixing brings both soluble and insoluble components to the bubble surface, the movement of which might be otherwise hindered by the structure and viscosity of the gluten–starch matrix of dough (Ornebro et al. 2000). It is thought that the liquid film is formed from the aqueous phase of dough, when dough is mixed above a certain critical moisture content of approximately 0.3 g g^{-1} dry flour (Gan et al. 1995).

The ability of a liquid film to maintain bubble integrity in dough systems is determined by its composition of surface active components. These are likely to include small molecule surfactants, such as lipids and other secondary metabolites (Gan et al. 1995; Keller et al. 1997), as well as proteins such as nonspecific lipid transfer proteins, puroindolines (Douliez et al. 2000), and certain α-amylase-trypsin inhibitors (Gilbert et al. 2003). Arabinoxylans can also mediate interactions and cross-links between proteins in an adsorbed layer, increasing the viscosity of thin liquid films, thereby enhancing the stability of the foam structure. Emulsifiers are lipid-like compounds that are routinely included as ingredients in bread making for the improvement of loaf volume and the stability of the fragile foam structure during proofing (Carson and Sun 2000).

The mechanical forces on the bubbles during mixing are much greater than the surface tension forces that have a less significant effect on the process of bubble creation in dough. Nonetheless, bubble size and number can be affected by gas pressure applied during mixing (Campbell et al. 1998, 2001). If a partial vacuum is later applied during mixing, the number of bubbles is reduced, resulting in a finer crumb structure in the finished bread.

7.3.2 Dough Dividing and Molding

To generate the shape and size of product required, the bulk dough has to be first divided from the mixer into individual pieces and then shaped. Different dividers need to be matched to different dough types to give optimum dividing accuracy with minimal compression damage. The success of dividing depends on the homogeneity of the dough, which is largely determined by the distribution of gas bubbles. The constant density of the dough, composed of bubbles of uniform size and even distribution, results in more accurate dividing by volume. Mechanical molding subjects the dough to stresses and strains that may lead to damage to the existing gas bubble structure. Some bread-making processes require the rounder to have a degassing effect, however, which is unnecessary in no-time bread-making processes (e.g., the Chorleywood bread-making process) due to the low gas content in the dough at the end of mixing (Cauvain 2003). In dough-making processes in which the first proof time is short, the rounder adds little or nothing to the structural properties of the final product and, for a given first proof time, limits the extensibility of the dough piece during final molding. The intermediate or first proofing process is a period of rest between the works carried out by dividing and rounding and may be used before final shaping. The duration of this process is related to the dough rheology required for final molding. The longer the dough rests, the greater the changes are in its rheology. Especially in no-time dough-making processes (e.g., the Chorleywood bread-making process), the first dough proof can have a considerable effect on final bread quality. The elimination of the first proofing could lead to a reduction in loaf volume and poorer crumb cell structure because of damage to the bubble structure in the dough.

The basic functions of the final molding stage are to shape the dough to fit the product concept and to reorientate the cell structure. Final molding clearly requires the dough to have the appropriate rheology, particularly when starting from a rounded dough ball. The dough pieces should have low resistance to deformation and minimal elasticity; otherwise the high pressures required to change the dough's shape can cause loss of product quality (Cauvain and Young 2003). Dough that lacks water is "tight" and offers greater resistance to deformation, exhibiting greater elasticity and reduced extensibility and is more susceptible to damage during dividing and shaping. This increases the importance of judging flour water absorption capacity and optimizing added water levels during mixing (Cauvain and Young 2003). The degree to which dough may be degassed during the sheeting stages of final molding depends on its rheology and interactions with the equipment. A further problem that may be encountered during dough sheeting is the rupture of gas-stabilizing films and the subsequent coalescence of two gas bubbles to form one of larger size. Such damage to dough bubble in structures is thought to be a major factor in the formation of large, unwanted holes in bread crumb (Cauvain and Young 2003).

7.3.3 Proofing

After mixing and forming, dough is subject to proofing and its structure is expanded to a suitable size before it goes into the oven. Bakers' yeast is the most active at 32°C to 38°C and so running the proofer at approximately 40°C minimizes the time required for proofing. Successful proofing demands that the relative humidity of the air in the cabinet should be close to but slightly lower than that of the dough. In practice, this means approximately 80% to 85% relative humidity at approximately 40°C (Cauvain and Young 2003). During proofing, the starch from the flour is progressively converted into dextrins and sugars by the actions of the enzyme. Yeast feeds on the sugars to produce carbon dioxide and alcohol. The carbon dioxide diffuses into the gas bubbles in the dough, causing them to grow and the dough to expand

(Whitworth and Alava 1999). The rheology of the dough itself was not thought to influence bubble expansion significantly during the early stages of proofing, although it clearly becomes critical later. The structure of the dough is now dependent on the foam's stability to maintain the overall desired bubble structure. The bubbles will undergo instability processes, for example, coalescence and disproportionation, as they expand. However, disproportionation in bread dough is not simply a matter of mass transport of gas from the surrounding dough into the bubbles. Below a certain critical radius, the bubbles will not grow and will eventually disappear altogether, with only larger bubbles increasing in size during proofing (Kumagai et al. 1991; Shimiya and Nakamura 1997). During the latter stages of proofing and in the early stage of baking, the bubbles undergo a further, more rapid expansion phase as a result of both increased CO_2 production by the yeast and the formation of steam. This increases the tendency for bubbles to coalesce as a consequence of failure in the gluten–starch matrix forming the bubble walls. The failure of the starch–gluten films separating some bubbles has been implicated in the instability of the foam structure of bread at the end of proofing (Gan et al. 1995). It has been postulated that the bubbles are lined by a thin aqueous film, which ensures that gas is retained, even when discontinuities form in the starch–gluten matrix, thus preventing coalescence (Sahi 1994; Gan et al. 1995). The drainage, surface properties, and composition of such a thin film play an important role in determining the stability of adjoining bubbles at this stage of the process, and hence the crumb structure of baked bread (Mills et al. 2003). The limit of expansion of these bubbles is related directly to their stability due to coalescence and the eventual loss of gas when the bubbles fail.

7.3.4 BAKING

After proofing, the dough must be baked. A series of physical, chemical, and biological changes such as evaporation of water, formation of porous structure, volume expansion, protein denaturation, starch gelatinization, crust formation, etc., take place during bread baking. Baking temperatures will vary depending on the oven and the product, but typically range from 220°C to 250°C, aiming to achieve a core temperature of approximately 92°C to 96°C by the end of the baking period (Cauvain 2003). The driving force for heat transfer is the temperature gradient from regions near the crusts (where the temperature is limited to the boiling point of water) to the center of loaf. The heat transfer mechanism is conduction along the cell walls, and the center temperature rises independently of the oven temperature and approaches boiling point asymptotically (Cauvain and Young 2003). The production of CO_2 by yeast continues at an increased rate during the first stage of baking until yeast is destroyed at a temperature of approximately 55°C. According to the Gay–Lussac law, occluded gas expands when temperature increases from 25°C to 70°C (Bloksma 1986).

The rising temperature during baking causes thermal expansion of vapor and raises the saturation pressure of water within the dough. This leads to a local expansion but at the same time forces compression and higher densities elsewhere. CO_2 also plays an important role in the expansion of bubbles during bread baking. It is also released from the bread when the bubble walls start to break under pressure, making the porous structure more continuous and open to the outside of the bread. Depending on rheological properties such as elasticity and viscosity, the closed cell membranes in the dough may resist expansion (Zhang et al. 2007b). Starch gelatinization in the cell membranes occurs at more than 65°C, increases dough viscosity, and impairs the extensibility of the dough, which results in increased pressure in closed gas cells leading to rupture of the cell membranes. As a result, the gas molecules will exchange between adjacent cells and ultimately be transported to the outside of the dough, resulting in a loss of gas and presumably limited capacity for expansion.

During baking, a crust forms and restricts the expansion of the dough (Zhang et al. 2007b). Zanoni et al. (1994) suggested that the crust could restrict water vapor flow from the pores to the dough's surface. The crust is a hard, vitreous surface layer that is a result of dried starch gel with dispersed protein and lipid aggregates (Eliasson and Larsson 1993). The formation of crust and browning during baking seem to be primary contributors to the formation of bread flavor. In bread crust, the higher temperature and lower water content activate nonenzymatic browning reactions including Maillard reactions (sugar-amine) and caramelization (Gogus et al. 2000). Flavor is another quality attribute developed during the baking process in the form of n-heterocycles via Maillard reactions. During baking, the flavor compounds formed are adsorbed by pore curvatures. Crust structure also provides a barrier against the loss of flavor (Eliasson and Larsson 1993). Most breads are characterized by a crispy crust. To obtain gloss, it is essential that vapor condenses on the surface to form a starch paste that will gelatinize, form dextrins, and eventually caramelize to give both color and shine. If there is excess water, paste-type gelation takes place, whereas with insufficient water, crumb-type gelation occurs. To deliver the necessary water, steam is introduced into the oven at the beginning of the baking (Cauvain 2003). The crust, being less flexible, does not contract as fast or to the same degree as the crumb, and the strains placed on the crust by attached portions of crumb become such that splits begin to occur in the surface. In the production of part-baked breads in which proofed dough is baked just sufficiently to inactivate the yeast and enzymes, and to set the structure with a minimum of crust coloration and moisture loss, appropriate steaming is crucial (Cauvain and Young 2003).

7.4 MEASURING RHEOLOGICAL PROPERTIES OF DOUGH

Rheology is the study of the deformation of matter (Steffe 1996). Rheological tests measure a well-defined property, such as stress, strain, stiffness, or viscosity. A small piece of a material is usually deformed in a controlled way, normally on a motor-driven machine, and the force is measured as well as the distance moved or displacement of the object. The force is then usually plotted against the displacement to give a force–displacement curve (Dobraszczyk 2003). Rheometry is a powerful technique for explaining the quality of cereal foods as well as for simulating and predicting responses to the deformation conditions that are present in practical processing but inaccessible to normal rheological measurement (Ahmed 2010).

The rheology of bread dough and particularly wheat dough is complex due to its viscoelastic behavior and the fact that it experiences a wide range of conditions of stress states and strain rates during processing and baking (Ahmed 2010). Therefore, the rheological properties of dough are dependent both on time and strain, and because rheological measurements use only one deformation force, there is often a discrepancy between single point tests and actual performance in the plant, where conditions of strain and strain rate may be poorly defined and very different from those in the laboratory tests.

The main techniques used for measuring cereal properties have traditionally been divided into descriptive empirical techniques and fundamental measurements (Dobraszczyk 2003), and they are categorized according to the type of strain imposed: compression, extension, shear, torsion, and also the relative magnitude of the imposed deformation (small or large deformation). The baking process is usually divided into the cold (mixing, fermenting) and hot (oven) stages. During the cold stage, doughs can be tested for their strain-dependent, time-dependent, and concentration-dependent behaviors, whereas during the hot phase, only the time-dependent and temperature-dependent behaviors can be recorded (Weipert 1990).

7.4.1　FUNDAMENTAL RHEOLOGICAL TESTS

Fundamental rheological tests (Table 7.1) measure physical properties independent of shape, size, and how they are measured (Dobraszczyk 2003). In most tests, the deformation or the stress is held constant or is made to vary as a simple function of time, and therefore the main types of fundamental rheological tests used in cereal testing are dynamic oscillation (deformation or stress varies sinusoidally with time), creep (stress constant) and stress relaxation (deformation constant), extensional measurements, and flow viscometry (deformation increases linearly with time; Bloksma and Bushuk 1988). The type of deformation applied can be stretch or shear.

Dynamic oscillatory rheometers simultaneously measure the elastic as well as the viscous components of a material's complex viscosity by the application

---

TABLE 7.1
Fundamental Rheological Tests

Fundamental Methods	Products	Property Measured
Dynamic oscillation concentric cylinders, parallel plates	Fluids, pastes, doughs	Dynamic shear moduli, dynamic viscosity
Tube viscometers: capillary pressure, extrusion, pipe flow	Fluids, sauces, pastes, dough	Viscosity, in-line viscosity
Transient flow: concentric cylinders, parallel plates	Semisolid (viscoelastic materials)	Creep, relaxation, moduli, and time
Extension: uniaxial, biaxial, TAXT2 dough inflation system, lubricated compression	Solid foods, doughs	Extensional viscosity, strain hardening

Source:　Dobraszczyk, B. J. 2003, Measuring the rheological properties of dough. In: *Bread making: Improving quality*, ed. S. P. Cauvain, 375–395. Cambridge: Woodhead Publishing Limited.

of sinusoidally oscillating stress or strain with time and measuring the resulting response (Dobraszczyk 2003).

Creep–recovery is a static rheological method in which an instantaneous stress is applied to the sample and the change in strain is measured over time. A creep phase is usually followed by a recovery phase in which the applied stress is removed (Steffe 1996). Creep–recovery measurements can be performed in compression and uniaxial tension or shear.

There are several methods that have been used to measure the rheological properties of dough in extension: simple uniaxial extension, in which dough is stretched in one direction; and biaxial extension, in which dough is stretched in two opposing directions, which can be achieved either by compression between lubricated surfaces or by bubble inflation (Bagley and Christianson 1986; Dobraszczyk 1997). Uniaxial extension of doughs can be measured by many methods, some of which are the Simon Research extensometer, Brabender's extensigraph, and Stable Micro Systems' Kieffer dough and gluten extensibility rig.

The most widely used methods for measuring biaxial extension properties of food materials have been inflation techniques and compression between flat plates using lubricated surfaces, which produce purely extensional flow provided no friction occurs (Chatraei and Macosko 1981). Bubble inflation and lubricated compression are the most appropriate methods for measuring the rheological properties of dough because the deformation in these tests resembles the ones within the dough during the proofing and oven rise stages.

The aim of viscometry is to determine the relation between the rate of shear and shear stress in a material in a steady state. Viscosimeters record the torque and the shear rate in a geometrically defined gap during steady state shear flow. Based on calculations of its stress and shear rate, the viscosity of Newtonian (ideal) and non-Newtonian liquids can be calculated from analysis of the flow field between coaxial cylinders, plate and cone, or plate and plate devices (Weipert 1990). Dough is a viscoelastic material with explicit nonlinear behavior.

7.4.2 Descriptive Rheological Measurements

The properties of dough are objectively assessed by numerous instruments (Table 7.2) to evaluate performance during processing and for quality control. Although the physical properties measured (such as consistency, hardness, texture, viscosity, etc.) can be classified as rheological properties (Bloksma and Bushuk 1988), these instruments are descriptive and depend on the specific conditions under which the test was performed because the sample geometry is variable and the stress and strain states are not precisely defined (Dobraszczyk and Morgenstern 2003). It is also impossible to compare results between different testing machines (e.g., between a farinograph and an extensigraph) or to extrapolate the results to other deformation conditions, such as during baking. For example, it is difficult to access dough during mixing, sheeting, proving, and baking without interrupting the process or disturbing the structure of the material. Prediction of the range of conditions the dough experiences during a given process can be extrapolated from rheological measurements made under laboratory conditions, often with modeling simulations using computational fluid dynamics software (Scott and Richardson 1997).

TABLE 7.2

Physical Dough Testing Instruments

Instrument	Product	Property Measured	Standard Methods No.
		Recording Dough Mixers	
Farinograph (Brabender)	Dough	Mixing time/torque Apparent viscosity	AACC Methods (54-21, 54.22; 38-20.01); ICC (115/1; 109/1); ISO 55301
Do-Corder (Brabender)			
Valorigraph (Labor-MIM)			
Resistograph (Brabender)			
Mixograph (National Mfg. Corp., Lincoln, NE)			AACC Method 54-40.02; 54-60.01
Consistograph			AACC Method 54-50.01; ICC 171
		Load-Extension Meters	
Extensigraph	Dough	Extensibility	AACC 54-10.01; ICC 114/1; ISO 5530-2
Research extensometer (The Halton)			
TAXT2/Kiffer rig	Dough, gluten	Biaxial	AACC 54-30.02; ICC121
Chopin Alveograph		extensibility	
Dobraszczyk/Roberts attachment (TA-XT2-BUBBLE)			
		Pasting Tests	
Amylograph (Brabender)	Pastes, suspensions	Apparent viscosity, Gelatinization temp	AACC 22-10.01: 22-12.01; 61-01.01; ICC 126/1; ISO 7973
Rapid Visco Analyzer (The Newport Scientific)			ICC 161; ICC 162
Falling number			AACC 56-81.03
Texturometer	Solid foods	Texture profile, firmness	
Flow cup	Fluids, sauces, batters	Apparent viscosity	
Falling ball	Fluids		
Flow viscometers	Fluids, pastes		
Fermentometers	Dough	Height, volume	
Penetrometers	Semisolid foods, gels	Firmness, hardness	

Source: Modified from Bloksma, A. H., and Bushuk, W. 1988, Rheology and chemistry of dough. In *Wheat chemistry and technology*, ed. Y. Pomeranz, 131–218. St. Paul: AACC; Kulp, K., and Ponte, J. G. 2000, Breads and yeast-leavened bakery foods. In *Handbook of cereal science and technology*, eds. K. Kulp and J. G. Ponte, 516–525. New York: Marcel Dekker; Dobraszczyk, B. J. 2003, Measuring the rheological properties of dough. In: *Bread making: Improving quality*, ed. S. P. Cauvain, 375–395. Cambridge: Woodhead Publishing Limited.

Recording mixers measure viscosity and are used to measure and record the changes in the resistance to mixing with time. Numerous studies have shown that rheological measurements after mixing parallel changes in mixer torque and power consumption (Mani et al. 1992; Anderssen et al. 1998; Zheng et al. 2000), especially if rheological measurements are made under large, nonlinear deformation conditions closer to those experienced in the mixer (Mani et al. 1992; Hwang and Gunasekaran 2001). Stress–strain instruments provide information on the potential behavior of dough during its rise due to the development and expansion of gas during fermentation and early baking stages. Load-extension instruments are applied to measure the resistance of dough to extension. Unlike the extensigraph or extensometer, which both stretch the test dough piece in only one direction, the alveograph subjects dough to extension into two dimensions by blowing a molded and rested sheet into a bubble. From the physical viewpoint, such an extension mode is well linked with the gas cell expansion in rising dough.

Pasting tests concentrate on changes in physical properties taking place during the baking stage, which are a result of starch gelatinization and its degradation by amylolytic enzymes. Continuously measuring the sample's viscosity while it is being heated in a controlled manner is well suited for this purpose (Kulp and Ponte 2000).

7.4.3 PHYSICOCHEMICAL TESTS

Physicochemical tests predict flour's potential bread baking strength from its gluten behavior. They measure the quality and quantity of gluten (swelling ability of wet gluten, change in viscosity) as well as tests evaluating a dough's gas production and retention capacity. Standard methods for gluten determination are the AACC Methods 56-50, 56-51A, and 56-63, whereas AACC Method 22-11 is for gas production (Kulp and Ponte 2000). To produce a loaf of bread with a light and even crumb, the yeast must produce sufficient CO_2, and the dough must retain the gas produced over a sufficiently long period. Most methods do not distinguish between the gas that is retained in the dough and the gas that escapes from it; therefore, they measure gas production. Exceptions are methods that measure the height of fermenting dough, such as the maturograph, or the buoyant force on a test piece of dough (oven rise recorder). Measurements of gas production can be converted into measurements of gas retention if the method is modified to absorb, in a solution of alkali, the escaped CO_2. The Chopin rheofermentometer is an instrument that allows fully automated monitoring of the fermentation process in dough. This is done through the continuous measurement and recording of dough rise, gas formation, and gas retention.

7.5 INFLUENCE OF BREAD-MAKING PROCESS ON BREAD QUALITY

Bread quality is reflected in its sensory attributes, physical properties, chemical composition, safety, and shelf life. Quality, together with eating habits and price, influences consumers in purchase choices. Consumers of bakery products are heterogeneous, and they differ on their geographical and demographical backgrounds. Besides consumer's concerns, the producers need to fulfill legislative regulations.

The number of features in a type of bread is assessed during product development and quality control using sensory and instrumental methods. Bread volume is a result of crumb structure. The required volume depends on the bread variety. The shape of the bread should be symmetric and rounded. It is brought about by correct dough fermentation and molding. Crust should be of the proper thickness, color, and glossiness typical for that bread type, without cracks and detachment from the crumb. Crust color needs to be even throughout the surface. It is influenced by the baking time and temperature, and the amount of sugar in dough. Consumers prefer a fine uniform crumb structure accompanied with thin cell walls. The type of flour used and the structure of the crumb influence the color. The texture of the crust and crumb is completely different. Most consumers prefer crispy crust but soft, elastic, and moist crumb. Crumb texture is strongly influenced by the crumb porosity and bread volume. The flavor of the bread is the most important. It is composed of approximately 200 compounds, among which are alcohols, aldehydes, ketones, esters, and pyrazines (Chang et al. 1995). Bread volatiles are products of proteolysis, lipid oxidation, and Maillard reactions that occur during fermentation and baking.

Bread quality is the result of a good recipe and the quality of the ingredients, appropriate processing, and a skillful baker. Take the baguette, for example. Baardseth et al. (2000) demonstrated how process variations accounted for 40% of the variation in product quality whereas flour quality accounted for 16% of the variation. They showed that baguettes produced using the traditional French process got a higher score for porosity, elasticity, crust crispiness, and volume compared with breads produced using industrial processes. Nonetheless, industrial processes can ensure consistent quality if each bread-making step is well controlled and adjusted to the recipe.

The energy input into dough mixing and baking volumes are interrelated. The importance of mixing on bread quality is the most obvious in the Chorleywood bread-making process, which enables it to gain the same bread volume and crumb softness even with lower protein flour. The higher the level of energy transferred to the dough during mixing, the better the bread volume and crumb softness will be after baking. Fine, uniform cell structures are assured by using pressures at less than atmospheric values, whereas open, more random structures can be obtained using pressures greater than atmospheric values (Cauvain et al. 1999; Campbell et al. 2001).

During the final shaping of dough pieces, mechanical defects, which are most commonly manifested as dark-colored and firm patches in the crumb, could occur as a result of the thick cell walls (Cauvain and Young 2000).

Fermentation is an essential step for bread's cellular structure, crust crispness, and aroma formation (Primo-Martin et al. 2010). In general, the longer the fermentation time, the better the aroma and texture of bread. Pagani et al. (2006) showed that after 24 h, the hardness of crumb produced by straight dough method is 50% higher than that of a sponge and dough loaf.

Shittu et al. (2007) showed that varying combination of baking temperature and time leads to significant differences in the quality of bread in terms of volume, crust color, crumb moisture, density, porosity, and softness. In general, breads baked at higher heating rates have a lower moisture content compared with breads baked at lower heating rates (Patel et al. 2005). During baking, bread dough is first expanded and then slightly contracted at the end of baking. During postbaking chilling, Ben-Aissa et al. (2010) determined that 24% of volume shrinkage occurred in the case of shorter baking (maximum core temperature of 75°C kept for 10 min) in comparison with 2.4% of volume contraction for longer baking (at 98°C for 10 min). Baking with adequate steaming increases bread volume compared to baking without steam. An increased amount of steam during baking leads to reduced color, failure force, and failure firmness, although it also leads to intensified glossiness (Altamirano-Fortoul et al. 2012).

Alternatives to conventional baking that are trying to offer reduced baking time and energy consumption could have a variety of effects on bread quality. Microwave-baked bread has unacceptable texture in terms of rubbery and tough crust, and a firm and dry crumb, rapid staling, and lack of crust color (Ovadia and Walker 1995). Therefore, microwaves are often combined with other baking techniques to obtain quality resembling that of conventionally baked breads (Table 7.3). Willyard (1998) reported slower staling of buns when baked with an additional application of microwave energy in combination with conventional baking. The combination of microwave heating with halogen (infrared) lamp or air jet impingement, and a combination of forced convection with microwave energy using a hybrid oven, can improve overall bread quality (Patel et al. 2005; Datta et al. 2007). Datta et al. (2007) showed that breads baked in jet impingement had the highest total porosity followed by microwave jet impingement and microwave infrared.

During cooling from 20°C to 4°C, bread volume is reduced by 1.4% and after freezing at −18°C, it is reduced by a further 2.6% (Ben-Aissa et al. 2010). Vacuum cooling offers several advantages to the bakery industry in terms of quality. The shape and texture of vacuum-cooled bakery products are superior to air-cooled products because less contraction and collapse occur during storage (Wang and Sun 2001).

TABLE 7.3
Nonconventional Bread Baking Technologies

Process	Effect on Bread Quality	Reference
Electrical resistance	Delayed staling	Martin et al. 1991
Jet impingement	Increased porosity	Datta et al. 2007
Jet impingement + infrared radiation	Thinner crust	Olsson et al. 2005
Conventional + microwave	Lower volume, unacceptable texture	Ovadia and Walker 1995
Forced convection + microwave	Better quality than conventional depending on the heating rate	Patel et al. 2005
Halogen lamp + microwave	Specific volume and color comparable with conventional breads but higher weight loss and crumb firmness	Keskin et al. 2004
Halogen lamp + microwave	Quality comparable with conventional breads at the optimum baking conditions	Demirekler et al. 2004

7.5.1 INFLUENCE OF SOURDOUGH FERMENTATION ON BREAD QUALITY

Sourdough has traditionally been used as a natural improver in rye baking. The knowledge gained in rye bread-making can be transferred to wheat or gluten-free baking. To achieve sourdough of consistent quality, the use of commercial starters of pure or mixed cultures is recommended because the microflora of spontaneous sourdough is variable depending on the microflora of the raw materials used, hygienic conditions, and technological parameters of the process. Contradictory findings on sourdough influence on bread volume, texture, and flavor imply that improved product quality is obtained only under the optimized conditions of sourdough fermentation (Katina et al. 2006). The amount of sourdough added has to be chosen according to its characteristics and bread type. Too much sourdough can flatten the bread because acidic conditions during sourdough fermentation favor the activity of cereal protease that hydrolyzes wheat gluten and weakens the dough (Corsetti et al. 2000). Sourdough breads have softer crumb due to the increased solubility of arabinoxylans and dextrin production (Courtin and Delcour 2002). During sourdough fermentation, some lactic acid bacteria strains such as *Leuconostoc* spp. produce exopolysaccharides that have a positive influence on crumb texture (Corsetti et al. 2000; Tieking et al. 2003).

Although most of the aroma compounds are produced during baking, sourdough fermentation is a key factor in obtaining desirable bread aroma. Aroma depends on the characteristics of the raw materials, conditions of sourdough fermentation, starter choice, and their interactions, as well as the final proofing and baking (Katina et al. 2004; Hansen and Schieberle 2005). Biochemical changes in cereal flour during sourdough fermentation are responsible for the desirable and undesirable changes in bread flavor. Fermentation conditions should assure the formation of mild acidity, an increased amount of free amino acids, and specific volatiles that will contribute to the balanced sensory profile of bread (Katina et al. 2004). Acetic acid improves bread aroma when present in small amounts, although in higher concentrations, it has a negative effect. It is assumed that the combination of *Lactobacillus plantarum* and *Saccharomyces cerevisiae* guarantees a balanced aroma profile in wheat sourdough bread (Hansen and Hansen 1994). During fermentation with mixed starter cultures containing yeast, more sweet, acidic, and bitter compounds of flavor are produced (Meignen et al. 2001). In sourdough from whole grains, the production of organic, amino, and phenolic acids is particularly intensive and contributes to the bread aroma profile (Katina et al. 2004).

Depending on the type and fermentation conditions, the addition of sourdough can shorten or extend the freshness and shelf life of bread (Corsetti et al. 2000). The addition of sourdough with low pH values and high molar ratio of lactic and acetic acid is the most efficient in delaying bread staling (Barber et al. 1992; Corsetti et al. 2000). Dough acidification favors the amylolitic hydrolysis of starch. As a result, an increased amount of soluble starch affects the retrogradation of swollen starch granules. As mentioned previously, exopolysaccharides produced by some lactic acid bacteria have a positive effect on bread's freshness (Corsetti et al. 2000). Acetic acid effectively suppresses the growth of molds and ropiness on bread (Gobbetti et al. 2005).

7.5.2 QUALITY OF FROZEN DOUGH BREAD

Bread is characterized by its short shelf life, a term that means that the aroma of bread is rapidly lost, crust is softened, and the crumb is hardened and loses elasticity. The need to provide a wide range of freshly baked bakery products for institutional and direct sales to consumers drove producers to become involved in refrigerated/ frozen dough processing. By choosing the right ingredients and process parameters, excellent bread can be produced from frozen doughs stored for up to 3 months or longer. Nonetheless, the freezing process and frozen storage of dough can significantly affect the quality of the resulting bread. Yeast cells and the gluten network are damaged during freezing (Havet et al. 2000). This results in reduced CO_2 production and retention, which is reflected in decreased porosity, uneven distribution of bubbles, reduced specific volume of breads, and harder crumb (Phimolsiripol et al. 2008). It has been noticed that ice recrystallization and water redistribution result in losses in yeast activity and dough strength (Varriano-Marston et al. 1980). Furthermore, the starch granules that are damaged by the formation of large ice crystals during freezing may favor moisture retention and, in turn, an increased amount of freezable water (Berglund et al. 1991; Lu and Grant 1999).

During frozen storage, the changes in the dough are extended. During the first 2 months of frozen storage, samples undergo their main changes, reaching an asymptotic behavior after 2 to 3 months and remaining stable thenceforth, having satisfying quality characteristics even after 9 months of storage (Giannou and Tzia 2007). The specific loaf volume of frozen dough bread is further decreasing whereas crumb firmness is increasing as the duration of frozen dough storage is extended (Phimolsiripol et al. 2008). Also, the color of the resultant bread is darker and less uniform as frozen dough is stored for longer periods (Yi and Kerr 2009). This may be due to an increase in leached amylose and degraded dextrins, which contribute to Maillard reactions. Water redistribution and availability may also play a role in color darkening and nonuniformity.

To preserve yeast activity, it is advisable to apply slow freezing rates and to use freeze-tolerant yeast in increased levels of up to 6% (Neyreneuf and Delpuech 1993). Takano et al. (2002) demonstrated that the damaged gluten can be partially repaired by the mechanical force of punching, molding, and resheeting after thawing, which increases bread volume by 10% to 110% if freeze-tolerant yeast is used. With extended frozen storage, the proofing time has to be prolonged. A strong, high-quality wheat flour or extra gluten flour is to be used for frozen dough (Lallemand Baking Update 1996). At the same time, the addition of water should be reduced to approximately 2% to 4% lower than the optimum to avoid stickiness of the dough after thawing and to minimize ice crystals growth. Oxidation requirements may increase during prolonged frozen dough storage because reducing compounds may leak from dead yeast cells. The level of oxidants, such as ascorbic acid, that are compatible with a no-time dough process should be increased. The negative effects of dough freezing can be minimized by the inclusion of suitable improvers—concentrated proteins, hydrocolloids, emulsifiers, and enzymes (Selomulyo and Zhou 2007; Table 7.4). Minervini et al. (2011) demonstrated that sourdough in combination with cryoprotectant (skimmed milk, sucrose, and trehalose), conventional additives (guar

TABLE 7.4

Baking Improvers for Frozen Dough and Par-Baked Frozen Bread

Improver	Function in Bread	Application	Dosage in Flour	Reference
		Hydrocolloids		
Hydroxypropyl-methylcellulose	Improves volume and shelf life	FD, PBF	0.3%–0.5%	Bárcenas and Rosell 2007; Polaki et al. 2010
Carboxymethyl-cellulose	Improves volume, structure, texture, and color	FD	1%–3%	Sharadanant and Khan 2006
Xanthan gum	Improves volume and crumb elasticity	FD, PBF	0.2%	Mandala et al. 2008; Matuda et al. 2008
Guar gum	Improves volume and texture	FD	0.5%	Ribotta et al. 2004; Matuda et al. 2008
Locust bean gum	Improves volume, moisture content, and shelf life	FD, PBF	0.4%–3%	Sharadanant and Khan 2006; Polaki et al. 2010
Gum arabic	Improves volume, structure, texture, and color	FD	1%–3%	Asghar et al. 2005; Sharadanant and Khan 2006
		Proteins		
Carrot protein	Improves dough stability to freezing	FD	6.2%	Zhang et al. 2007a
Whey protein	Improves dough rheology	FD	2.5%–5%	Asghar et al. 2009
Egg yolk	Improves dough expansion and bread volume	FD	3%	Hosomi et al. 1992
		Lipids and Emulsifiers		
Shortening	Improves texture	FD, PBF	1%–10%	Inoue et al. 1995; Carr and Tadini 2003
DATEM	Improves volume, shape, and texture	FD	0.5%–0.6%	Kenny et al. 1999; Ribotta et al. 2003
Sodium or calcium stearoyl lactylate	Improves volume and texture	FD	0.5%	Inoue et al. 1995; Kenny et al. 1999
Sucrose esters	Improve crumb structure, volume, stability, and shelf life	FD	0.5%	Hosomi et al. 1992

(*continued*)

TABLE 7.4 (Continued)
Baking Improvers for Frozen Dough and Par-Baked Frozen Bread

Improver	Function in Bread	Application	Dosage in Flour	Reference
		Enzymes		
Xylanase	Improves volume and crumb softness, delays staling	PBF	100 ppm	Jiang et al. 2008
Hemicellulase	Improves volume, prevents crust flaking	PBF	300 ppm	Ribotta and Le Bail 2007
Sulfhydryl oxidase (+ ascorbic acid)	Improves dough rheology and fermentation features	FD	1–100 U/g	Faccio et al. 2012
		Miscellaneous		
Glycerol	Prevents the formation of ice crystals during freezing and frozen storage	FD	2%	Huang et al. 2011
Trehalose (+ gluten flour 2%–4%)	Yeast cryoprotectant	FD	10%	Salas-Mellado and Chang 2003
Sourdough (+ cryoprotectant and/or honey)	Improves volume, porosity, crumb softness, and color	FD	30% on dough weight	Minervini et al. 2011

Note: FD, frozen dough; PBF, par-baked frozen bread.

gum, diacetyl tartaric acid esters of monoglycerides, and ascorbic acid), and honey or fructose and glucose decrease the hardness of frozen dough bread even when compared with unfrozen dough bread.

Prefermentation of dough before freezing is an attractive but challenging alternative to unfermented frozen dough. The advantages of prefermented frozen dough include a reduced need for yeast viability after frozen storage, less time-consuming production after freezing (ready to bake), and improved heat transfer. Nonetheless, there is a danger that a fragile dough structure will collapse due to gas contraction during freezing. The degree of fermentation before freezing and the freezing rate of dough are crucial factors for obtaining high-quality breads. Räsänen et al. (1995) recommended a fermentation ratio of approximately half of the full fermentation time needed for nonfrozen dough with a high freezing rate to avoid dough deterioration. Gabric et al. (2011) observed that chilling at 4°C (for 120 min) before freezing the prefermented dough minimizes the loss in bread volume. Otherwise, the volume that is lost has to be recovered during the thawing and baking phases. During frozen storage of prefermented frozen dough, similar changes occur as in the unfermented frozen dough (Baier-Schenk et al. 2005).

7.5.3 QUALITY OF BREADS PRODUCED BY PARTIALLY BAKED FROZEN TECHNOLOGY

Partially baked bread technology is gaining significant shares in the bakery market. The par-baked frozen process involves initial baking encompassing approximately 75% of the total baking time, followed by freezing and storage. Final baking can be done in baking stations or households by nonskilled personnel, making freshly baked bread available at any time. This is a concept of "brown-and-serve." Fik and Surowka (2002) concluded that, after rebaking, par-baked frozen bread has close sensory and texture properties to the breads obtained from conventional baking. Freezing does not change bread's taste and aroma. In contrast, some studies found that the par-baked frozen breads after rebaking have a lower volume, denser structure, harder crumb, flaking crust, and a limited shelf life compared with directly baked breads. Baking at lower temperature, applied in par-baking, results in lower volume gain during baking (Ben-Aissa et al. 2010). Moreover, during cooling and freezing, gluten proteins in the crumb are contracted, which cannot be fully recovered during final baking. Par-baked bread is subjected to mechanical stress due to the compression of the gas phase during cooling (Lucas et al. 2005). To produce more aerated crumb and satisfying volume, the amount of yeast for part-baked frozen bread can be increased (Carr and Tadini 2003). Hamdami et al. (2007) showed that rapid freezing (5 m/s and 233 K) tends to minimize the weight loss and the freezing time of par-baked bread. On the other hand, slow freezing (0.5 m/s and 253 K) applied at the beginning of the freezing process tends to minimize the ice content at the crust–crumb interface and can be followed by a rapid freezing at the end (0.5 m/s and 233 K).

In relation to volume, par-baked bread has a harder crumb compared with conventionally baked bread. This is because par-baked bread is exposed twice to temperatures of –7°C to 4°C, at which starch retrogradation is the highest (Cauvain and Young 2000). Lowering the initial baking temperature with the prolongation of baking time decreases crumb firmness of the rebaked bread (Park and Baik 2007). An increased amount of shortening can soften the crumb (Carr and Tadini 2003; Table 7.4). For producers of par-baked breads, vacuum-based cooling systems provide several advantages such as reduced baking time and maintained bread volumes (Wang and Sun 2001).

Another problem after rebaking the par-baked bread is the crust flaking as the result of the crumb contraction and excessive drying of the surface by the end of freezing (Le Bail et al. 2005; Lucas et al. 2005; Hamdami et al. 2007). This has been ascribed to the concentration of ice under the crust due to the presence of the freezing front and thermomechanical differences between crust and crumb (Lucas et al. 2005). Le Bail et al. (2005) found that the chilling condition after partial baking is the most important parameter for crust flaking. Par-baked breads should be cooled to room temperature after baking and before entering the freezer. Flaking can be minimized by placing the product in the freezer at –35°C rather than at a higher temperature and by applying higher air humidity during fermentation and chilling.

Proper steam injections are necessary to help par-baked bread form a soft and thin crust to retain the bread's moisture.

Par-baked breads can be stored in a freezer for up to 12 months without microbial spoilage. During frozen storage, bread volume is progressively reduced, especially during the first month of storage (Rosell and Santos 2010). Changes in bread texture occur after 6 weeks of storage (Vulicevic et al. 2004). The moisture content of crust and crumb, crumb springiness, and mouthfeel are the most sensitive attributes of par-baked bread (Vulicevic et al. 2004). After 3 weeks of frozen storage, the amount of total phenolics in par-baked bread decreases due to lipid oxidation (Novotni et al. 2011). Thus, the specific improvers that act as antioxidants and protective packaging are needed for the production of bread with extended shelf life. Packaging material for frozen dough and par-baked bread must keep the product's moisture loss at a minimum and must have good oxygen barrier characteristics, physical strength against breakage at low temperature, good heat sealability, and a low cost (Cauvain and Young 2000).

For shorter storage periods, refrigeration or modified atmosphere packaging of par-baked breads can be recommended. Lainez et al. (2008) showed that par-baked bread stored at 1°C has a shelf life of up to 28 days in which the specific volume, shape, and moisture content of bread are preserved.

7.6 CONCLUSION

The bakery industry is constantly changing due to the demand for convenience, diversity, health, and environmental issues. The growing market for easily prepared and functional bakery items will continue to drive technological advances. The formulation may need adaptation from traditional to modern baking production by using different flours from new–old cereals and pseudocereals, defined prefermentation starters, and various ingredients for enrichment. Each bread-making step, such as mixing, fermentation, baking, cooling, and storage conditions as well as the quality of the raw materials, affects bread's quality. Substantial knowledge and understanding of dough rheology allow for the bread-making process to be conducted efficiently and enable the bakery industry to manufacture a large variety of products.

REFERENCES

Ahmed, J. 2010. Effect of high pressure on structural and rheological properties of cereal and legume proteins. In *Novel food processing: Effect on rheological and functional properties*, eds. J. Ahmed, H. S. Ramaswamy, S. Kasapis, J. I. Boye, 225–231. Boca Raton: CRC Press, Taylor & Francis Group.

Altamirano-Fortoul, R., Le-Bail, A., Chevallier, S., and Rosell, C. M. 2012. Effect of the amount of steam during baking on bread crust features and water diffusion. *J. Food Eng.* 108:128–134.

Anderssen, R. S., Gras, P. W., and MacRitchie, F. 1998. The rate-independence of the mixing of wheat flour dough to peak dough development. *J. Cereal Sci.* 27:167–177.

Asghar, A., Anjum, F. M., Allen, J. C., and Daubert, C. R. 2009. Effect of modified whey protein concentrates on empirical and fundamental dynamic mechanical properties of frozen dough. *Food Hydrocoll.* 23:1687–1692.

Asghar, A., Anjum, F. M., Tariq, M. W., and Hussain, S. 2005. Effect of carboxy methyl cellulose and gum arabic on the stability of frozen dough for the bakery products. *Turk. J. Biol.* 29:237–241.

Baardseth, P., Kvaal, K., Lea, P., Ellekjaer, M. R., and Faegestad, E. M. 2000. The effects of bread making process and wheat quality on French baguettes. *J. Cereal Sci.* 32:73–87.

Bagley, E. B., and Christianson, D. D. 1986. Response of commercial chemically leavened doughs to uniaxial compression. In *Fundamentals of dough rheology*, eds. H. Faridi and J. M. Faubion, 27–36. St. Paul: AACC.

Baier-Schenk, A., Handschin, S., and Conde-Petit, B. 2005. Ice in prefermented frozen bread dough—An investigation based on calorimetry and microscopy. *Cereal Chem.* 82:251–255.

Baker, J. C., and Mize, M. D. 1946. Gas occlusion during dough mixing. *Cereal Chem.* 23:39–51.

Barber, B., Ortola, C., Barber, S., and Fernandez, F. 1992. Storage of packaged white bread. III. Effects of sourdough and addition of acids on bread characteristics. *Z. Lebensmittel Unter Forsch* 194:442–449.

Bárcenas, M. E., and Rosell, C. M. 2007. Different approaches for increasing the shelf life of partially baked bread: Low temperatures and hydrocolloid addition. *Food Chem.* 100:1594–1601.

Belton, P. S. 2003. The molecular basis of dough rheology. In *Bread making—Improving quality*, ed. S. P. Cauvain, 273–287. Cambridge: Woodhead Publishing.

Ben-Aissa, M. F., Monteau, J. Y., Perronnet, A., Roelens, G., and Le-Bail, A. 2010. Volume change of bread and bread crumb during cooling, chilling and freezing, and the impact of baking. *J. Cereal Sci.* 51:115–119.

Berglund, P., Shelton, D., and Freeman, T. 1991. Frozen bread dough ultra structure as affected by duration of frozen storage and freeze-thaw cycles. *Cereal Chem.* 68:105–107.

Bloksma, A. H. 1986. Rheological aspects of structural changes during baking. In *Chemistry and physics of baking*, eds. J. M. Blanshand, P. J. Frazier, T. Galliard, 170–178. London: The Royal Society of Chemistry.

Bloksma, A. H., and Bushuk, W. 1988. Rheology and chemistry of dough. In *Wheat chemistry and technology*, ed. Y. Pomeranz, 131–218. St. Paul: AACC.

Campbell, G. M., Rielly, C. D., Fryer, P. J., and Sadd, P. A. 1998. Aeration of bread dough during mixing: Effect of mixing dough at reduced pressure. *Cereal Food World* 43:163–167.

Campbell, G. M., Herrero-Sanchez, R., Payo-Rodriguez, R., and Merchan, M. L. 2001. Measurement of dynamic dough density and effect of surfactants and flour type on aeration during mixing and gas retention during proofing. *Cereal Chem.* 78:272–277.

Campbell, G. M., and Shah, P. 1999. Entrainment and disentrainment of air during bread dough mixing and their scale-up of dough mixers. In *Bubbles in food*, eds. G. M. Campbell, C. Webb, S. S. Pandiella, K. Niranjan, 11–20. Minnesota: Eagan Press.

Carr, L. G., and Tadini, C. C. 2003. Influence of yeast and vegetable shortening on physical and textural parameters of frozen part baked French bread. *Lebensm-Wiss Technol* 36: 609–614.

Carson, L. C., and Sun, X. S. 2000. Breads from white grain sorghum: Effect of SSL, DATEM and xanthan gum on sorghum bread volume. *Appl. Eng. Agric.* 16:431–436.

Cauvain, S. P., Whitworth, M. B., and Alava, J. M. 1999. The evolution of bubbles structure in bread doughs and its effects on bread structure. In *Bubbles in food*, eds. G. M. Campbell, C. Webb, S. S., Pandiella, K. Niranjan, 85–88. St. Paul: Eagan Press.

Cauvain, S. 2003. Breadmaking: an overview. In *Bread making: Improving quality*, ed. S. P. Cauvain, 8–28. Cambridge: Woodhead Publishing.

Cauvain, S. P., and Young, L. S. 2003. Water control in baking. In *Bread making: Improving quality*, ed. S. P. Cauvain, 447–464. Cambridge: Woodhead Publishing.

Cauvain, S. P., and Young, L. S. 2000. Strategies for extending bakery product shelf-life. In *Bakery food manufacture and quality*. Water control and effects, eds. S. P. Cauvain and L. S. Young, 188–205. Oxford: Blackwell Science.

Chamberlain, N. 1979. Gases–The neglected bread ingredients. In *Proceedings of the 49th Conference of the British Society of Baking*, 12–17. Stratford-on-Avon, UK.

Chang, C.-Y., Seitz, L. M., and Chambers, E. 1995. Volatile flavor components of breads made from hard red winter wheat and hard white winter wheat. *Cereal Chem.* 72:237–242.

Chatraei, S. H., and Macosko, C. W. 1981. Lubricated squeezing flow: A new biaxial extensional rheometer. *J. Rheol.* 25:433–443.

Corsetti, A., Gobbetti, M., De Marco, B., Balestrieri, F., Paletti, F., and Rossi, J. 2000. Combined effect of sourdough lactic acid bacteria and additives on bread firmness and staling. *J. Agric. Food Chem.* 48:3044–3051.

Courtin, C., and Delcour, J. A. 2002. Arabinoxylans and endoxylanases in wheat flour breadmaking. *J. Cereal Sci.* 35:225–243.

Datta, A. K., Sahin, S., Sumnu, G., and Keskin, S. O. 2007. Porous media characterization of breads baked using novel heating modes. *J. Food Eng.* 79:106–116.

Demirekler, P., Sumnu, G., and Sahin, S. 2004. Optimization of bread baking in halogen lamp–microwave combination oven by response surface methodology. *Eur. Food Res. Technol.* 219:341–347.

Dobraszczyk, B. J. 2004. The physics of baking: Rheological and polymer molecular structure–function relationships in breadmaking. *J. Non-Newtonian Fluid Mech.* 124:61–69.

Dobraszczyk, B. J. 1997. Development of a new dough inflation system to evaluate doughs. *Cereal Food World* 42:516–519.

Dobraszczyk, B. J. 2003. Measuring the rheological properties of dough. In: *Bread making: Improving quality*, ed. S. P. Cauvain, 375–395. Cambridge: Woodhead Publishing Limited.

Dobraszczyk, B. J., and Morgenstern, M. P. 2003. Review: Rheology and the breadmaking process. *J. Cereal Sci.* 38:229–245.

Douliez, J. P., Michon, T., Elmoranji, K., and Marion, D. 2000. Structure, biological and technological functions of lipid transfer proteins and indolines, the major lipid binding proteins from cereal kernels. *J. Cereal Sci.* 32:1–20.

Eliasson, A. C., and Larsson, K. 1993. *Cereals in breadmaking: A molecular colloidal approach*. New York: Marcel Dekker.

Fik, M., and Surowka, K. 2002. Effect of prebaking and frozen storage on the sensory quality and instrumental texture of bread. *J. Sci. Food Agric.* 82:1268–1275.

Gabric, D., Ben-Aissa, F., Le-Bail, A., Monteau, J. Y., and Curic, D. 2011. Impact of process conditions on the structure of pre-fermented frozen dough. *J. Food Eng.* 105:361–366.

Gan, Z., Ellis, P. R., and Schofield, J. D. 1995. Gas cell stabilization and gas retention in wheat bread dough. *J. Cereal Sci.* 21:215–230.

Giannou, V., and Tzia, C. 2007. Frozen dough bread: Quality and textural behavior during prolonged storage—Prediction of final product characteristics. *J. Food Eng.* 79:929–934.

Gilbert, S. M., Burnett, G. M., Mills, E. N. C., Belton, P. S., Shewry, P. R., and Tatham, A. S. 2003. Identification of the wheat seed protein CM3 as a highly active emulsifier using a novel functional screen. *J. Agric. Food Chem.* 51:2019–2025.

Gira 2007. European BVP & BVP Companies Panorama 2006–2011 Mini Market Report. www.ebookbrowse.com/gira-bake-off-mini-market-report-2011-pdf.d169918254.

Gobbetti, M., De Angelis, M., Corsetti, A., and Di Cagno, R. 2005. Biochemistry and physiology of sourdough lactic acid bacteria. *Trends Food Sci. Technol.* 16:57–69.

Gogus, F., Duzdemir, C., and Eren, S. 2000. Effect of some hydrocolloids and water activity on nonenzymic browning of concentrated orange juice. *Nahrung*. 44:438–442.

Hamdami, N., Pham, Q. T., Le-Bail, A., and Monteau, J. Y. 2007. Two stage freezing of part-baked breads: Application and optimization. *J. Food Eng.* 82:418–426.

Hansen, Å., and Hansen, B. 1994. Influence of wheat flour type on the production of flavour compounds in wheat sourdoughs. *J. Cereal Sci.* 19:185–190.

Hansen, Å., and Schieberle, P. 2005. Generation of aroma compounds during sourdough fermentation: applied and fundamental aspects. *Trends Food Sci. Technol.* 16:85–94.

Havet, M., Mankai, M., and Le-Bail, A. 2000. Influence of the freezing condition on the baking performances of French frozen dough. *J. Food Eng.* 45:139–145.

Hosomi, K., Nishio, K., and Matsumoto, H. 1992. Studies on frozen dough baking. I. Effects of egg yolk and sugar ester. *Cereal Chem.* 69:82–92.

Huang, L., Wan, J., Huang, W., Rayas-Duarte, P., and Liu, G. 2011. Effects of glycerol on water properties and steaming performance of prefermented frozen dough. *J. Cereal Sci.* 53:19–24.

Hwang, C. H., and Gunasekaran, S. 2001. Determining wheat dough mixing characteristics from power consumption profile of a conventional mixer. *Cereal Chem.* 78:88–92.

Inoue, Y., Sapirstein, H., and Bushuk, W. 1995. Studies on frozen doughs. IV. Effect of shortening systems on baking and rheological properties. *Cereal Chem.* 72:221–226.

Katina, K., Poutanen, K., and Autio, K. 2004. Influence and interactions of processing conditions and starter culture on formation of acids, volatile compounds, and amino acids in wheat sourdoughs. *Cereal Chem.* 81:598–610.

Katina, K., Heiniö, R.-L., Autio, K., and Poutanen, K. 2006. Optimization of sourdough process for improved sensory profile and texture of wheat bread. *LWT—Food Sci. Technol.* 39:1189–1202.

Keller, R. C. A., Orsel, R., and Hamer, R. J. 1997. Competitive adsorption behaviour of wheat flour components and emulsifiers at an air–water interface. *J. Cereal Sci.* 25:175–183.

Kenny, S., Wehrle, K., Dennehy, T., and Arendt, E. K. 1999. Correlations between empirical and fundamental rheology measurements and baking performance of frozen bread dough. *Cereal Chem.* 76:421–425.

Keskin, S. O., Sumnu, G., and Sahin, S. 2004. Bread baking in halogen lamp–microwave combination oven. *Food Res. Int.* 37:489–495.

Kilborn, R. H., and Tipples, K. S. 1972. Factors affecting mechanical dough development. I. Effect of mixing intensity and work input. *Cereal Chem.* 49:34–47.

Kulp, K., and Ponte, J. G. 2000. Breads and yeast-leavened bakery foods. In *Handbook of cereal science and technology*, eds. K. Kulp and J. G. Ponte, 516–525. New York: Marcel Dekker.

Kumagai, H., Lee, B.-H., Kumagai, H., and Yano, T. 1991. Critical radius and time course of expansion of an isolated bubble in wheat flour dough under temperature rise. *Agric. Biol. Chem.* 55:1081–1087.

Lainez, E., Vergara, F., and Bárcenas, M. E. 2008. Quality and microbial stability of partially baked bread during refrigerated storage. *J. Food Eng.* 89:414–418.

Lallemand Baking Update 1996. Frozen Dough. Vol. 1, no. 18 www.lallemand.com/ BakerYeastNA/eng/PDFs/LBU%20PDF%20FILES/1_18FROZ.PDF. Accessed date December 3, 2013.

Le Bail, A., Monteau, J.-Y., Margerie, F., Lucas, T., Chargelegue, A., and Reverdy, Y. 2005. Impact of selected process parameters on crust flaking of frozen part baked bread. *J. Food Eng.* 69:503–509.

Levine, H., and Slade, L. 1990. Influences of the glassy and rubbery states on the thermal, mechanical and structural properties of doughs and baked products. In *Dough rheology and baked product texture*, eds. J. Faubion and H. Faridi, 157–330. New York: Van Nostrand Reinhold.

Lu, W., and Grant, L. A. 1999. Effects of prolonged storage at freezing temperatures on starch and baking quality of frozen doughs. *Cereal Chem.* 76:656–662.

Lucas, T., Quellec, S., Le Bail, A., and Davenel, A. 2005. Chilling and freezing of part-baked breads. II. Experimental assessment of water phase changes and of structure collapse. *J. Food Eng.* 70:151–164.

Mandala, I., Kapetanakou, A., and Kostaropoulos, A. 2008. Physical properties of breads containing hydrocolloids stored at low temperature. II. Effect of freezing. *Food Hydrocoll.* 22:1443–1451.

Mani, K., Tragardh, C., Eliasson, A. C., and Lindahl, L. 1992. Water content, water soluble fraction, and mixing affect fundamental rheological properties of wheat flour doughs. *J. Food Sci.* 57:1198–1200.

Martin, M. L., Zeleznak, K. J., and Hoseney, R. C. 1991. A mechanism of bread firming. I. Role of starch swelling. *Cereal Chem.* 68:498–507.

Matuda, T. G., Chevallier, S., Filho, P. A. P., Le Bail, A., and Tadini, C. C. 2008. Impact of guar and xanthan gums on proofing and calorimetric parameters of frozen bread dough. *J. Cereal Sci.* 48:741–746.

Meignen, B., Onno, B., Gélinas, P., Infantes, M., Guilois, S., and Cahagnier, B. 2001. Optimization of sourdough fermentation with *Lactobacillus brevis* and baker's yeast. *Food Microbiol.* 18:239–245.

Millar, S. J., Bar, L'Helgouac'h, C., Massin, C., and Alava, J. M. 2005. Flour quality and dough development interactions—The first crucial steps in bread production. In *Using cereal science and technology for the benefit of consumers*, eds. S. P. Cauvain, S. E. Salmon, L. S. Young, 132–136. Cambridge: Woodhead Publishing.

Mills, E. N. C., Wilde, P. J., Salt, L. J., and Skeggs, P. 2003. Bubble formation and stabilization in bread dough. *Food Bioprod Process* 81:189–193.

Minervini F., Pinto, D., Di Cagno, R., De Angelis, M., and Gobbetti, M. 2011. Scouting the application of sourdough to frozen dough bread technology. *J. Cereal Sci.* 54:296–304.

Neyreneuf, O., and Delpuech, B. 1993. Freezing experiments on yeasted dough slabs. Effects of cryogenic temperatures on the baking performance. *Cereal Chem.* 70:109–111.

Novotni, D., Ćurić, D., Galić, K., Škevin, D., Neđeral, S., Kraljić, K., Gabrić, D., and Ježek, D. 2011. Influence of frozen storage and packaging on oxidative stability and texture of bread produced by different processes. *LWT—Food Sci. Technol.* 44:643–649.

Olsson, E. E. M., Trägårdh, A. C., and Ahrné, L. M. 2005. Effect of near-infrared radiation and jet impingement heat transfer on crust formation of bread. *J. Food Sci.* 70:484–491.

Ornebro, J., Nylander, T., and Eliasson, A. C. 2000. Interfacial behaviour of wheat proteins. *J. Cereal Sci.* 31:195–221.

Ovadia, D. Z., and Walker, C. E. 1995. Microwave baking of bread. *J. Microwave Power Electromag. Energy* 30:81–89.

Pagani, M. A., Lucisano, M., and Mariotti, M. 2006. Italian Bakery. In *Bakery products science and technology*, ed. Y. H. Hui, 536. Ames: Blackwell Publishing.

Park, C. S., and Baik, B.-K. 2007. Influences of baking and thawing conditions on quality of par-baked French bread. *Cereal Chem.* 84:38–43.

Patel, B. K., Waniskab, R. D., and Seetharamana, K. 2005. Impact of different baking processes on bread firmness and starch properties in breadcrumb. *J. Cereal Sci.* 42:173–184.

Phimolsiripol, Y., Siripatrawan, U., Tulyathan, V., and Cleland, D. J. 2008. Effects of freezing and temperature fluctuations during frozen storage on frozen dough and bread quality. *J. Food Eng.* 84:48–56.

Polaki, A., Xasapis, P., Fasseas, C., Yanniotis, S., and Mandala, I. 2010. Fiber and hydrocolloid content affect the microstructural and sensory characteristics of fresh and frozen stored bread. *J. Food Eng.* 97:1–7.

Primo-Martin, C., van Dalen, G., Meinders, M. B. J., Don, A., Hamer, R. H., and van Vliet, T. 2010. Bread crispness and morphology can be controlled by proving conditions. *Food Res. Int.* 43:207–217.

Räsänen, J., Harkonen, H., and Autio, K. 1995. Freeze-thaw stability of prefermented frozen lean wheat doughs: Effect of flour and fermentation time. *Cereal Chem.* 72:637–642.

Ribotta, P. D., and Le Bail, A. 2007. Thermo-physical and thermo-mechanical assessment of partially baked bread during chilling and freezing process. Impact of selected enzymes on crumb contraction to prevent crust flaking. *J. Food Eng.* 78:913–921.

Ribotta, P. D., Pérez, G. T., León, A. E., and Añón M. C. 2004. Effect of emulsifier and guar gum on micro structural, rheological and baking performance of frozen bread dough. *Food Hydrocoll.* 18:305–313.

Roos, Y. H. 2003. Molecular mobility in dough and bread quality. In *Bread making: Improving quality*, ed. S. P. Cauvain, 289–305. Cambridge: Woodhead Publishing.

Rosell, C. M., and Santos, E. 2010. Impact of fibers on physical characteristics of fresh and staled bake off bread. *J. Food Eng.* 98:273–281.

Sahi, S. S. 1994. Interfacial properties of the aqueous phases of wheat flour doughs. *J. Cereal Sci.* 20:119–127.

Salas-Mellado, M. M., and Chang, Y. K. 2003. Effect of formulation of quality of frozen bread dough. *Braz. Arch. Biol. Technol.* 46:461–468.

Scott, G., and Richardson, P. 1997. The application of computational fluid dynamics in the food industry. *Trends Food Sci. Technol.* 8:119–124.

Selomulyo, V. O., and Zhou, W. 2007. Frozen bread dough: Effects of freezing storage and dough improvers. *J. Cereal Sci.* 45:1–17.

Shah, P., Campbell, G. M., McKee, S. L., and Rielly, C. D. 1998. Proving of bread dough: Modelling the growth of individual bubbles. *Food Bioprod. Process* 76:73–79.

Sharadanant, R., and Khan, K. 2006. Effect of hydrophilic gums on the quality of frozen dough: Electron microscopy, protein solubility, and electrophoresis studies. *Cereal Chem.* 83:411–417.

Shimiya, Y., and Nakamura, K. 1997. Changes in size of gas cells in dough and bread during bread-making and calculation of critical size of gas cells that expand. *J. Texture Stud.* 28:273–288.

Shittu, T. A., Raji, A. O., and Sanni, L. O. 2007. Bread from composite cassava-wheat flour: I. Effect of baking time and temperature on some physical properties of bread loaf. *Food Res. Int.* 40:280–290.

Steffe, J. F. 1996. *Rheological methods in food engineering.* East Lansing: Freeman Press.

Takano, H., Naito, S. Ishida, N., Koizumi, M., and Kano, H. 2002. Fermentation process and grain structure of baked breads from frozen dough using freeze-tolerant yeasts. *J. Food Sci.* 67:2725–2733.

Tieking, M., Korakli, M., Ehrmann, M. A., Gänzle, M. G., and Vogel, R. F. 2003. In situ production of exopolysaccharides during sourdough fermentation by cereal and intestinal isolates of lactic acid bacteria. *Appl. Environ. Microbiol.* 69:945–952.

Varriano-Marston, E., Hsu, K. H., and Mahdi, J. 1980. Rheological and structural changes in frozen dough. *Bakers Dig.* 54:32–34.

Vulicevic, I. R., Abdel-Aal, E. S. M., Mittal, G. S., and Lu, X. 2004. Quality and storage life of part-baked frozen breads. *LWT—Food Sci. Technol.* 37:205–213.

Wang, L., Sun, and D.-W. 2001. Rapid cooling of porous and moisture foods by using vacuum cooling technology. *Trends Food Sci. Technol.* 12:174–184.

Weipert, D. 1990. The benefits of basic rheometry in studying dough rheology. *Cereal Chem.* 67:311–317.

Whitworth, M. B., and Alava, J. M. 1999. The imaging and measurements of bubbles in bread doughs. In *Bubbles in food*, eds. G. M. Campbell, C. Webb, S. S. Pandiella, 221–231. St. Paul: AACC.

Willyard, M. R. 1998. Conventional browning and microwave baking of yeast raised dough. *Cereal Food World* 43:131–138.

Yi, J., and Kerr, W. L. 2009. Combined effects of dough freezing and storage conditions on bread quality factors. *J. Food Eng.* 93:495–501.

Zanoni, B., Pierucci, S., and Peri, C. 1994. Study of bread baking process—II. Mathematical modeling. *J. Food Eng.* 23:321–336.

Zhang, C., Zhang, H., and Wang, L. 2007a. Effect of carrot (*Daucus carota*) antifreeze proteins on the fermentation capacity of frozen dough. *Food Res. Int.* 40:763–769.

Zhang, L., Lucas, T., Doursat, C., Flick, D., and Wagner, M. 2007b. Effects of crust constraints on bread expansion and CO_2 release. *J. Food Eng.* 80:1302–1311.

8 Confectionary Baking

Servet Gulum Sumnu and Ozge Sakiyan Demirkol

CONTENTS

8.1 INTRODUCTION

Cakes are defined as aerated, chemically leavened bakery products. Essentially, there are two types of cakes. Sponge cakes are very airy batters that turn into cakes with a rather open structure. Layer cakes, on the other hand, contain a solid fat, which results in an aerated batter with distinct flow properties, and a cake that has a fine grain and relatively small air cells when creamed with sugar (McWilliams

1989). There are three steps on which the production of layer and sponge cakes is based. The first step is the aeration of a liquid batter to form foam. The second step is the expansion of air bubbles within the foam during baking. The last step is the transformation of the foam to a sponge structure, caused by a large increase in viscosity, as the starch granules gelatinize and swell (Guy and Sahi 2006).

Biscuits are cereal-based products, baked to a moisture content of less than 5%. The cereal component is variously enriched with two major ingredients, namely, fat and sugar (Manley 1991). The quality of biscuits is governed by the nature and quality of the ingredients used. A variety of shapes and textures may be produced by varying the proportions of these ingredients (Indrani and Rao 2008). The name "cookie" can be regarded as synonymous with biscuit, but the former is more comprehensive in meaning in the United States, and the latter in the United Kingdom (Manley 1991). On the other hand, the word cookie means little cake. They are mostly made from soft and weak flours, and their formula is high in sugar and fat but low in water content (Indrani and Rao 2008).

8.2   FUNCTIONS OF RAW MATERIALS

The ingredients of cakes, biscuits, and cookies can be listed as flour, sugar, shortening, eggs, nonfat dry milk, salt, leavening agent, water, additives, and so forth. The percentage and the role of each ingredient vary from one confectionary to another.

8.2.1   FUNCTIONS OF FLOUR

Flour gives unique textural and appearance characteristics to the product in which it is used. There are some important parameters to specify flour. The first one is the moisture content. The moisture content of the flour is generally approximately 14% and should not vary by more than 1%. Another important specification parameter is the protein content. The protein content varies depending on the final product. For baking, if highly elastic dough is needed, flour with high protein content would be preferred. The ash content of flour is used as one of the main quality criteria to indicate the suitability of flour for baking. Total α-amylase and falling number are other important parameters for flour specification. Determination of damaged starch, water absorption, and rheological properties of flour are also important for specification. Flour treatment can be achieved through the addition of enzymes or gluten supplementation (Bennion and Bamford 1997).

Cake flours are generally made from wheat with lower protein levels. Other than protein, it mainly contains starch, lipids, some minerals, vitamins, and ash. When wheat starch is heated in water, the granules begin to absorb water and expand to many times their original size. The crystalline structure melts, amylose leaches out of the granules, and the granules become deformed. This gelatinization occurs over a wide temperature range and is influenced by the presence of other ingredients such as sucrose or emulsifiers (Bennion and Bamford 1997).

Special cake flours are from specially selected soft wheat that is milled to a finer particle size. The protein content of cake flour is approximately 8%. General-purpose cake flours have slightly higher protein contents than the high-ratio types and are given heavy treatments of chlorine bleach, producing flour with a pH value of approximately 5.2. This treatment affects the gelation properties of the starch, which is necessary for making very soft batters containing high percentages of sugar, as it is the case in high-ratio cakes (Bennion and Bamford 1997).

There are many types of biscuits for which special types of flour are required. Flour for short and semisweet biscuits require protein contents of 8.0% to 9.5%, respectively. It is typically milled from weak wheat with a low protein content. The rheological properties of the dough for biscuits are the most important. Flour that produces dough with a higher extensibility but less spring (resistance) than that of bread dough is required. The extensibility of biscuit flour dough may be increased by treating the flour with a proteolytic enzyme or with a reducing agent (Kent 1983).

8.2.2 FUNCTIONS OF SUGAR

Depending on the quantity used, sugar can affect not only the taste but also the texture and appearance of the baked cake. It plays a role in controlling batter viscosity, the degree of gelatinization of starch, and the heat setting temperature of proteins. Sugar also serves as a tenderizing agent by retarding gluten development during mixing. In addition, sugar elevates the coagulation temperature of egg and milk proteins, causing increased expansion of cake batter (McWilliams 1989). Almost all commercially available sugars contain more than 99.8% sucrose with less than 0.05% moisture, approximately 0.05% invert sugar and other carbohydrates besides sucrose, and a trace of ash (Matz 1972). Sugar in the high-ratio cake formulation results in good air incorporation, leading to a more viscous and stable foam (Paton et al. 1981). In addition, sugar affects the physical structure of baked products by regulating the gelatinization of starch. Sucrose is known to delay the gelatinization of the granules. Delay in starch gelatinization during baking allows air bubbles to expand properly due to vapor pressure buildup by carbon dioxide and water vapor before the cake sets. This allows the formation of desired cake structure (Kim and Setser 1992; Kim and Walker 1992).

Common sugar is chemically known as sucrose and is derived almost exclusively from sugar cane or sugar beet. In its pure state, it is normally available as white crystals, but it can also be bought as liquid sugar, which is a solution in water (Manley 1991).

Sucrose is a disaccharide and a nonreducing sugar. Functionally, it is a very important component of many types of biscuits because it not only imparts sweetness but also contributes to the eating texture. During the mixing of dough, increasing levels of sugar in the formulation result in softening of the gluten and hence a reduction in the level of water needed (Manley 1991). The limited amount of water used in biscuit formulation, and also its nonavailability to protein and starch, particularly contributes to the crispness of biscuits (Indrani and Rao 2008).

Vetter (1984) studied the effect of sugar quality and its grain size on biscuit spreading. It was concluded that fine grain size and a high concentration of sugar contribute to significant spreading of the biscuit.

Because the quantity of sugar is relatively high in cookie recipes, the quality of sugar has a great importance. Sugar adds sweetness, acts as a tenderizing agent, and affects cookie spread. Sugar makes the cooked product fragile because it controls hydration and tends to disperse the protein and starch molecules thereby preventing the formation of a continuous mass (Bean and Setser 1992). The water levels in most cookie dough preclude the total solution of the sugar present, so the crystal size of the sugar is an important quality factor for the texture of the baked cookies. Smaller crystals will dissolve preferentially in the dough (Manley 1991).

8.2.3 FUNCTIONS OF FAT

Fat or shortening is one of the important ingredients in a cake's formulation. Fat helps to entrap air during the creaming process, resulting in aeration and hence leavens the product. It also imparts desirable flavor and softer texture to the cakes. Most types of cakes require fairly high levels of shortening for the development of their characteristic crumb structure. In a cake system, shortening serves three major functions: to entrap air during the creaming process, to physically interfere with the continuity of starch and protein particles, and to emulsify the liquid in the formulation. Thus, the shortening affects the tenderness and moisture of the cake (Freeland-Graves and Peckham 1987). In addition, fats are also known to delay gelatinization by deferring the transport of water into the starch granule. This is due to the formation of complexes between the lipid and amylose during baking (Elliasson 1985; Ghiasi et al. 1982; Larsson 1980). The effect of four different fat types on the rheology of the cookie dough and subsequently their effect on the quality of cookies was studied by Jacob and Leelavathi (2007). They found that the quality of these cookies was significantly affected by the fat type.

In biscuit dough, fat functions as a textural agent that softens the final product. During the mixing of dough, there is a competition between the aqueous phase and the fat for the flour surface. The water or sugar solution interacts with the flour protein to create gluten, which forms a cohesive and extensible network. When some fat coats the flour, this network is interrupted and eating properties after baking are less hard and more inclined to melt in the mouth. If the fat level is high, the lubricating function in the dough is so pronounced that little or no water is required to achieve the desired consistency. Starch swelling and gelatinization are also reduced, resulting in a very soft texture. By combining with a sugar solution, fat also prevents the hard vitreous mass upon cooling (Manley 1991). The addition of fat influences the texture and taste of cookies, making the cookies crispier because of its spreading effect on cookie dough.

8.2.4 FUNCTIONS OF EGG

Eggs are a highly functional food ingredient, and it has three primary functions: foaming, emulsification, and coagulation. Foaming is the incorporation of air into a product, usually achieved by whipping. Although many food ingredients are capable of forming foams, egg and egg products are especially good foaming agents. They produce a large foam volume, which is relatively available for cooking, and coagulate upon heating to maintain a stable foam structure. The second function is emulsification, which is the stabilization of the suspension of one liquid in another.

Egg yolk contains an excellent food emulsifier, lecithin. Coagulation is the last function. It is the conversion of the liquid egg to a solid or semisolid state that is usually accomplished by heating. This property of egg is difficult to duplicate with any other food ingredient (Bennion and Bamford 1997).

Eggs can be examined in two parts as egg yolk and egg white. Egg yolk is a dispersion of particles in a continuous phase. This system contains egg lipids, 70% of which are triglycerides. The particles that make up 25% of the dry matter of the yolk are phosvitin and lipovitellin. The continuous phase contains 75% of the dry matter of the yolk in the form of lipovitellin and globular protein. Cholesterol and lecithin are also present in egg yolk. The color of the yolk is determined by the amount of xanthophyll, a yellow coloring pigment.

Egg white or albumen is made up of a complex structure of proteins, such as ovalbumin and conalbumin. It contains (in dry matter) about 85% of the total protein content of an egg. The egg white is very viscous and alkaline in a fresh egg and contains natural inhibitors, such as lysozyme, which form a chemical protection against invading microorganisms (Bennion and Bamford 1997).

Eggs affect the texture of cakes as a result of their emulsifying, leavening, tenderizing, and binding actions. They add color and nutritional value, and in many cases, desirable flavor. They are essential for obtaining the characteristic organoleptic qualities (Matz 1972).

Eggs help in puffing, emulsifying the dough, and bringing the water and fat phases together to result in a creamier and smoother texture in cookies. Egg whites have a drying effect and contribute to structure or shape (Indrani and Rao 2008).

8.2.5 Functions of Water

The quality of water used as an ingredient can have greater effects on bakery products than is generally recognized. The amount and types of dissolved minerals and organic substances present in the water can affect the flavor, color, and physical attributes of the finished baked goods. The hardness of water is an important characteristic for bakery products. Soft waters may result in sticky dough, which require less than the normal amount of ingredient water. However, this may be overcome by using more salt in the formulation (Matz 1972). Water serves as a solvent in the cake batter to dissolve dry ingredients. It is also necessary for gelatinization of starch (McWilliams 1989).

Water has a complex role in biscuits because it determines the conformational state of biopolymers, affects the nature of interactions between the various constituents of the formula, and contributes to dough structuring (Elliasson and Larsson 1993).

8.2.6 Functions of Baking Powder

Baking powder is the leavening agent produced by the mixing of an acid-reacting material and sodium bicarbonate, with or without starch or flour. It yields not less than 12% of available carbon dioxide. The acid-reacting materials in baking powders are tartaric acid or its acid salts, acid salts of phosphoric acid, and compounds of aluminum (Matz 1972). Chemical leavening involves the action of an acid on bicarbonate to release carbon dioxide gas for aeration of a dough or batter during mixing and baking. The aeration provides a light, porous cell structure, fine grain, and a texture with desirable appearance along with palatability to baked goods.

There are essentially two components in a chemical leavening system: bicarbonate and acid. Bicarbonate supplies carbon dioxide gas, and acid triggers the liberation of carbon dioxide from bicarbonate upon contact with moisture. Sodium bicarbonate is the primary source of carbon dioxide gas in practically all chemical leavening systems. It is stable and obtainable as a highly purified dry powder at relatively low production costs. The prevalent baking acids in modern chemical leavening systems are sodium or calcium salts of ortho, pyro, and complex phosphoric acids.

Basically, there are two mechanisms of decomposition of sodium bicarbonate (Bennion and Bamford 1997). The first one is thermal decomposition and the second one is acid-activated decomposition. Thermal decomposition of sodium bicarbonate takes place at high temperatures (90°C and above) and is not of particular benefit in cake baking. In its simplest form, it can be represented by the following chemical reaction:

$$2NaHCO_3 + heat \xrightarrow{\Delta} Na_2CO_3 + CO_2 + H_2O$$

Acid-activated decomposition involves the reactions of hydrogen ions in aqueous solution and can be represented by the following general chemical reaction:

$$NaHCO_3 + H^+ \xrightarrow{\Delta} Na^+ + CO_2 + H_2O$$

8.2.7 RECENT STUDIES ON THE EFFECTS OF INGREDIENTS ON QUALITY

One can find a large number of studies about the effects of ingredients on the final quality of cakes, biscuits, and cookies in literature. In this section, some of the recent studies will be discussed.

As previously mentioned, fat is the major ingredient of cookie formulation. It affects dough's properties, causes variations in dough dimensions during baking, and (in the end) has a significant effect on the baked product. In light of this information, changing the fat level of recipes was found to be effective on dough and cookie properties. Pareyt et al. (2010) reported that reducing fat levels increased dough elasticity and dough intrinsic hardness. Moreover, they declared that reducing the fat level in the recipe increased the intrinsic cookie break strength, which can be related to more gluten cross-linking.

The use of erythritol as a sweetener to replace sugar in Danish cookies was investigated by Lin et al. (2010). They found that the erythritol was stable during baking and the cookies had lighter surface color. It was also reported that the cookies tasted less sweet as erythritol levels increased and sucrose levels decreased.

Biscuits prepared from different flour composites were evaluated for their dough characteristics and final product quality. The best formulation was chosen in terms of biscuit quality. It was reported that the higher the level of wheat flour, the more adhesive the dough that was obtained (Saha et al. 2011).

Sakiyan et al. (2011) investigated the effects of different cake formulations on the degree of starch gelatinization during baking. Fat-free, 25% fat, and 25% fat replacer–containing cake samples were used. The results showed that the addition of fat reduced the degree of starch gelatinization in conventional baking (Figure 8.1). On the other hand, fat enhanced gelatinization in microwave and infrared–microwave combination ovens (Figures 8.2 and 8.3). This difference can be explained by the higher dielectric properties of the samples with higher fat contents (Sakiyan et al. 2007). Although fat is known to have lower dielectric properties, fat-containing cakes were shown to have higher dielectric properties than fat-free cakes because fat-containing cakes had lower porosity (Sakiyan et al. 2007). The higher temperatures achieved in these samples resulted in higher gelatinization degree. On the other hand, for conventional baking, because the dielectric properties did not affect the heating mechanism, the nonfat samples experienced higher gelatinization degree. This may be because fats are known to delay starch gelatinization by deferring heat transport and the transport of water into the starch granule due to the formation of complexes between the lipid and amylose during baking (Elliasson 1985; Ghiasi et al. 1982; Larsson 1980).

Another study about the effects of fats on cake quality is the evaluation of different types of fats in high-ratio layer cakes. Zhou et al. (2011) compared plastic shortening, liquid shortening, liquid oil, and liquid oil plus emulsifier combinations in terms of

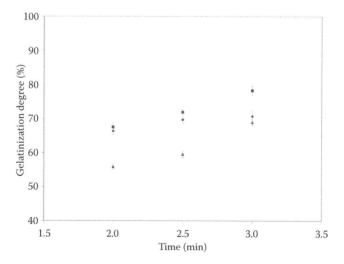

FIGURE 8.1 Gelatinization degree of cakes with different formulations baked in microwave oven for different baking times: (♦), fat free; (■), 25% fat; and (▲), 25% fat replacer. (With kind permission from Springer Science+Business Media: *Food Bioprocess and Technology*, A study on degree of starch gelatinization in cakes baked in three different ovens, 4, 2011, 1237–1244, Sakiyan, O. et al.)

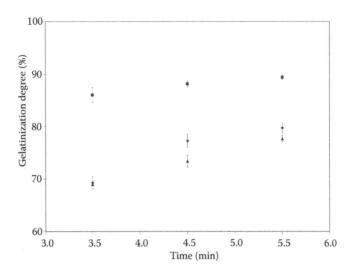

FIGURE 8.2 Gelatinization degree of cakes with different formulations baked in infra-red–microwave oven for different baking times: (♦), fat free; (■), 25% fat; and (▲), 25% fat replacer. (With kind permission from Springer Science+Business Media: *Food Bioprocess and Technology*, A study on degree of starch gelatinization in cakes baked in three different ovens, 4, 2011, 1237–1244, Sakiyan, O. et al.)

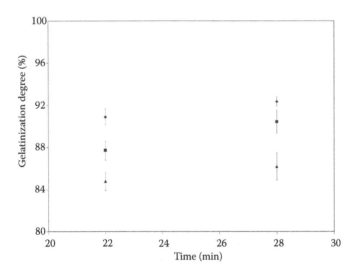

FIGURE 8.3 Gelatinization degree of cakes with different formulations baked in a conventional oven for different baking times: (♦), fat free; (■), 25% fat; and (▲), 25% fat replacer. (With kind permission from Springer Science+Business Media: *Food Bioprocess and Technology*, A study on degree of starch gelatinization in cakes baked in three different ovens, 4, 2011, 1237–1244, Sakiyan, O. et al.)

cake performance and firming over time. They concluded that the liquid shortening provided the best fresh cake characteristics and cake firmness performance.

Finally, Cheong et al. (2011) evaluated the baking performance of palm diacyl-glycerol (PDG)-enriched fats and compared with that of commercial bakery fats. They reported that PDG-enriched shortenings produced cakes with significantly higher specific volume. On the other hand, PDG-enriched margarines led to less spread of cookies as compared with that of commercial margarines.

8.3 TECHNOLOGY OF CONFECTIONARY BAKING

The common stages of production of baked products involve mixing, dividing/scaling/depositing, forming/molding/shaping, expansion, baking, cooling, and packaging (Cauvain and Young 2006). Detailed information on the production steps of cakes, biscuits, and cookies is given in the works of Tireki (2008a,b) and Manley (1998). This section will be focused on baking technologies. Baking is a thermal process in which a liquid dough or batter is transferred to a solid baked form. The reactions that take place during baking are the evolution of gases, expansion, setting of the structure, reduction in moisture content, and formation of crust and surface color (Cauvain and Young 2006).

8.3.1 CONVENTIONAL BAKING TECHNOLOGIES

Cakes are generally baked in ovens using pans or containers. A few types of cakes, such as Swiss rolls, may be baked as free-standing forms and formed as thin sheets because of their low viscosity and tendency to flow. Thus, they should be transferred to the oven quickly. Expansion of the product, moisture loss, and crust and color formation are the major changes that take place during cake baking. Biscuits and cookies are baked directly on the oven band. The major changes during baking of biscuits are darkening of the surface color and a reduction of the moisture content by up to 1% to 4% (Cauvain and Young 2006).

The oven can be considered as a hot box or a tunnel wherein the desired conditions of heat and moisture transfers are achieved. Heat is provided by burning fuel, such as gas, oil, or electricity, and is transferred to the product by conduction, convection, and radiation heating mechanisms (Manley 1998). The temperature and humidity of the oven depend on the type of the product being baked. If the oven used in cake baking is too hot, darker crust color, low volume, and close or irregular crumb will be obtained. On the other hand, a too cold oven causes poor crust color, very high volume, and weak crumb (Tireki 2008a).

The production rate of a baked product is a function of the length of the oven and baking time. In general, baking speed is determined by the drying time of the baked product. Tunnel ovens consist of independently heated and controlled zones. For the shortest ovens, there is only one zone, but for longer ovens, there are more than two zones (Manley 1998). The conventional types of ovens in baking are direct-fired, indirect-fired, and hybrid ovens.

In direct gas-fired ovens, burners are located above and below the baking band. There are different arrangements to adjust the flame in the oven so that uniform heating can be achieved across the band (Tireki 2008b). The radiation mechanism is dominant in these ovens. In direct-fired ovens, the burner may be supplied with carbureted gas and air or electricity. In indirectly fired, forced convection ovens, a heat exchanger near the zone burner heats the air passing through plenum chambers in the baking chamber.

Hybrid ovens are a combination of direct-fired and indirect-fired ovens. Most of the radiant heat is present in the early baking period while the convective heat takes place in the drying period.

8.3.2   Alternative Baking Technologies

The energy used for baking ovens becomes relatively more expensive, and there is a tendency to search for alternative baking technologies to save time and energy during baking and to improve product quality. Alternative baking technologies include jet impingement ovens, microwave ovens, and hybrid technologies that are a combination of microwave–jet impingement and microwave–infrared ovens. There are various studies in the literature that show the effects of alternating technologies on the quality of baked products.

8.3.2.1   Jet Impingement Baking Technology

Jet impingement ovens were first designed by Donald Smith (1975) for the purpose of improving the uniformity of heated air over the surface of the product as compared with natural convection in conventional type ovens. Jet impingement ovens have been used in the food industry for the baking of confectionary products such as cookies and cakes (Li and Walker 1996).

In jet impingement ovens, heated air is sent to the food by nozzles so that the air reaches a very high velocity. Hot air with high velocity removes the cold air on the product's surface and replaces it with hotter and drier air. Thus, the rates of heat and mass transfer increase at the product's surface and uniform baking is achieved (Walker and Li 1993; Yin and Walker 1995). The advantages of air impingement over conventional baking are less processing time, lower processing temperatures, energy efficiency, more uniform heating, and less moisture loss (Kocer et al. 2008).

The heat transfer coefficient in the jet impingement oven is very high. Typical heat transfer coefficient values are 6 to 12 W/m^2 K for natural convection, 13 to 30 W/m^2 K for forced convection to flat surfaces, and 40 to 200 W/m^2 K for air impingement (Kocer et al. 2008). The important factors that should be considered in jet impingement oven design include the distance between the impingement nozzle and the product surface, nozzle diameter/width, and spacing between nozzles (Ovadia and Walker 1998).

8.3.2.2   Microwave Baking Technology

The use of microwave baking brings energy, time, and space savings with high nutritional product quality as compared with conventional baking. The use of microwave baking in the industry is limited. This is due to the failures in product quality such as lack of color and crust formation, high moisture loss, and tough and firm texture (Sumnu 2001). APV Baker uses a multimedia oven, which combines the simultaneous application of conventional heat and microwaves inside the baking chamber (Sumnu and Sahin 2005a). This helps to achieve high product quality by conventional baking, and baking times are significantly reduced by microwave technology. The advantages of this combination technology include the retention of distinctive flavor, color, and texture of the baking process, increased throughput, elimination of the need for checking, and instantaneous response to changes in control settings. A recent trend in the field of baking is the combination of microwave heating with other heating methods such as jet impingement or infrared heating.

The mechanisms that explain how microwaves interact with foods can be categorized into two groups: ionic conduction and dipolar rotation. In ionic conduction, the positively charged ions will accelerate in the direction of the electric field in the microwave oven, whereas negatively charged ions will move in the opposite direction (Sahin and Sumnu 2006). Because the electric field is changing its direction depending on the frequency, the directions of the motion of the positive and negative ions will change. Moving particles collide with the adjacent particles and set them into more agitated motion. Thus, the temperature of the particle increases. In the presence of polar molecules, heat generation in the microwave oven takes place through the mechanism of dipolar rotation. When polar molecules interact with the electric field, they try to orient themselves in the direction of the field and they collide with other molecules. The change in direction of the field results in further collision because they try to line up with the reversed directions. Thus, thermal agitation and heating take place.

The important factors of foods that affect the magnitude and uniformity of power absorption during microwave baking are shape, size, and dielectric properties, whereas oven size and geometry, turntables and mode stirrers, and position of the load are the oven parameters that are responsible for achieving uniform microwave heating (Zhang and Datta 2001).

8.3.2.3 Hybrid Baking Technology

As mentioned previously, microwave baking results in unacceptable product quality. To minimize inferior baked product quality, microwaves can be combined with other heating mechanisms such as jet impingement or infrared, which are known as hybrid technologies. The hybrid (combination) ovens have the advantages of energy efficiency, energy saving, and improvement of product quality (Kocer et al. 2008). These ovens are energy efficient because of the faster rate of heat transfer. Use of these ovens results in energy savings because baking takes place at lower temperature in a shorter time.

The quality of products baked in a hybrid oven is improved because crust, color, and flavor can be developed. In a microwave oven, the air inside is not heated by microwaves. Thus, the surfaces of the product do not reach the required temperatures for Maillard, and caramelization reactions cannot be reached. Moreover, the high amount of moisture that comes into contact with the ambient air condenses and a soggy surface is obtained. Thus, no color or crust formation can be achieved in microwave-baked products. In the presence of microwave–impingement combination baking, hot air with high velocity strikes the surface and replaces the cool stagnant air with hot and dry air. This increases the heat transfer rate and removes the surface moisture, which leads to quick crust formation and color development. Moisture loss will also be reduced because the rapidly formed surface crust will keep the moisture inside and thus prevent excessive drying of the product (Kocer et al. 2008). When microwave heating is combined with infrared heating, baking will be focused on the surface because the penetration depth of infrared heating is small. Thus, browning reactions and crust formation can be achieved in baked products.

Li and Walker (1996) showed that the desired color can be achieved in a shorter time when microwaves are combined with impingement. The quality of different products baked in conventional, impingement, hybrid jet impingement, and microwave ovens was also compared. It was found that although the baking time was the shortest in a hybrid oven, the quality of the products was similar for the three ovens (Walker and Li 1993).

Combining microwaves with infrared to improve food quality dates back to 1999 with the invention of the Advantium Oven by General Electric Company (Louisville, KY). This oven includes three halogen lamps in addition to a classic microwave oven design. Two 1500-W halogen lamps are located at the top and one 1500-W halogen lamp is at the bottom. The halogen lamp is the near-infrared source used in this oven.

The application of microwave–infrared combination has been applied on confectionery products such as cakes, biscuits, and gluten-free cakes. In these studies, it was seen that product quality was almost comparable with conventionally baked products.

The microwave–infrared combination reduces the baking time of cakes by 79%. The quality of cakes baked in microwave–infrared combination ovens was similar to that of the cakes baked in conventional ovens in terms of firmness, volume, and weight loss (Sevimli et al. 2005).

When microwave–infrared combination heating was applied to cookie baking, the quality of the cookies was similar to that of conventionally baked cookies, and the baking time was reduced by 50% (Keskin et al. 2005).

Microwave–infrared combination baking was recommended to be used for gluten-free rice cakes (Turabi et al. 2008a). Cakes baked by using microwave–infrared combination ovens had similar color and firmness values to those of conventional ovens. In addition, microwave–infrared combination baking decreased conventional baking time by 77%. The only disadvantage of combination baking was that it led to higher moisture loss.

8.4 HEAT AND MASS TRANSFER DURING BAKING

Baking is a complex process, which includes simultaneous heat and mass transfer. The biochemical changes, such as starch gelatinization and protein denaturation as well as volume increase and shrinkage of cakes during baking, complicate the modeling of heat and mass transfer during baking.

8.4.1 HEAT TRANSFER DURING BAKING

Different heat transfer mechanisms (conduction, convection, and radiation) take place during the baking of the cookie, biscuit, or cake in an oven. Heat is transferred by convection and radiation to the surface of the baked product and by conduction within the product. Because the products are placed on the baking pan or band, there is no convection through the bottom surface.

Standing (1974) showed that in biscuit baking, radiation is the dominant heat transfer mechanism (43% of total heat transfer). Convection and conduction were responsible for 37% and 20% of the total heat transfers, respectively.

Heat transfer during baking can be expressed by the following equation (Sablani et al. 1998):

$$\rho_B C_{pB} \frac{\partial T}{\partial t} = \nabla(k\nabla T) + \rho_L \lambda \frac{\partial C}{\partial t} \tag{8.1}$$

where ρ_B is the density of the baked product (kg/m³), C_{pB} is the heat capacity of the baked product (J/kg K), T is the temperature (°C), t is time (s), k is the thermal conductivity of the product (W/m K), ρ_L is the density of the liquid (kg/m³), λ is the latent heat of vaporization (J/kg), and C is the moisture content (kg/kg).

The boundary conditions at the surface to solve the heat transfer equation can be written as

$$k\nabla Tn = h_t(T_a - T_s) + \varepsilon\sigma(T_a^4 - T_s^4) \tag{8.2}$$

where n is the unit normal vector, k is the thermal conductivity of the product (W/m K), h_t is the effective convective surface heat transfer coefficient (W/m² K), T_a is the temperature of heating medium and enclosure (K), T_s is the temperature of the product's surface (K), ε is the emissivity, and σ is the Stefan Boltzman constant (W/m² K⁴). Equation 8.2 can also be expressed as

$$k\nabla Tn = (h_t + h_r)(T_a - T_s) \tag{8.3}$$

where h_r is the radiation heat transfer coefficient (W/m² K) and expressed as

$$h_r = \frac{\varepsilon\sigma(T_a^4 - T_s^4)}{(T_a - T_s)} \tag{8.4}$$

The heat transfer equation and boundary conditions used in microwave baking are different from conventional baking. In microwave baking, heat is generated inside the food by the energy equation

$$\frac{\partial T}{\partial t} = \alpha\nabla^2 T + \frac{Q}{\rho C_p} \tag{8.5}$$

where T is the temperature (°C), t is time (s), α is the thermal diffusivity (m²/s), ρ is the density (kg/m³), C_p is the specific heat capacity of the material (J/kg K), and Q is the heat generated per unit volume of material (W/m^3), which represents the conversion of electromagnetic energy.

The relationship between Q and electric field intensity (E) at that location can be derived from Maxwell's equations of electromagnetic waves, as shown by Metaxas and Meredith (1983):

$$Q = 2\Pi\varepsilon_0\varepsilon''fE^2 \tag{8.6}$$

where ε_0 is the dielectric constant of free space, ε'' is the dielectric loss factor of the food, f is the frequency of the oven, and E is the root-mean-squared value of the electric field intensity.

Radiative heat transfer is often neglected in microwave baking, except for the cases when the surfaces of packaging material act as susceptors (Sumnu and Sahin 2005b). The generalized surface boundary conditions for microwave baking are shown as

$$k\nabla Tn = h_t(T_s - T_a) + m_w\lambda \tag{8.7}$$

where n is the unit normal vector, k is the thermal conductivity of the product (W/m K), h_t is the effective convective surface heat transfer coefficient (W/m^2 K), T_a is the temperature of the heating medium (K), T_s is the temperature of the product's surface (°C), λ is the latent heat of vaporization (J/kg), and m_w is the rate of moisture evaporated (kg/s).

8.4.2 MASS TRANSFER DURING BAKING

Moisture diffuses from inside of the product to the surface. Moisture diffusion during baking can be modeled as

$$\frac{\partial C}{\partial t} = \nabla(D\nabla C) \tag{8.8}$$

where C is the moisture content (kg/kg), t is time (s), and D is the effective moisture diffusivity (m^2/s).

The surface boundary condition (Sablani et al. 1998) can be written as

$$D\nabla Cn = h_m(C_a - C_s) \tag{8.9}$$

where h_m is the effective surface convective mass transfer coefficient (m/s), C_a is the moisture content in air (kg/kg), and C_s is the moisture content at the surface (kg/kg).

When food is heated in a microwave oven, moisture evaporates and moves out of the food material through the surface. The equation to express moisture transport for a porous body can be written as a combination of concentration, pressure, and temperature gradients in which flow due to temperature gradient is generally ignored in microwave heating (Datta 1990):

$$\frac{\partial C}{\partial t} = \nabla(D\nabla C) + D\delta_p \nabla^2 P \qquad (8.10)$$

where C is the moisture content (kg/kg), t is time (s), D is the effective moisture diffusivity (m^2/s), δ_p is the pressure gradient coefficient, and P is pressure (Pa).

8.4.3 Thermophysical Properties of Cakes, Cookies, and Biscuits

The thermophysical properties of foods, such as thermal conductivity, specific heat, mass diffusivity, and density, are important for the mathematical modeling of heat and mass transfer during baking. Baik et al. (2001) and Sablani (2008) presented comprehensive reviews on the thermophysical properties of confectionary foods.

8.4.3.1 Thermal Conductivity

Thermal conductivity is an important physical property of batter or dough. Although it is kept constant in some modeling studies, it is known to be a function of composition and temperature (Sahin and Sumnu 2006). The thermal conductivity of cake was shown to decrease during baking, whereas the thermal conductivity of cupcake was found to be one-third of the batter (Baik et al. 1999). Sumnu et al. (2007) also showed that the thermal conductivity of breads decreased sharply during the initial stages of baking and remained constant afterward. This was attributed to an increase in specific volume of bread as more air was entrapped. Baik et al. (1999) showed that thermal conductivity of cupcake was a function of moisture content, density, and temperature. The prediction model for thermal conductivity of cupcake was found as

$$k = 0.00263T - 0.831m - 0.000910\rho + 0.00422m\rho \qquad (8.11)$$

where k is the thermal conductivity (W/m K), m is the moisture content (g water/g sample), ρ_B is the density of the baked product (kg/m^3), C_{pB} is the heat capacity of the baked product (J/kg K), and T is temperature (°C).

The thermal conductivity of various bakery products was modeled as functions of moisture content, temperature, and density of the product by using an artificial neural network (Sablani et al. 2002). The model was able to predict the thermal conductivity of various baked products with a mean relative error of 10%.

8.4.3.2 Specific Heat Capacity

Kulacki and Kennedy (1978) measured the specific heat of two commercial biscuit doughs (AACC dough and hard-sweet dough), and the specific heat of both types of dough was shown to increase with increasing temperature. The specific heat capacity of cupcakes was shown to be dependent on moisture content and temperature (Baik et al. 1999) and is expressed by the following equation:

$$C_p = 7107m + 18.77T - 45.3mT \qquad (8.12)$$

8.4.3.3 Thermal Diffusivity

In most studies, the thermal diffusivity of baked products increased during baking, whereas thermal conductivity decreased and the specific heat changed slightly (Baik et al. 2001). This can be related to the porosity and moisture changes in the baked product. In general, the thermal diffusivity of crust ($3-37 \times 10^{-8}$ m²/s) was found to be lower than the crumb ($15-53 \times 10^{-8}$ m²/s). This was explained by the lower heat capacity and density of crust compared with the crumb (Zanoni et al. 1995).

8.4.3.4 Moisture Diffusivity

Baik and Marcotte (2003) examined moisture diffusivity of cake batter and found that moisture diffusivity was a function of porosity and temperature. The moisture diffusivity of biscuits was shown to increase with baking time and temperature (Demirkol et al. 2006). Effective moisture diffusivities of cake batter were shown to be in the range of 10^{-10} to 10^{-8} m²/s during baking (Sakin et al. 2007a). It was shown to increase with oven temperature (Table 8.1).

In the presence of microwave baking (100–900 W), the effective moisture diffusivity of Madeira cake batter ranged from 14.3×10^{-9} to 84.4×10^{-9} m²/s at the start of the falling rate, whereas it ranged from 6.50×10^{-9} to 35.6×10^{-9} at the end of baking time (Megahey et al. 2005). These values were higher than the effective diffusivity values of cake during conventional baking (0.328×10^{-9} and 0.178×10^{-9} m²/s at the start of the falling rate period and at the end of baking time, respectively).

8.4.3.5 Density/Specific Volume

The volume of the baked products is generally expressed as a specific volume, which is in fact the reciprocal of density. The generation of gas inside the batter or dough during baking and in the early stages of baking results in the expansion of the baked product. As a result, an increase in volume or a decrease in density of the product is observed.

Specific gravity is a commonly used physical property for cake batter. Low specific gravity is desired in cake batter because it indicates that more air is incorporated

TABLE 8.1
Effective Diffusivities of Cake Batter during Baking at Different Temperatures

Oven Temperature (°C)	Effective Diffusivity (m²/s)
50	0.0167
80	0.102
100	0.608
140	2.43
160	2.66

Source: Reprinted from *J. Food Eng.*, 80, Sakin, A. et al., Modeling the moisture transfer during baking of white cake, 822–831, Copyright 2007, with permission from Elsevier.

into the batter. Turabi et al. (2008b) showed that in gluten-free cake formulations, the addition of gum and emulsifier decreased the specific gravity, which showed that more air was incorporated in the presence of gum and emulsifier blends.

Baking time and method are known to affect the volume of cakes. It was shown that the specific volume of cakes increased as baking time increased (Sumnu et al. 2005). Modification of the starch–gluten matrix and structure development needs sufficient crumb temperature and baking time for the occurrence of complex reactions required for optimum cake volume. Cakes baked in a microwave oven or infrared oven had lower specific volumes than conventionally baked cakes (Sumnu et al. 2005). The lower volume of microwave cakes can be explained by the failure of starch to reach its set point during the microwave baking period. In the presence of infrared heating, radiation is focused at the surface and a thick crust is formed, which retards the expansion of the crumb.

The different formulations of cakes, cookies, and biscuits are known to affect the density/volume of biscuits and cakes. Manohar and Rao (1999) showed that the density of biscuits decreased with increasing fat level. Replacement of inulin with fat in muffin formulations increased crumb density but decreased muffin volume (Zahn et al. 2010). Use of egg alternatives (collagen, cryogel gelatin, solugel collagen hydrolysates, gelatin, whey protein concentrate, fish protein, whey protein isolate, hydrolyzed whey protein isolate, pea protein, rice protein concentrate, soy protein, corn zein, and casein) instead of eggs was shown to decrease the volume of angel cakes (Abu-Ghoush et al. 2010). Muffin volume decreased significantly when wheat flour was replaced with resistant starch at concentrations of 15% (by weight of total formulation) or higher (Baixauli et al. 2008).

Mamat et al. (2010) investigated the physicochemical properties of commercial biscuits and reported that the density of semisweet biscuits in the market ranged from 328.7 ± 11.5 to 402.4 ± 21.9 g/L.

8.4.4 Modeling of Heat and Mass Transfer during Baking

Most of the baking experiments used trial and error procedures to achieve high-quality baked products. Modeling heat and mass transfer during baking reduces the effort, time, and money involved in these baking experiments.

Lostie et al. (2002) used a one-dimensional model for heat and mass transfer within a sponge cake batter during baking. The heat was given only over the upper part of the cylindrical cake, whereas bottom and side surfaces were thermally insulated so that one-dimensional heat transfer assumption was valid. This study was a modification of other baking models in a way that cake batter expansion, gas phase internal convective flux, and vapocondensation heat flux were also considered in the model. As a numerical solving method, the Newton–Raphson algorithm was used.

Lostie et al. (2004) used lumped state variables model for heat and mass transfer during the baking of a sponge cake. The researchers divided the baking period into two groups: the heating up period and the crumb/crust period. They proposed a model for the "crumb/crust" period, in which the main resistances for heat and mass transfer were located in the dry crust where heat was transferred by conduction and water vapor was migrated by convection under a gradient of gaseous phase total pressure. The crust acted as a porous rigid shell, inducing significant temperature and pressure gradients, whereas the temperature and pressure distributions within the crumb were found to be almost uniform. The advantage of the model used was its mathematical simplicity. On the other hand, the disadvantage of the model was the use of an adjustable parameter, which depends on the product's properties and on the processing conditions, and which has to be numerically estimated from experimental baking curves.

Sakin et al. (2007b) modeled heat and mass transfer of cupcakes by an implicit finite difference method. The heat and mass transfer mechanisms were defined by Fourier's law and Fick's second law, respectively, in a two-dimensional cylindrical system. The Fourier equation was modified by adding a phase change term to consider the evaporation of water during baking. The simulated temperature and moisture profiles of cakes were in good agreement with the experimental data.

There are few studies in the literature that introduce air circulation in industrial baking ovens into the models. A steady state mathematical model was used to predict heat and mass transfer during the baking of biscuits in a continuous, indirect gas-fired oven (Broyart and Trystram 2002). Radiation, convection, conduction, and product–water phase change were considered in the heat transfer model. For mass transfer, the condensation of steam from the baking atmosphere to the product's surface and product drying were taken into consideration. The predicted temperature of the product and moisture content variations along the length of the oven were in good agreement with the experimental data. Mirade et al. (2004) used three-dimensional computational fluid dynamics approach to predict air temperature and velocity profiles during the baking of biscuits in an industrial, indirect gas-fired tunnel oven. Comparison of the results with experimental data showed that the numerical results were in good agreement with temperature profiles within the baking chamber, although they did not align well with air velocity profiles.

Flow field and heat transfer during the baking of a cookie-shaped object in a jet impingement oven were simulated in a three-dimensional system (Kocer and Karwe 2005). It was found that heat transfer coefficient was a significant function of air velocity, but air temperature had no effect on the heat transfer coefficient. Flow field close to the surface of the model cookie was different from that of a flat surface because of the presence of a suction fan in the oven.

Contact baking is a process in which the product is heated at high temperature (140°C–300°C) by contact with a hot surface and is used in the baking of pizzas, tortillas, pancakes, and pita bread (Feyissa et al. 2011). Feyissa et al. (2011) have recently modeled coupled heat and mass transfer by taking multiphase water transport and local evaporation into consideration during contact baking of a pancake. A good agreement was found between the experimental temperature data and the predicted temperature profile.

8.5 INFLUENCE OF PROCESSING ON BAKERY PRODUCT QUALITY

Quality is the level at which a specific product meets the expectations or receives the approval of consumers. The characteristics of a food product determine the approval of the consumer. Because quality is very important, the factors affecting quality gain added importance. Baking type and conditions are reported as the key factors affecting the quality of baked products.

In the conventional baking temperature and humidity of an oven, air velocity and time are the parameters affecting the quality of baked products (Lara et al. 2011; Xue and Walker 2003; Xue et al. 2004).

Zareifard at al. (2009) reported the effect of convective heat flux on quality parameters (volume expansion, texture properties, color, and moisture content) of cake samples. Convective heat flux was changed by modifying air characteristics and wall temperatures. It was found that cakes baked in extreme oven conditions had a 10% reduction in volume expansion and a 30% increase in texture properties compared with those baked in standard conditions. Surface color was always darker but more uniform for the ovens with less convection. In addition, the moisture content of the middle part of the cake was always higher than that of the top, bottom, and sides. This study proved that air and oven wall temperatures and air velocity affected the convective heat transfer and hence the quality of the product.

In alternative baking technologies, microwave power, infrared power, and air velocity are the critical factors that affect the quality. Li and Walker (1996) showed that increasing baking time, air velocity, and temperature in jet impingement ovens increased both crust color and cake firmness. The effect of humidity on crust color and volume of cake samples baked in an air jet impingement oven was studied by Xue and Walker (2003). It was declared that high humidity lightened cake crust color, increased volume, and increased the final moisture content of the cake (Xue and Walker 2003).

It was found that the spread ratio of the cookies was significantly affected by infrared and microwave power when they were baked in an infrared–microwave combination oven. Additionally, the moisture content of the cookies was found to decrease as infrared or microwave power increased (Keskin et al. 2005). It was reported that the rate of weight loss and color change values of cakes increased as infrared power increased, which can be explained by subjecting the samples to more radiation in the presence of high infrared powers, and, consequently, higher surface temperatures might be obtained (Sumnu et al. 2005; Turabi et al. 2008a). The baking method was found to be important in affecting the quality of baked products such as volume, texture, color, porosity, and moisture content.

When the textural properties of biscuits baked in a conventional oven and a microwave oven were compared, it was seen that the structure of conventional biscuits weakened at a faster rate than microwaved biscuits. Microwaved biscuits were also found to be less susceptible to checking upon exposure to high ambient humidity (Ahmad et al. 2001).

The effects of infrared–microwave combination baking on the quality of cookies in terms of texture, color, and spread ratio were studied by Keskin et al. (2005). They concluded that the best baking condition in a combination oven was baking at 70% halogen lamp and 20% microwave power levels for 5.5 min. It was also found that the baking time of these cookies was half that of those that were conventionally baked.

In another study, Bernussi et al. (1998) compared conventional and microwave baking in terms of cookie quality. As the major conclusion of the study, it can be said that microwave baking significantly reduced the moisture gradient within the cookies and total moisture content of the cookies.

It has been recently shown that cakes baked in infrared–microwave combination oven were more porous than those baked in a conventional oven (Figure 8.4; Turabi et al. 2010). This may be due to the significant differences in the baking mechanisms. As can be seen in scanning electron microscopy images, more uniform structures were obtained when xanthan and xanthan–guar gums were added into the cake formulations.

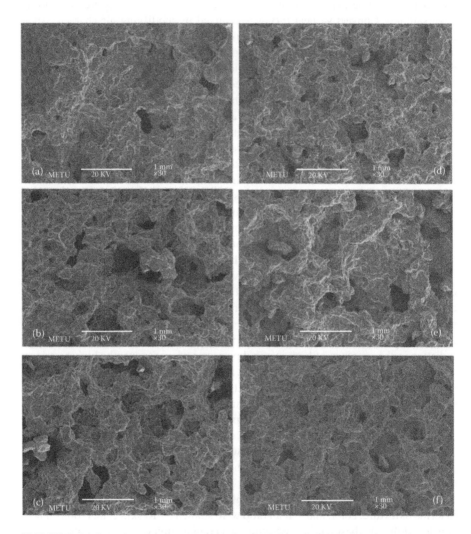

FIGURE 8.4 Scanning electron micrographs (30×) for cakes baked in conventional and infrared–microwave oven: (a) conventional oven–control, (b) conventional oven–xanthan gum, (c) conventional oven–xanthan–guar gums, (d) infrared–microwave oven–control, (e) infrared–microwave oven–xanthan gum, and (f) infrared–microwave oven–xanthan–guar gums. (Reprinted from *Food Hydrocolloids*, 24, Turabi, E. et al. Quantitative analysis of macro and micro-structure of gluten free rice cakes containing different types of gums baked in different ovens, 755–762, Copyright 2010, with permission from Elsevier.)

With regard to cake quality, the most important quality parameters to be investigated are volume, texture, color, and moisture loss. Many researchers have studied this topic (Sakiyan 2007; Sumnu et al. 2005). Sumnu et al. (2005) studied the variation of these parameters during baking in microwave, infrared, and infrared–microwave combination ovens. They found that cakes baked in a microwave oven had the lowest quality. In addition, they concluded that infrared–microwave combination baking reduced weight loss and firmness and increased the volume of cake samples. Combination baked cakes were found to have comparable firmness and color results with conventional baked samples.

Sakiyan (2007) declared that cakes baked in a microwave oven had the lowest specific volume as compared with other baking methods. This can be explained by the failure of starch to reach its set point during the microwave baking period. Similarly, Li and Walker (1996) showed that cakes baked by microwave had lower volumes than conventionally baked cakes.

Acrylamide formation is closely related to the development of the desired sensory properties (color, flavor, and texture) of baked products. On the other hand, it is important to decrease acrylamide formation without changing the quality parameters of the product. Anese et al. (2008) investigated the effect of radio-frequency (RF) heating on acrylamide formation in cake and biscuit samples. The heating modes used in the study were conventional convection heating and different combinations of conventional and RF heating. The samples examined were leavened cakes and short dough biscuits baked to a final moisture of 3.5% and 3.0%, respectively. They concluded that applying RF heating in the last stages of the baking process was a promising strategy to keep the acrylamide level in the product at a minimum. They also found that RF-assisted baking should be preferred for thin bakery products, such as biscuits, compared with thick products, such as leavened cakes, because of excessive browning in the internal portion.

The control of starch gelatinization in cake systems has been emphasized to be important in obtaining a noncollapsing, porous cake structure (Kim and Walker 1992). One of the most important drawbacks of microwave baking is the insufficient starch gelatinization in baked products. This can be because of the short baking period of microwave systems. Less gelatinization in microwave-baked cakes was reported (Sánchez-Pardo et al. 2008; Sakiyan et al. 2011). The degree of gelatinization in microwave-baked cakes ranged from 55% to 78%, depending on the formulation (Sakiyan et al. 2011). On the other hand, it ranged from 85% to 93% in conventionally baked cakes. This may be due to the limited availability of water during the microwave heating process. Besides, Sakiyan et al. (2011) reported that combining infrared with microwaves increased the degree of gelatinization of samples to a level (70%–90%) comparable to that of conventionally baked cakes (85%–93%).

REFERENCES

Abu-Ghoush, M., Herald, T.J., and Aramouni, F.M. 2010. Comparative study of egg white protein and egg alternatives used in an angel food cake system. *Journal of Food Processing and Preservation* 34: 411–425.

Ahmad, S.S., Morgan, M.T., and Okos, M.R. 2001. Effects of microwave on the drying, checking and mechanical strength of baked biscuits. *Journal of Food Engineering* 50: 63–75.

Anese, M., Sovrano, S., and Bortolomeazzi, R. 2008. Effect of radiofrequency heating on acrylamide formation in bakery products. *European Food Research and Technology* 226: 1197–1203.

Baik, O.D., and Marcotte, M. 2003. Modeling the moisture diffusivity in a baking cake. *Journal of Food Engineering* 56: 27–36.

Baik, O.D., Sablani, S.S., Marcotte, M., and Castaigne, F. 1999. Modeling the thermal properties of a cup cake during baking. *Journal of Food Science* 64: 295–299.

Baik, O.D., Marcotte, M., Sablani, S.S., and Castaigne, F. 2001. Thermal and physical properties of bakery products. *Critical Reviews in Food Science and Nutrition* 41: 321–352.

Baixauli, R., Sanz, T., Salvador, A., and Fiszman, S.M. 2008. Muffins with resistant starch: Baking performance in relation to the rheological properties of the batter. *Journal of Cereal Science* 47: 502–509.

Bean, M.M., and Setser, C.S. 1992. *Polysaccharide Sugars and Sweeteners.* New York. Macmillan Publishing Company.

Bennion, E.B., and Bamford, G.S.T. 1997. Chemical aeration. In *The Technology of Cake Making,* ed. A.J. Bent, 100–107, Sixth Edition. London: Blackie Academic and Professional.

Bernussi, A.L.M., Chang, Y.K., and Martinez-Bustos, E. 1998. Effects of production by microwave heating after conventional baking on moisture gradient and product quality of biscuits (cookies). *Cereal Chemistry* 75: 606–611.

Broyart, B., and Trystram, G. 2002. Modeling heat and mass transfer during continuous baking of biscuits. *Journal of Food Engineering* 51: 47–57.

Cauvain, S., and Young, L. 2006. *Baked Products: Science, Technology and Practice.* Oxford: Blackwell Publishing.

Cheong, L.Z., Tan, C.P., Long, K., Idris, N.A., Yusoff, M.S.A., and Lai, O.M. 2011. Baking performance of palm diacylglycerol bakery fats and sensory evaluation of baked products. *European Journal of Lipid Science and Technology* 113: 253–261.

Datta, A.K. 1990. Heat and mass transfer in the microwave processing of food. *Chemical Engineering Progress* 86: 47–53.

Demirkol, E., Erdoğdu, F., and Palazoğlu, T.K. 2006. Analysis of mass transfer parameters (changes in mass flux, diffusion coefficient and mass transfer coefficient) during baking of cookies. *Journal of Food Engineering* 72: 364–371.

Elliasson, A.C. 1985. Starch gelatinization in the presence of emulsifiers: A morphological study of wheat starch. *Starch* 37: 411–415.

Elliasson, A.C., and Larsson, K. 1993. *Cereal in Bread Baking: A Molecular Colloidal Approach.* New York: Marcel Dekker.

Feyissa, A.H., Gernaey, K.V., Ashokkumar, S., and Adler-Nissen, J. 2011. Modelling of coupled heat and mass transfer during a contact baking process. *Journal of Food Engineering* 106: 228–235.

Freeland-Graves, J.H., and Peckham, G.C. 1987. *Foundations of Food Preparation.* New York: Macmillan Publishing Company.

Ghiasi, K., Hoseney, R.C., and Varriano-Marston, E. 1982. Effects of flour components and dough ingredients on starch gelatinization. *Cereal Chemistry* 60: 58–61.

Guy, C.E.R., and Sahi, S.S. 2006. Application of lipase in cake manufacture. *Journal of the Science of Food and Agriculture* 86: 1679–1687.

Indrani, D., and Rao, G.V. 2008. Functions of ingredients in baking of sweet goods. In *Food Engineering Aspects of Baking Sweet Goods,* eds. S.G. Sumnu and S. Sahin, 31–48. Boca Raton: CRC Press (Taylor and Francis group).

Jacob, J., and Leelavathi, K. 2007. Effect of fat-type on cookie dough and cookie quality. *Journal of Food Engineering* 79: 299–305.

Kent, N.L. 1983. *Technology of Cereals.* London: Pergamon Press.

Keskin, S.O., Ozturk, S., Sahin, S., Koksel, H., and Sumnu, G. 2005. Halogen lamp–microwave combination baking of cookies. *European Food Research and Technology* 220: 546–551.

Kim, S.S., and Setser, C.S. 1992. Wheat starch gelatinization in the presence of polydextrose or hydrolyzed barley β-glucan. *Cereal Chemistry* 69: 447–451.

Kim, C.S., and Walker, C.E. 1992. Effects of sugars and emulsifiers on starch gelatinization evaluated by differential scanning calorimetry. *Cereal Chemistry* 69: 212–217.

Kocer, D., and Karwe, M.V. 2005. Thermal transport in a multiple jet impingement oven. *Journal of Food Engineering* 28: 378–396.

Kocer, D., Karwe, M.V., and Sumnu S.G. 2008. Alternative baking technologies. In *Food Engineering Aspects of Baking Sweet Goods*, eds. S.G. Sumnu and S. Sahin, 215–244. Boca Raton: CRC Press (Taylor and Francis group).

Kulacki, F.A., and Kennedy, S.C. 1978. Measurement of the thermo-physical properties of common cookie dough. *Journal of Food Science* 43: 380–384.

Lara, E., Cortés, P., Briones, V., and Perez, M. 2011. Structural and physical modifications of corn biscuits during baking process. *Lebensmittel Wissenschaft und Technologie—Food Science and Technology* 44: 622–630.

Larsson, K. 1980. Inhibition of starch gelatinization by amylase-lipid complex formation. *Starch* 32: 125–126.

Li, A., and Walker, C.E. 1996. Cake baking in conventional, impingement and hybrid ovens. *Journal of Food Science* 61: 188–191, 197.

Lin, S.D., Lee, C.C., Mau, J.L., Lin, L.Y., and Chiou, S.Y. 2010. Effect of erythritol on quality characteristics of reduced-calorie Danish cookies. *Journal of Food Quality* 33: 14–26.

Lostie, M., Peczalski, R., Andrieu, J., and Laurent, M. 2002. Study of sponge cake batter baking process. II. Modeling and parameter estimation. *Journal of Food Engineering* 55: 349–357.

Lostie, M., Peczalski, R., and Andrieu, J. 2004. Lumped model for sponge cake baking during the "crust and crumb" period. *Journal of Food Engineering* 65: 281–286.

Mamat, H., Abu, Hardan, M.O., and Hill, S.E. 2010. Physicochemical properties of commercial semi-sweet biscuit. *Food Chemistry* 121: 1029–1038.

Manley, D.J.R. 1998. *Biscuit, Cookie and Cracker Manufacturing Manuals*. Cambridge, England: Woodhead Publishing.

Manley, M. 1991. *Technology of Biscuits, Crackers and Cookies*. London: Ellis Horwood.

Manohar, R.S., and Rao, P.H. 1999. Effect of emulsifiers, fat level and type on the rheological characteristics of biscuit dough and quality of biscuits. *Journal of the Science of Food and Agriculture* 79: 1223–1231.

Matz, S.A. 1972. *Bakery Technology and Engineering*. Westport, Connecticut: The AVI Publishing Company, Inc.

McWilliams, M. 1989. *Food Experimental Perspectives*. New York: Macmillan Publishing Company.

Megahey, E. K., McMinn, W.A.M., and Magee, T.R.A. 2005. Experimental study of microwave baking of Madeira cake batter. *Food and Bioproducts Processing* 83: 277–287.

Metaxas, A.C., and Meredith, R.J. 1983. *Industrial Microwave Heating*. London: Peter Peregrinus.

Miradea, P.S., Daudina, J.D., Ducept, F., Trystramb, G., and Clément, J. 2004. Characterization and CFD modelling of air temperature and velocity profiles in an industrial biscuit baking tunnel oven. *Food Research International* 37: 1031–1039.

Ovadia, D.Z., and C. E. Walker. 1998. Directing jets of fluid such as air against the surface of food provides advantages in heating, drying, cooling, and freezing. *Food Technology* 52: 46–50.

Pareyt, B., Brisj, K., and Delcour, J.A. 2010. Impact of fat on dough and cookie properties of sugar-snap cookies. *Cereal Chemistry* 87(3): 226–230.

Paton, D., Larocque, G.M., and Holme, J. 1981. Development of cake structure: influence of ingredients on the measurement of cohesive force during baking. *Cereal Chemistry* 58: 527–529.

Sablani, S.S. Physical and thermal properties of sweet goods. 2008. In *Food Engineering Aspects of Baking Sweet Goods*, eds. S.G. Sumnu and S. Sahin, 191–213. Boca Raton: CRC Press (Taylor and Francis group).

Sablani, S.S., Marcotte, M., Baik, O.D., and Castaigne, F. 1998. Modeling of simultaneous heat and water transport in the baking process. *Lebensmittel Wissenschaft und Technologie* 31: 201–209.

Sablani, S.S., Baik, O.D., and Marcotte, M. 2002. Neural networks for predicting thermal conductivity of bakery products. *Journal of Food Engineering* 52: 299–304.

Saha, S., Gupta, A., Singh, S.R.K., Bharti, N., Singh, K.P., Mahajan, V., and Gupta, H.S. 2011. Compositional and varietal influence of finger millet flour on rheological properties of dough and quality of biscuits. *LWT—Food Science and Technology* 44: 616–621.

Sahin, S., and Sumnu, S.G. 2006. *Physical Properties of Foods*. New York: Springer.

Sakin, M., Kaymak-Ertekin, F., and Ilicali, C. 2007a. Modeling the moisture transfer during baking of white cake. *Journal of Food Engineering* 80: 822–831.

Sakin, M., Kaymak-Ertekin, F., and Ilicali, C. 2007b. Simultaneous heat and mass transfer simulation applied to convective oven cup cake baking. *Journal of Food Engineering* 83: 464–474.

Sakiyan, O. 2007. Investigation of physical properties of different cake formulations during baking with microwave and infrared-microwave combination. PhD thesis, Ankara, Turkey: Middle East Technical University.

Sakiyan, O., Sumnu, G., Sahin, S., and Meda, V. 2007. Investigation of dielectric properties of different cake formulations during microwave and infrared–microwave combination baking. *Journal of Food Science* 72: 205–213.

Sakiyan, O., Sumnu, G., Sahin, S., Meda, V., Köksel, H., and Chang, P. 2011. A study on degree of starch gelatinization in cakes baked in three different ovens. *Food and Bioprocess Technology* 4: 1237–1244.

Sánchez-Pardo, M., Ortiz-Moreno, A., Mora-Escobedo, R., Bello-Pérez, A., Yee-Maderia, H., and Ramos-López, G. 2008. Effect of baking method on some characteristics of starch in pound cake crumbs. *Journal of the Science of Food and Agriculture* 88: 207–213.

Sevimli, M., Sumnu, G., and Sahin, S. 2005. Optimization of halogen lamp–microwave combination baking of cakes: A response surface methodology study. *European Food Research and Technology* 221: 61–68.

Smith, D.P. 1975. Cooking apparatus. U.S. Patent # 3,884,213.

Standing, C.N. 1974. Individual heat-transfer modes in band oven biscuit baking. *Journal of Food Science* 39: 267–271.

Sumnu, G. 2001. A review on microwave baking of foods. *International Journal of Food Science and Technology* 36: 117–127.

Sumnu, G., Datta, A.K., Sahin, S., Keskin, S.O., and Rakesh, V. 2007. Transport and related properties of breads baked using various heating modes. *Journal of Food Engineering* 78: 1382–1387.

Sumnu, G., and Sahin, S. 2005a. Baking using microwave processing. In *Microwave Processing of Foods*, eds. M. Regier and H. Schubert, 119–142. Cambridge: Woodhead Publishing Ltd.

Sumnu, G., and Sahin, S. 2005b. Recent developments in microwave processing. In *Emerging Technologies for Food Processing*, ed. D. Sun, 419–444. London: Elsevier.

Sumnu, G.S., Sahin, S., and Sevimli, M. 2005. Microwave, infrared and infrared-microwave combination baking of cakes. *Journal of Food Engineering* 71: 150–155.

Tireki, S. 2008a. Technology of cake production. In *Food Engineering Aspects of Baking Sweet Goods*, eds. S. Sahin and S.G. Sumnu, 149–158. Boca Raton: CRC Press (Taylor and Francis group).

Tireki, S. 2008b. Technology of cookie production. In *Food Engineering Aspects of Baking Sweet Goods*, eds. S. Sahin and S.G. Sumnu, 159–172. Boca Raton: CRC Press (Taylor and Francis group).

Turabi E., Sumnu, G., and Sahin, S. 2008a. Optimization of baking of rice cakes in infrared–microwave combination oven by response surface methodology. *Food and Bioprocess Technology* 1: 64–73.

Turabi, E., Sumnu, G., and Sahin, S. 2008b. Rheological properties and quality of rice cakes formulated with different gums and an emulsifier blend. *Food Hydrocolloids* 22: 305–312.

Turabi, E., Sumnu, S., and Sahin, S. 2010. Quantitative analysis of macro and micro-structure of gluten free rice cakes containing different types of gums baked in different ovens. *Food Hydrocolloids* 24: 755–762.

Vetter, J.L. 1984. Technical Bulletin. VI. Manhattan. KS: American Institute of Baking.

Walker, C.E., and Li, A. 1993. Impingement oven technology—Part III—Combining impingement with microwave (hybrid oven). *AIB Research Department Technical Bulletin*, Volume XV, Issue 9, September.

Xue, J., and Walker, C.E. 2003. Humidity change and its effects on baking in an electrically heated air jet impingement oven. *Food Research International* 36: 561–569.

Xue, J., Lefort, G., and Walker, C.E. 2004. Effects of oven humidity on foods baked in gas convection ovens. *Journal of Food Processing and Preservation* 28: 179–200.

Yin, Y., and Walker, C.E. 1995. A quality comparison of breads baked by conventional versus nonconventional ovens: a review. *Journal of the Science of Food and Agriculture* 67: 283–291.

Zahn, S., Pepke, F., and Rohm, H. 2010. Effect of inulin as a fat replacer on texture and sensory properties of muffins. *International Journal of Food Science and Technology* 45: 2531–2537.

Zanoni, B., Peri, C., and Gianotti, R. 1995. Determination of the thermal diffusivity of bread as a function of porosity. *Journal of Food Engineering* 26: 497–510.

Zareifard, M.R., Boissonneault, V., and Marcotte, M. 2009. Bakery product characteristics as influenced by convection heat flux. *Food Research International* 42: 856–864.

Zhang, H., and Datta, A.K. 2001. Electromagnetics of microwave heating. In *Handbook of Microwave Technology for Food Applications*, eds. A.K. Datta and R.C. Anatheswaran, 33–67. New York: Marcel Dekker Inc.

Zhou, J., Faubion, J.M., and Walker, C.E. 2011. Evaluation of different types of fats for use in high-ratio layer cakes. *LWT—Food Science and Technology* 44: 1802–1808.

9 Breakfast Cereals

Maria João Barroca

CONTENTS

9.1 INTRODUCTION

Cereal processing technology constitutes a strategic industry delivering a range of products worldwide from finished items to raw materials used in baking, brewing, etc. Breakfast cereals production is a strategic activity that has evolved from traditional cereals that require further cooking or heating before consumption to ready-to-eat (RTE) cereals that can be consumed without further cooking at home.

Breakfast has been described as one of the most important meals of the day, which usually includes breakfast cereals. Thus, RTE foods are quite convenient and are constituted by cereal grains (mostly oats, rice, or wheat), which have been processed in such a way that makes them suitable for human consumption without requiring further processing or cooking.

Even though the use of grains (raw, dried, or flour) as well as their cooked or baked forms dates back to before Christ's time, the RTE industry only really started in the late nineteenth century and originated from pioneering attempts to manage health in the United States, and is based on the teachings of the Seventh-Day Adventist Church (Whalen et al. 2000). In the middle of the 1800s, the Seventh-Day Adventist Church established its national headquarters in Battle Creek, Michigan. Later on, one of the Kellogg brothers, John Harvey, became the physician-in-chief of the Seventh-Day Adventist Church's Battle Creek Sanitarium and filed a patent (in 1894) that included flakes of wheat, corn, oats, barley, and other grains (Jackson et al. 2004). Also on that year, a former patient of the sanitarium, C. W. Post, invented a coffee substitute from a mixture of roasted wheat, bran, and molasses (he called it *Postum*), and would go on to form The Postum Cereal Co. Ltd. Four years later, he made a second substitute, in granular form, called Grape-Nuts. Despite its name, the cereal contained neither grapes nor nuts. Since the beginning, this product has received poor acceptance as a breakfast food, yet it is still on the market more than 100 years later (Fast 2001).

By the 1900s, a change occurred in how corn flakes were being promoted—from being a health food to a breakfast food that has "good taste" following the addition of barley malt extract and sugar to enhance the flavor of the basic toasted corn flakes.

In fact, the enhancement of flavor with sugar, syrup glaze, or salt, and the loss of nutrients during the RTE processing, which are compensated for by the addition of minerals and vitamins, namely, B vitamins, thiamin, riboflavin, niacin, and mineral fractions, are playing a big role in the nutritional improvement and marketing tactics of breakfast cereals.

During the next 40 years, marketing became the real explosive propelling force of breakfast cereals through the widespread development of radio and the use of premiums, or "gimmicks" inserted in each cereal box. However, during and after World War II, the breakfast cereal industry truly blossomed into maturity as food scientists from the US government decided that breakfast cereals were food, along with bread, and would be good vehicles for vitamin restoration (Fast 2001).

Other major changes and innovations took place in the 1950s and 1960s with the use of continuous cooking extruders. The extruders, classified according to the method of operation (cold or extruder-cookers) and the method of construction (single-screw or twin-screw extruders), have similar operating principles, in which the raw materials are fed into the extruder barrel and the screw(s) then convey the food along it. The food mix is thermomechanically cooked at high temperature under pressure, and undergo shear stress generated in the screw barrel assembly. The material is compressed and deformed in the screw to form a semisolid mass. This is forced through a restricted opening (the die) at the discharge end of the screw, allowing the production of a significant range of products with different shapes (including rods, spheres, doughnuts, tubes, strips, or shells), textures, colors, and flavors from basic ingredients (Midden 1989; Fellows 2000).

Extrusion cooking, classified as a high-temperature/short-time process, is an important food processing technique used worldwide for the production and modification or improvement of the quality of various products, namely, RTE cereals. The thermomechanical action during extrusion brings about starch gelatinization, protein denaturation, and inactivation of enzymes, microbes, and many antinutritional factors. All this occurs in a shear environment, resulting in a plasticized continuous mass (Battacharya and Prakash 1994).

The development of extruded cereals led to the development of low bulk density extruded/expanded breakfast cereals in many grain combinations, shapes, flavors, and colors. This continuous cooking associated with the automation of the gun puffing process greatly increased the output per hour and improved the technology employed to form intricate shapes of pastas. In addition, oven toasting was improved and introduced to increase the output of the classic rotary drum of toasting ovens for flakes (Fast 2001).

The developments in shredding and rotary cutting techniques allowed the introduction of bite-sized products produced from grains other than wheat, such as rice and corn, and products with added flavors, such as sugar, salt, and malt.

In the 1970s and 1980s, market forces put the emphasis on more nutritious cereals, leading the manufacturers and marketers to work together in marketing strategies and to introduce protein-fortified RTE beyond vitamin and mineral supplements. Moreover, new varieties in shapes, flavors, colors, textures, resistance to sagging in milk, and additives were being introduced almost every day. Additives included nuts of all kinds, raisins, and other dried fruits, even freeze-dried fruits and ice cream coatings.

Since 1990, great technical developments in the RTE manufacturing processes of weighing, blending, cooking, drying, tempering, forming, and quality control allowed many cereal manufacturing companies to move to a higher level of total plant control (Fast 2001).

A new generation of cereal products prepared with sugar, honey, and more sophisticated flavors, including texture additives and reinforcement with nutrients and minerals in accordance with the tastes and fulfillment of consumers' requirements, have been developed, constituting a complete meal rather than part of a well-rounded breakfast. In many cases, these breakfast cereals are made from an existing product as the base (flakes, puffed grains, or shredded biscuits) and with evolved production lines that include coating systems and spray systems of minerals and vitamins.

Nowadays, this industry has been dominated by a relatively few large producers. The worldwide leaders are Kellogg, General Mills, Nestlé, Weetabix, Uncle Toby's, Sanitarium, and General Foods. However, there are smaller players who have market shares in various regions of the world.

The processing of RTE cereals typically involves the cooking of the grain with flavoring materials, sweeteners, and fortifying agents if they are heat stable. Two general cooking methods are employed in this industry: direct steam injection into the grain mass using rotating batch vessels and continuous extrusion cooking, classified as a high-temperature/short-time process. Due to its high versatility and peculiar ability to modify the structural physicomechanical properties of products, extrusion cooking technology has been largely used for manufacturing breakfast cereals (Paradiso et al. 2008). The rheological properties of the product are influenced by the extruder's operating parameters, such as barrel and die temperature, screw speed, screw configuration, and die shape, as well as raw material formulation, namely, moisture, protein, starch, and lipid contents (Moraru and Kokini 2003). Furthermore, ingredients such as sugar, salt, protein, and fiber can affect the extrusion system's variables, as well as product characteristics, such as texture, structure, expansion, and sensory attributes (Jin et al. 1994).

After the cooking process, which in some manner may gelatinize the starch of raw grains, these are then flattened (flaked), formed (extruded), shredded, or expanded (puffed; Culbertson 2004).

Texture in RTE cereals is fundamental for product acceptance by consumers. Texture characteristics such as crispness and crunchiness are generally expected, which can be imparted to raw materials through several ways. Crispness is perceived through a combination of tactile, kinesthetic, visual, and auditory sensations and represents the key texture attribute of dry snack products (Heidenreich et al. 2004). Texturizing processes will provide crispness or crunchiness by changing, basically, the surface/volume ratio and thickness. Texture alteration can be achieved by different operations such as flaking, puffing, or extruding. However, as RTE snacks are usually low-moisture foods, a small change in moisture content directly affects their texture. As is generally known, less crispy cereals and snacks are not acceptable.

Most RTE cereals may be grouped into general categories that are related with their manufacturing process: flaked cereals (cornflakes, wheat flakes, and rice flakes), puffed whole cereals, shredded whole grains, and granola cereals (Valentas et al. 1991; Fast 1993; Hoseney 1994). Each of these categories can be produced through a traditional process or through an extruded process that, basically, differs from the way in which the grit for flaking is formed by extruding the mixed ingredients through a die hole and cutting off pellets of the dough in the desired size.

To maintain the level of moisture, to preserve the nutritional and organoleptic properties, and to extend their shelf life, special packaging is required and antioxidants are often used, namely, natural mixed tocopherols. These can be used as effective antioxidants that slow down the development of off-flavors during the extrusion cooking process of breakfast cereals (Paradiso et al. 2008).

9.2 FLAKED CEREALS

Flaked cereals include those made directly from whole-grain kernels or parts of kernels of corn, wheat, or rice and extruded formulated flakes. The main objective of flaked cereal production is to process the grain in such a way that particles that form one flake each are obtained. Grain selection is therefore very important to the finished characteristics of flaked cereals and size reductions, and size screening operations may be necessary to provide uniform flakes, known as flaking grits.

Corn flakes and wheat flakes are typically made from whole-grain kernels or parts of them. In the final step, the cereal is treated to restore vitamins lost through cooking and is often coated with sweet flavorings to make them more attractive (Sumithra and Bhattacharya 2008).

Corn flakes are the best example of a cereal made from parts of whole and are possibly the most common form of breakfast cereals (Sumithra and Bhattacharya 2008). To produce traditional cornflakes, the corn is dried to remove the germ and bran from the kernel and what is left are essentially chunks of the endosperm. The size needed for cornflakes is from one-half to one-third of the whole kernel. These pieces retain their identity throughout the process, each producing a single flake, although sometimes two small grits stick together and wind up as one flake.

A typical formula for flaked cereals includes corn grits (to corn flakes), bumped wheat (to wheat flakes) and head rice (whole grain) or second heads (broken pieces of whole kernels), granulated sugar or liquid sucrose, malt syrup, salt, and water sufficient to yield cooked grits with a moisture content (% w/w) of no more than 32% (cornflakes), 28% to 30% (wheat flakes), and about 28% (rice flakes) after allowing for steam condensate (Fast 1993). The processing of wheat flakes is different from that of cornflakes given the differences between the grains. For cornflakes, the starting raw material is broken chunks of kernel endosperm from which the bran has been removed. On the other hand, for wheat flakes, the starting material is whole kernels with all seed parts intact (germ, bran, and endosperm). In the case of rice flakes, one can use either the whole grain or broken pieces, although, from an economic perspective, the latter is preferred.

The processing of the different traditional flaked cereals (cornflakes, wheat flakes, and rice flakes) only differs in some minor specifications due to differences between the grains. Typically, the unit operations to produce the traditional flaked cereals include mixing, cooking, dumping, delumping, drying, cooling and tempering, flaking, and toasting, which are described in detail by Fast (1993).

Toasting is an important processing step in the manufacture of breakfast cereals that dictates the attributes of RTE corn flakes, which are usually characterized by their unique crispness and the maintenance of its integrity while being eaten with milk. A number of parameters affect the degree of expansion of these cereals, which are related both with the compositional characteristics of the raw material (Chen and Yeh 2001; Jones et al. 2000) and with the processing conditions (Chandrasekhar and Chattopadhyay 1990), including the effects of important toasting variables such as moisture content, temperature, and time of toasting of flakes. Such parameters also influence quality attributes such as the thickness of puffed flakes, bulk density, color parameters, and maximum puncture force. Properly toasted flakes possess appropriate moisture content and desired texture and color. The toasted flakes have a moisture content from 1% to 3% (w/w) depending on the grain used, the process time, and temperature (Sumithra and Bhattacharya 2008).

The extruded flakes differ from those made using traditional processes because the extrusion method combines several unit operations including mixing, cooking, kneading, shearing, shaping, and forming (Fellows 2000). After mixing and extruding the ingredients, the pellets should have a moisture content in the range of 17% to 18% (w/w) to eliminate an otherwise necessary drying step before tempering and toasting the extruded pellets.

The ingredients to produce extruded flakes are similar to the ones previously described for flaked cereals, but in different proportions and with sufficient water for pellet formation and cooking. However, natural or artificial colorings are usually added to compensate for the dull, slightly gray, appearance of the cereals, due to the high mechanical work or shearing that the ingredients are submitted to during the processing steps. Thus, the grit for flaking is formed by extruding the mixed ingredients through a die hole and cutting off pellets of the dough in the desired size. With the cooking extruders, finer materials, such as flours, can be used because the flakable size particles are attained by mechanical means. By cooking dough and by forming flakes via extrusion, much of the equipment for preflaking size reduction and screening in the traditional process can be eliminated. The major advantage of the cooker-extruder is that it can handle cereal flours at relatively low moisture content (Hoseney 1994).

Even though wheat and maize flour are the most commonly used raw materials (Guy 2001), other grains are beginning to be used, such as the *Andean quinoa* (pseudo cereal) with the objective of improving the nutritional value of extruded flakes (Medina et al. 2011).

9.3 PUFFED WHOLE GRAINS

Puffed cereals are commonly used as RTE breakfast foods or as ingredients in snack formulations. They are appreciated mainly for their lightness and crispness, qualities related to their cellular structure (Peleg 1997), and degree of expansion (Owusu-Ansah et al. 1984).

Rice and wheat are the two main cereals whose granules can be processed individually to expand them or to inflate them, obtaining RTE products known as puffed cereals (Fast 2001). However, some other grain formulations can also be expanded to obtain puffed products using extrusion technology (Sevatson and Huber 2000).

The puffing cereals that involve decreasing the bulk density can be carried out using two general technological processes (Guy 1995): oven puffing, which is the sudden application of heat at atmospheric pressure (the water is vaporized before it has time to diffuse to the surface), and gun puffing, which is the sudden transfer of the grain containing superheated steam from a high pressure to a low pressure, thereby allowing the water to suddenly vaporize and cause expansion (Matz 1991; Kent and Evers 1994). The driving force in both methods is the transfer of the water to a vapor phase, and the degree of puffing is the sudden change in temperature or pressure (Hoseney 1994).

Oven-puffed cereals are almost always made from rice or corn, or mixtures of these two grains, because both inherently puff with appropriate moisture content and in the presence of high heat. In the gun-puffed process, the puffed cereals originate from rice and wheat as whole kernel, but if the material is pressured or extruded puffed, the products can be made from a cereal flour and dough mixed and moistened with steam.

The grains must be cooked and then submitted to different processes of puffing, thus promoting different degrees of puffing. For instance, the sudden application of heat (oven-puffed cereals) allows an expansion of 2 to 5 times, but with the pressure-puffed process (also referred to as the gun-puffed process), the degree of expansion obtained is 15 to 20 times greater than that of oven puffing.

Extrusion cooking is another method of expanding cereal products. This method uses both temperature and pressure to expand kernels or flours. In general, the extruded gun-puffed cereals originate from flours and not from whole grains. In this process, the basic ingredients are mixed and cooked in the extruder and the dough obtained is fed to a forming extruder, which generates extruded cooked shapes with moisture contents in the range of 20% to 24% (w/w). The next step is drying and tempering, usually to a moisture content of 9% to 12% (w/w) and then the shapes are gun-puffed (Fast 1993).

The effect of the puffing treatment (degree of expanding) is strongly influenced by the morphology and composition characteristics of raw material. Puffed rye and rice have a very porous matrix that is made up of numerous cavities of different sizes and is separated by a very thin "wall." On the other hand, puffed wheat, emmer wheat, and barley show a much more compact, homogeneous, and less porous structure; puffed buckwheat (pseudo cereal) is characterized by a large number of small and regular cavities. Moreover, puffing induces significant changes in the structure and physical properties of the starch and an increased water-holding capacity of both the grains and the flours (Mariotti et al. 2006).

Other cereal grains, such as maize and oats, are also used for puffing but are commonly used as flours extruded at high temperatures rather than as whole grains (Fast 1993).

After puffing the grains, the product is screened to remove the unpuffed and broken puffed kernels, bran, and dust. Then, it is dried to keep the moisture content at less than 3% (w/w). To maintain this level of product moisture and their crispiness, special packaging is also required.

Many puffed cereals, particularly the fun-puffed extruded products, are nutritionally fortified as well as sugarcoated before packaging. The sugar coating protects the product from moisture and hence gives longer shelf life for certain products. Because the grain is very porous, it takes moisture very rapidly and easily, so that packaging materials with good moisture barrier qualities are needed.

9.4   SHREDDED BISCUITS

Shredded biscuits are one of the oldest RTE cereals. The primary form used is whole kernel wheat. For example, white wheat produces shredded wheat biscuits that are light in color with a golden brown top crust and bottom when properly baked (Fast 1993). However, whole kernels of red wheat, rice, corn, and other grains, as well as parts of kernels and flours can be used alone or in mixtures to extrude shredded cereals.

The traditional process for shredding biscuits begins by cooking with steam injected directly into the water inside the cooker. After the cooking cycle, water is drained from the cooker and the grains are dumped and conveyed to cooling units. The objective of this step is to surface-dry the grain and cool it down to room temperature to stop the cooking process. After cooling, the grains are placed in large holding bins for up to 24 h before shredding to equilibrate the moisture in the kernels. With this step, the kernels become more firm, probably because of retrogradation of the starch (Jankowski and Rha 1986). This property is crucial for obtaining shreds of good strength for cutting and for handling of the unbaked biscuits.

Precooking, which takes place before shredding by extrusion, allows better control over the size of individual pieces, as determined by the cut at the die face of the extruder. Furthermore, the moisture content of the cooked dough pieces for shredding is 25% to 32% (w/w), which is much lower than that of grain moisture atmospheric pressure cooked (45%–50%).

In the shredding operation, the kernels are squeezed between two rolls with surface temperatures in the range of 35°C to 45°C. Nowadays, many different variations in rolls and grooving, related with the shape and dimensions of the grooves, are used in the cereal industry. After this step, the layered strands of dough are separated into biscuits by passing them under blunt knives to form individual biscuits. The individual biscuits are then baked in a band, continuous conveyor belt oven, or fluid bed reaching final moisture of approximately 3% (w/w).

To avoid rancid odors from developing during storage and to ensure a reasonable shelf life, an antioxidant treatment is usually required before packaging. The product is normally packaged in breather-type boxes with no inner or outer gas barriers. Another system that is quite effective in increasing the shelf life of cereals is the use of packaging materials that contain antioxidants such as butylated hydroxyanilose and butylated hydroxytoluene (Hoseney 1994).

9.5   GRANOLA CEREALS

The names *Granula* and *Granola* were trademarked terms in the late nineteenth century in the United States for foods consisting of whole-grain products crumbled and then baked until crisp. Granula was composed of whole flour and was similar to Grape-Nuts.

The major raw material used to make a granola cereal is rolled oats, either regular whole-rolled or quick-cooking oats. Nut pieces, coconut, brown sugar, honey, malt extract, dried fruits, dried milk, water, and vegetable oil are all mixed with the oats. Spices such as cinnamon and nutmeg can be also added (Fast 1993).

Oats are blended with other dry materials, and then the water, oil, and other liquid flavorings are mixed together in the proper amounts. The wetted mass is then spread in a uniform layer on the band of a continuous dryer or oven. The next step is baking, which takes place at a temperature range of 149°C to 218°C until the mat is uniformly toasted to a light brown and moisture reduced to approximately 3%. Finally, the mat is broken up into chunky pieces. Most of the granolas on the American market are sold as "natural" and, because of this, do not contain antioxidants and neither artificial flavors nor colors (Fast 1993).

ACKNOWLEDGMENTS

The author thanks the reviewers of the present chapter: Cristina M. C. Laranjeira, MD, AdI expert (Department of Food Technology, Biotechnology and Nutrition, ESAS, Polytechnic Institute of Santarém); Elsa C. D. Ramalhosa, PhD (Department of Production and Technology of Vegetable Products, ESA, Polytechnic Institute of Bragança); Goreti Botelho, PhD (CERNAS/Department of Food Science and Technology, Polytechnic Institute of Coimbra); and Maria Gabriela O. L. B. Lima, MD (Department of Food Technology, Biotechnology and Nutrition, ESAS, Polytechnic Institute of Santarém).

REFERENCES

Battacharya, S. and Prakash, M. 1994. Extrusion of blends of rice and chick pea flours: a response surface analysis. *Journal of Food Engineering*, 21: 315–330.
Chandrasekhar, P.R. and Chattopadhyay, P.K. 1990. Studies on microstructural changes of parboiled and puffed rice. *Journal of Food Processing and Preservation*, 14: 27–37.
Chen, C.H. and Yeh, A.I. 2001. Effect of amylose content on expansion of extruded rice pellet. *Cereal Chemistry*, 78: 261–266.
Culbertson, J.D. 2004. Grain, cereal: ready-to-eat breakfast cereals. In: *Food Processing: Principles and Applications*, eds. Smith, J.S and Hui, Y.H. Blackwell Publishing, Oxford, UK.
Fast, R.B. 1993. Manufacturing technology of ready-to-eat cereals. In: *Breakfast Cereals and How They Are Made*, eds. Fast, R.B. and Caldwell, E.F. American Association of Cereals Chemists, Inc, St. Paul, Minnesota, USA.
Fast, R.B. 2001. Cereal production methods. In: *Cereal Processing Technology*, ed. Owens, G. Woodhead Publishing Ltd., Cambridge, UK.
Fellows, P. 2000. *Food Processing Technology—Principles and Practice*, 2nd ed., Woodhead Publishing Limited, USA.
Guy, R.C.E. 1995. Breakfast cereals and snackfoods. In: *Physico-Chemical Aspects of Food Processing*, ed. Beckett, S.T. Chapman & Hall, London, England.
Guy, R.C.E. 2001. Raw materials for extrusion cooking. In: *Extrusion Cooking: Technologies and Applications*, ed. Guy, R. Woodhead Publishing Limited, Cornwall, England.

Heidenreich, S., Jaros, D., Rohm, R. and Ziems, A. 2004. Relationship between water activity and crispness of extruded rice crisps. *Journal of Texture Studies*, 35: 621–633.

Hoseney, C.R. 1994. *Principles of Cereal Science and Technology*, 2nd ed., American Association of Cereal Chemists, St. Paul, Minnesota, USA.

Jackson, S.M., Dudrick, S.J. and Sumpio, B.E. 2004. John Harvey Kellogg; surgeon, inventor, nutritionist (1852–1943). *Journal of the American College of Surgeons*, 199: 817–821.

Jankowski, T. and Rha, C.K. 1986. Retrogradation of starch in cooked wheat. *Strarch/Staerke*, 38: 1–6.

Jin, Z., Hsieh, F. and Huff, H.E. 1994. Extrusion of corn meal with soy fibre, salt and sugar. *Cereal Chemistry*, 71: 227–234.

Jones, D., Chinnaswamy, R., Tan, Y. and Hanna, M. 2000. Physiochemical properties of ready-to-eat breakfast cereals. *Cereal Foods World*, 45: 164–168.

Kent, N.L. and Evers, A.D. 1994. Breakfast cereals and other products of extrusion cooking. In: *Technology of Cereals: an Introduction for Students of Food Science and Agriculture*, eds. Kent, N.L. and Evers, A.D. Pergamon Press, New York, USA.

Mariotti, M., Alamprese, C., Pagani, M.A. and Lucisano, M. 2006. Effect of puffing on ultra-structure and physical characteristics of cereal grains and flours. *Journal of Cereal Science*, 43: 47–56.

Matz, S.A. 1991. *The Chemistry and Technology of Cereals as Food and Feed*, 2nd ed., Van Nostrand Reinhold/AVI, New York.

Medina, W.T., Andrés, A.L., Condori, J.L. and Aguilera, J.M. 2011. Physical properties and microstructural changes during soaking of individual corn and quinoa breakfast flakes. *Journal of Food Science*, 76: E254–E265.

Midden, T.M. 1989. Twin screw extrusion of corn flakes. *Cereal Foods World*, 34: 941–943.

Moraru, C.I. and Kokini, J.L. 2003. Nucleation and expansion during extrusion and micro-wave heating of cereal foods. *Comprehensive Reviews in Food Science and Food Safety*, 2: 147–165.

Owusu-Ansah, J., van de Voort, F.R. and Stanley, D.W. 1984. Textural and microstructural changes in maize starch as a function of extrusion variables. *Canadian Institute of Food Science and Technology Journal*, 17: 65–70.

Paradiso, V.M., Summo, C., Trani, A. and Caponio, F. 2008. An effort to improve the shelf life of breakfast cereals using natural mixed tocopherols. *Journal of Cereal Science*, 47: 322–330.

Peleg, M. 1997. Review: mechanical properties of dry cellular solid foods. *Food Science and Technology International*, 3: 227–240.

Sevatson, E. and Huber, G.R. 2000. Extruders in the food industry. In: *Extruders in Food Applications*, ed. Riaz, M.N. CRC Taylor & Francis, Boca Raton FL, USA.

Sumithra, B. and Bhattacharya, S. 2008. Toasting of corn flakes: product characteristics as a function of processing conditions. *Journal of Food Engineering*, 88: 419–428.

Valentas, K.J., Levine, L. and Clark, J.P. 1991. *Food Processing Operations and Scale-Up*. Marcel Dekker, New York.

Whalen, P.J., DesRochers, J.L. and Walker, C.E. 2000. Ready-to-eat breakfast cereals. In: *Handbook of Cereal Science and Technology*, eds. Kulp, K. and Ponte, J.G. Marcel Dekker, New York.

10 Pasta

Eleonora Carini, Elena Curti, Michele Minucciani,
Franco Antoniazzi, and Elena Vittadini

CONTENTS

10.1 INTRODUCTION

Pasta is one of the most prevalent Italian foods in the world and its consumption is globally widespread. Pasta owes its vast popularity to its versatility, simplicity of use, healthfulness, and low cost. According to the Italian Union of Pasta Industries (UNIPI 2009), the production of pasta in the world has increased from 7 to 12 million/year in the past decade due to the increase of pasta-based dishes present in fast foods and other restaurants. Italy is the largest pasta producer with 26% of the world production, followed by the United States (17%), Brazil (9%), Turkey (5%), and Russia (4%). The world production of pasta is summarized in Figure 10.1 (UNIPI 2009). Of the 3.2 million tons of pasta produced in Italy in 2009 (including dry and

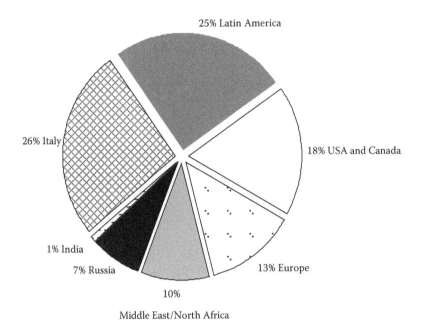

FIGURE 10.1   World production of pasta.

fresh pasta), 1.5 million tons was mainly exported to Germany (20%), France (16%), United Kingdom (15%), United States (7%), and Japan (5%).

Regarding pasta consumption (kilograms per capita per year), Italians are the largest consumers (~26 kg per capita/year), followed by Venezuela, Tunisia, and Greece (10–13 kg); United States, Canada, Brazil, and Argentina (7–9 kg); Russia and Turkey (6 kg); and Austria, Belgium, Panama, Finland, Colombia, Poland, Romania, Mexico, Ecuador, United Kingdom, Guatemala, Denmark, Libya, Japan, Egypt, Ireland, El Salvador, Estonia, Spain, and Slovak Republic (<6 kg; UNIPI 2009).

10.2   RAW MATERIALS AND LEGISLATION

Pasta is a very simple food in terms of ingredients, that is, wheat semolina or flour and water. Semolina (mainly from durum wheat) and water are mixed and kneaded together to form a dough, which is further processed to obtain fresh pasta with a large variety of shapes, and may eventually be dried in dry pasta products. Pasta products are subjected to diverse regulations in different parts of the world. Italy is the country with the strictest and most specific pasta regulation. According to Italian law (DPR 2001), "semolina pasta and low grade semolina pasta are the products obtained by extrusion, lamination and drying of the dough, exclusively prepared with semolina and low grade semolina of durum wheat and water." The legal requirements for Italian dry pasta products are shown in Table 10.1.

TABLE 10.1
Legal Requirements for Italian Dry Pasta Products

Type and Name	Maximum Humidity (%)	Ash (% Dry Matter) Min	Ash (% Dry Matter) Max	Proteins, Nitrogen ×5.7 (% Dry Matter)	Acidity
Durum wheat semolina pasta	12.50	—	0.90	10.50	4
Durum wheat low grade semolina pasta	12.50	0.90	1.35	11.50	5
Durum whole wheat semolina pasta	12.50	1.40	1.80	11.50	6

Special pasta products can be produced by adding ingredients (spinach, carrots, tomatoes, etc.), different from common wheat (*Triticum aestivum*) flour, as long as they are specified in the ingredients' list. Egg pasta products must be formulated only with semolina and at least four hen eggs (DL 1993) for a total weight of 200 g for each kilogram of semolina. Egg pasta must have a maximum humidity of 12.50%, a maximum ash content of 1.10% (dry matter), at least 12.50% of proteins (dry matter), and a maximum acidity of 5 degrees (a degree of acidity is equal to the number of cubic centimeters of normal alkaline solution required to neutralize 100 g of dry matter). The addition of eggs increases the lipid content in pasta, which is quantified in terms of ether extract (>2.8 g/100 g dry matter) and sterols (>0.145 g/100 g dry matter).

Fresh and stabilized pasta products must fulfill the legal requirements for dry pasta products (as reported in Table 10.1) except for humidity (at least 24% and 20% for fresh pasta products and stabilized products, respectively) and acidity (≤7 degrees). The addition of common wheat flour is also permitted. To be sold unpackaged, fresh pasta must be stored at temperatures equal to 4°C or lower. Prepackaged fresh pasta must have a water activity (a_w) in the range of 0.92 to 0.97, be subjected to a pasteurization treatment, and be stored at 4°C or below. Stabilized pasta must have a water activity (a_w) lower than 0.92 and must be subjected to a pasteurization treatment sufficient for allowing storage at room temperature.

Pasta production legislations in France, Portugal, and Greece are similar to that of Italy, as only durum wheat can be used to produce pasta. Other European countries (Austria, Belgium, Germany, United Kingdom, Netherlands, and Spain) and Russia allow for the use of common wheat in pasta manufacturing. The added wheat species are to be specified in the labels of pasta products according to European legislation (OJEC 1222/94).

In the United States, both durum semolina and common wheat flour (or any combination of the two) as well as other ingredients (i.e., eggs, vegetables, spices) are allowed in pasta production (FDA 2011). In Canada, pasta products are produced with durum wheat, whereas in Latin America (Venezuela and Mexico), the majority of pasta products are produced with durum wheat; on the contrary, in Brazil, only common wheat flour is used (Olivo 2005).

10.3 PASTA MANUFACTURING PROCESS

Pasta manufacturing is a relatively simple process consisting of the transformation of semolina and water into a coherent matter through the basic operations of dosing, mixing/kneading, and shaping (Dalbon et al. 1996; Feillet and Dexter 1996; Marchylo and Dexter 2001; Marchylo et al. 2004; Matsuo 1993; Milatovic and Mondelli 1991). Pasta manufacturing can be considered a "mature technology" given not only its pervasive use but also the very limited innovation of this process in the last 50 years. Many different types of pasta (Figure 10.2) can be obtained from these very simple operations. The large variety of pasta products are due to the different shapes obtained from different die cuts as well as the treatments following the shaping drying or pasteurization/sterilization steps.

10.3.1 DOSING

Raw materials used in pasta manufacturing are either solids with a granular consistency (i.e., semolina, flour, or powder additives) or liquids with different viscosities (i.e., water or liquid eggs). Solids are dosed either volumetrically or gravimetrically. Volumetric dosers are more commonly used in the continuous pasta process because of their lower cost as compared with gravimetric ones. Volumetric dosing works by controlling the movement of a constant volume that in turn determines the mass flow. Typically, there are two types of dosing: the star (Figure 10.3a) or screw feeders (Figure 10.3b). The star feeders consist of a capsular device that rotates at a specific rate depending on the desired flow, whereas in screw feeders, the flow is determined by a screw rotating into a steel cylinder. Gravimetric dosers are composed of a feeder (screw, Figure 10.3c; or belt, Figure 10.3d) mounted on a continuous weighing

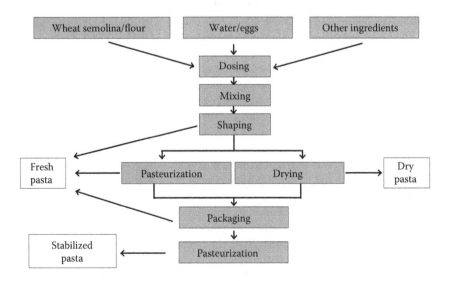

FIGURE 10.2 Pasta classification.

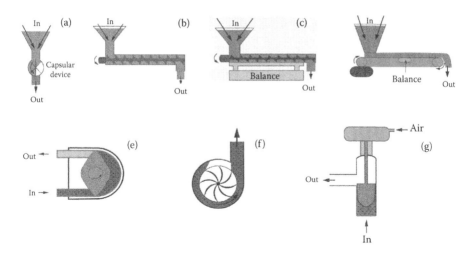

FIGURE 10.3 Typical dosing systems. Solids: volumetric (a, star feeder; b, screw feeder) and gravimetric (c, screw feeder; d, belt feeder). Liquids: peristaltic (e) and centrifugal (f) pumps, and modulating valve (g).

system, which is operated by an automatic controller that governs the speed of discharge and defines the mass transfer per time unit.

Liquid dosing is generally carried out by peristaltic pumps (Figure 10.3e) when dealing with high-viscosity fluids (i.e., liquid eggs), whereas centrifugal pumps (Figure 10.3f) are preferred with low-viscosity fluids (i.e., water). Modulating valves (Figure 10.3g) are an alternative for low-viscosity fluids (mainly water), as they are very reliable, efficient, and efficacious. Dosing liquid eggs with peristaltic pumps is preferred to a centrifugal pump to avoid egg foaming, which would occur in a centrifugal pump due to the high-speed rotation of the impeller. Foamed liquid eggs would not enable the correct flow measurement.

Dosing systems in mixers operating under vacuum conditions require the use of stellar valves for solid transfer and modular valves or peristaltic pumps for liquid transfer.

10.3.2 MIXING AND KNEADING

Mixing is a unit operation in which two or more components are dispersed within each other and a uniform mixture is obtained (Fellows 2001). The main purpose of the mixing stage in pasta making is to distribute water among semolina particles as evenly as possible (Dalbon et al. 1996).

A mixer is usually composed of one or two steel basins with one or two rotating shafts (Figure 10.4a). Each shaft is supplied with a sufficient number of paddles properly oriented to generate the mixing action and to continuously promote material advancement at a specific rate through the mixer toward the extrusion channel. The type of mixer, speed, and duration of mixing are variables that determine the

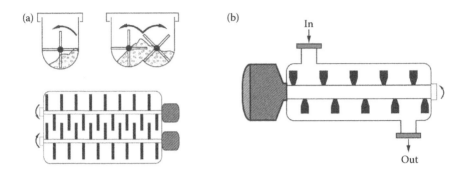

FIGURE 10.4 Mixing (a) and premixing (b) apparatus.

ingredients' dispersion and their interactions, as well as the processing efficiency and final product quality.

To obtain proper semolina hydration, a premixing stage may be used before mixing. Premixers consist of a centrifugal apparatus mounted on the top of the main mixer (Figure 10.4b). A large number of premixer designs are available and are all equipped with a high-speed rotational shaft (~900 rpm) that favors surface hydration of individual semolina grains in a very short time (a few seconds).

At the end of the mixing process, the semolina–water mixture has a granular structure characterized by particles with a diameter of approximately 0.1 to 0.5 cm, depending on the semolina/water ratio and the duration and intensity of the mixing process. In some cases, a proper and coherent dough is developed at this early stage of the process, but a kneading stage is often required to transform the hydrated semolina/flour particles into a dough. The premixing and kneading steps can be combined in a single operating unit (e.g., Bakmix, Storci, Parma Italy; Polymatic, Buhler, Uzwil, Switzerland). A "stabilization belt" can be used to improve the hydration of the semolina/water mix; after the premixing step, the granular material can be placed on a moving belt, allowing for a better hydration of the semolina particles before the shaping step.

Most modern pasta mixers operate in a vacuum (50–70 cm Hg) to remove oxygen and, therefore, reduce oxidation of yellow pigments and prevent the formation of air bubbles that might reduce pasta's brightness, translucence, and mechanical strength (Dexter 2004).

10.3.3 Shaping

Once the ingredients have been properly mixed, they undergo a shaping process that is carried out either by extrusion or lamination. In the first stages of the shaping process, the hydrated semolina/flour particles are kneaded into a dough that is then formed in the desired shape. An additional shaping step may be used in the production of filled pasta.

The extrusion process is preferably used for dry short (macaroni) or long pasta production, whereas the lamination is more suited for fresh, sheeted, and filled pasta products.

Extrusion is performed by single-screw extruders, which consist of a cylindrical stainless screw that rotates in a grooved cylindrical barrel ending with a head (usually conic) that includes and supports the plate die (Figure 10.5a). The pitch and the diameter of the screw, the number of flights, and the clearance between the flights and the barrel can be adjusted to optimize the extruder's performance (Fellows 2001). The shape of the pasta is defined by the geometry of the die through which the dough is forced. The main control parameters in the extruder are the screw speed and the barrel temperature. Screw rotational speed (generally 20–40 rpm) determines the extruder mass flow rate, which is, together with the open surface of the die, responsible for the extrusion speed. Extrusion barrels are equipped with water cooling jackets to dissipate heat and to maintain temperatures at less than 50°C to minimize pasta damage. The pressure of dough at the die plate (normally ~100 atm) is a consequence of the action and interaction of numerous factors, including die shape, screw speed, temperature, and hydration of semolina (Abecassis et al. 1994).

Plate dies are carefully designed to distribute pressure and product uniformly through the entire plate surface to keep a constant pasta extrusion speed. Long rectangular plate dies are generally used for long pasta, whereas round plate dies are

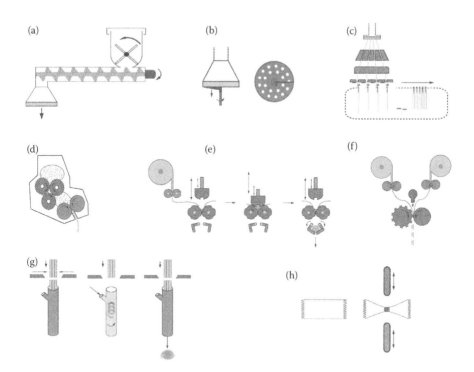

FIGURE 10.5 Shaping. Single screw extruder (a), short pasta equipment with a round plate die and rotary cutter (b), long pasta equipment with rectangular die and spreading machine (c), sheeter machine (d), details of single sheet filled pasta machinery (e), details of double sheet filled pasta machinery (f), details of nesting machinery (g), and details of bow tie pasta Bologna shaping system (h).

employed for short pasta and pasta sheets. Dies are either made fully in bronze or covered with Teflon; the material of choice has an effect on the pasta's appearance. The smoother Teflon surface produces a smoother pasta surface, whereas the coarser bronze leads to the formation of a rougher pasta surface. Bronze die–extruded pasta is believed to provide a better grip for the sauce, and it is perceived by consumers, at least in Italy, as being a higher quality product. From an industrial perspective, the use of Teflon dies is preferred because of their lower cost and lower wear during processing. At the exit of the plate die, the shaped pasta must be cut to the desired length. In short pasta production, round plate dies are followed by rotary cutters (Figure 10.5b), whereas in long pasta production, rectangular dies are generally associated with a spreading machine (Figure 10.5c). Long pasta is distributed onto sticks fixed on a chain by the spreading machine, cut to the desired length, and moved to the predrying chamber.

Lamination is performed by a sheeter (Figure 10.5d), which encompasses two or three corotating grooved rollers (called "gramole") that perform a kneading action, and a couple of smooth rolls, which press the dough into compact pasta sheets that are reduced to the desired thickness with multiple smooth calibrating rolls.

Both shaping processes (extrusion and lamination) can be performed at atmospheric pressure or under vacuum, which helps prevent oxidation that could negatively affect pasta quality (color in particular).

Other pasta shapes, for example, filled, nested, and "Bologna" bow ties pasta, can be obtained from extruded or laminated pasta sheets. Pasta products can be filled with mixtures of meat, cheese, or vegetables with bread crumbs and are produced from single or double pasta sheets. When a single sheet is used, it is reduced to the desired thickness by calibrating rolls, folded around the filling, and then pinched into the final shape (Figure 10.5e). In the case of double sheets filled pasta, two rolls of pasta, placed one in front of the other, feed two calibrating groups. The calibrated pasta sheets are then forced through molding and cutting rolls that define the product's shape (Figure 10.5f). The filling step for both types of filled pasta is carried out by a dosing system. In nested pasta production, a pasta sheet is guided through calibrating rolls to obtain the proper thickness, and it is then longitudinally cut into strands by rotative blades, conveyed into vertical tubes and cut to the desired length. The pasta strands are then hit by a tangential air spray and molded into nests, which are then conveyed on trays to the following processing step (Figure 10.5g). In "Bologna" pasta, pasta sheets are calibrated to the appropriate thickness and are then molded to the desired shape. For the production of bow ties, which are the most common Bologna shape, pasta is pinched in the center after the molding step (Figure 10.5h).

10.3.4 PASTEURIZATION

According to Italian law (DPR 2001), fresh pasta can be sold either loose or packaged. Packaged fresh pasta must undergo a pasteurization thermal treatment. Pasteurization can be a mild heat treatment applied to foods with the purpose of destroying selected vegetative microbial species (i.e., pathogens) and inactivating degradative enzymes. Because pasteurization does not eliminate all the vegetative microbial populations and almost none of the spore formers, pasteurized food must be held and stored at refrigerated conditions with, eventually, the addition of chemical additives or in modified atmosphere packaging (MAP) to minimize microbial growth (Francis 1999). The European Chilled Food Federation recommends a heat treatment at 90°C for 10 min ($z = 10$) for refrigerated products with pH >4 and $a_w > 0.85$ (i.e., pasta and filled pasta) to obtain a 6 log reduction of the most significant hazard contamination by psychotropic *Clostridium botulinum* type B (European Chilled Food Federation 1996). However, this heat treatment is not applied to pasta as it leads to a significant structural damage. Simultaneous control of multiple parameters allows for a better control of microbial growth (i.e., hurdle technology). In particular, the simultaneous adjustment of pH (<5), a_w (<0.97) and the use of MAP is preferred for these pasta products. The optimal time/temperature ratio to be applied for efficient and efficacious pasta pasteurization depends on formulation, physical–chemical properties (water activity, moisture content pH), initial bacterial load, size and shape of pasta, load of product (kg/cm²), and desired shelf life. In industrial practice, appropriate thermal treatments applied to fresh filled pasta (F_{10}^{70} 50 − 200) are sufficient to destroy the vegetative cells of pathogenic bacteria (Giavedoni et al. 1993; Lopez et al. 1998; Zardetto et al. 1999).

Different methods are available for pasta pasteurization. Belt pasteurizers (Figure 10.6a) are used for loose products and use nascent or injected steam (culinary steam). Nascent steam is generated from a tank containing boiling water located under the belt; water can be heated in different ways (electrical resistance system, heat exchangers, or direct flame). Injected steam is produced in a boiler located outside the plant and injected directly on the product in the pasteurizer or in the chamber (steam chamber). Superheated steam, produced with an electrical heat exchanger (to reach 120°C–130°C), is preferably used to avoid condensation on the pasta's surface, which could result in starch gelatinization, adhesion of pasta pieces, and a shiny pasta surface. After the pasteurization treatment (95°C–98°C for 2 min), pasta is dried in hot air tunnels (75°C–80°C for 1 min) to reduce humidity and a_w, as well as to prevent adhesion among pasta pieces. The product is then cooled, as fast as possible, to refrigeration temperatures (~4°C at the center of the product) to avoid bacterial growth before the packaging step. The air fluxed in the tunnels must be filtered, and the packaging operation is preferably carried out in white rooms. Pasta pasteurization can be also performed on packaged products in hot air static cells (100°C–120°C for 5 min).

(a)   (b)

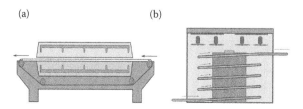

FIGURE 10.6 Pasteurization systems. Belt pasteurizer (a) and spiral ovens/static cell (b).

There are some pasta products ("stabilized pasta") that, after the first pasteurization and packaging steps, undergo a second pasteurization treatment with hot air (90°C–100°C for 10 min) in spiral ovens/static cells (Figure 10.6b) to reach a thermal treatment at the product core of 80°C for about 10 min (Batisti et al. 1995). These products are considered shelf-stable and can be stored at room temperature for 1 year.

10.3.5 DRYING

The objective of drying is to reduce the moisture content of pasta from approximately 30% to approximately 12% to 14%. Dehydration, or drying, is defined as the application of heat under controlled conditions to remove, by evaporation, the majority of the water present in a food. The pasta drying process involves four sequential steps: (1) air heating by electrical or hot water heat exchanger, (2) heating of pasta by ventilation through the heat exchanger, (3) pasta absorbing thermal energy (heat), and (4) water evaporation from the pasta core to its surface. The main technological parameters that define the drying conditions (temperature, humidity, and air velocity) depend on product shape, dimension, humidity, and load. Table 10.2 shows the most commonly used (and most frequently reported by the scientific community) pasteurization temperatures and time parameters. Low-temperature and long-time drying processes are preferred in artisanal production, whereas high-temperature and short-time drying conditions are suitable for larger industrial productions. Before being dried to the final humidity, pasta products undergo a fast (1–2 min) preliminary

TABLE 10.2
Temperature/Time Parameters in Pasta Drying Processing

Drying Technology	Drying Temperature (°C)	Drying Time (h)
Low temperature	40–60[a-d, g-l]	7.5–30[a-d, g-l]
High temperature	65–80[a, c-m]	5.5–12[a, c-m]
Very high temperature	80–110[a-c, f-m]	2.5–6.5[a-c, f-m]

[a] Malcolmson et al. 1993.
[b] Anese et al. 1999.
[c] Yue et al. 1999.
[d] Zweifel et al. 2000.
[e] Acquistucci 2000.
[f] Güler et al. 2002.
[g] Zweifel et al. 2003.
[h] Baiano et al. 2006.
[i] Lamacchia et al. 2007.
[j] Villeneuve and Gélinas 2007.
[k] Cubadda et al. 2007.
[l] Petitot et al. 2009.
[m] Fu et al. 2011.

drying (predrying) immediately after the shaping process to rapidly reduce pasta moisture from 30% to 17% to 18%; this leads to a decrease in the total drying time and the formation of a solid thin film on the surface of pasta that prevents a structural collapse and sticking phenomena. The equipment for short pasta predrying consists of a vibrating tray going through a chamber in which hot air (70°C–80°C) is fluxed from the bottom to the top. The vibration intensity allows proper pasta displacement and regulates the residence time. In long pasta production, the pasta proceeds in a tunnel in which hot air is fluxed.

In continuous pasta production, the drying step is carried out in a tunnel dryer: short pasta products are conveyed by several overlaid belts (Figure 10.7a) or a rotary drum (for smaller pasta shapes; Figure 10.7b), whereas long pasta is still hung on sticks and moved by the conveyer (Figure 10.7c).

During proper drying, the moisture content of pasta is reduced from approximately 17% to 18% to approximately 12.5% by alternate steps of hot air application and resting times to obtain a stable product. During the resting phases, diffusion of water molecules from the core to the surface of the pasta pieces occurs—minimizing internal structural tensions that could result in poor mechanical strength or cracking in dried pasta.

At the end of the drying step, pasta is "stabilized" by exposure to specific air temperature and humidity conditions to homogenize water distribution across the product, before being cooled and then packaged. After the cooling step, long pasta is automatically removed from the sticks and cut to the desired length. Production scraps are recovered and redirected to the mixing step.

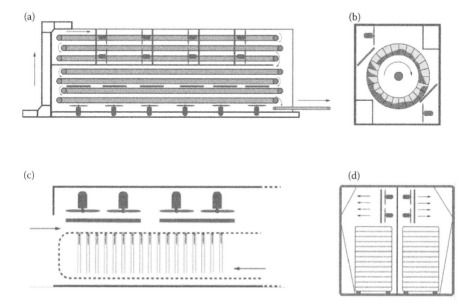

FIGURE 10.7 Drying systems. Overlaid belts for short pasta drying (a), rotary drum for smaller short pasta shapes (b), long pasta drying tunnel (c), and static drying cell (d).

In batch pasta production, pasta drying is carried out in large static and insulated cells (Figure 10.7d) and fluxed with hot air in discontinuous mode. After the predrying process, pasta is stacked on trays that are manually or automatically loaded into trolleys (~50 trays for each trolley), introduced into the drying cell (4 or 8 trolleys for each cell), and dried to the final desired moisture content.

10.3.6　Packaging

Packaging prevents pasta from contamination, protects it from damage during shipment and storage, and has an attractive function for the consumer (Walsh and Gilles 1977).

Dry pasta is generally packed in cardboard boxes or plastic wrappers, whereas prepacked fresh pasta is wrapped in multilayer plastic flow pack or boxes.

Cardboard boxes offer the best protection during handling and are generally made of recycled paperboard or solid unbleached/bleached sulfate cardboards. Plastic wrappers are commonly multilayer films of polyethylene terephthalate, ethylene vinyl alcohol, polyethylene, or polypropylene, and offer better protection from insects and moisture than cardboard. Plastic materials like polypropylene and polyethylene terephthalate (usually equipped with an oxygen barrier layer such as ethylene vinyl alcohol, polyvinylidene chloride, or polyamide) are manufactured into flexible pouches and semirigid containers and are used for heat-processed pasta products.

Fresh pasta, subjected to only one heat treatment, is packed immediately after pasteurization and cooling. The packaging must be kept at refrigerated temperatures ($T \leq 4°C$; DPR 2001), and the addition of chemical preservatives and bacteriostatic compounds is allowed to extend pasta shelf-life. Because the use of preservatives is prohibited in the European Union, MAP is therefore used. Different studies reported the implication of MAP to preserve the quality of fresh pasta (Del Nobile et al. 2009; Zardetto 2005). In particular, low O_2 and high CO_2 concentrations limit the development of microorganisms and reduce the growth and toxin production of different molds that are strictly aerobic and sensitive to high CO_2 concentrations (Zardetto 2005). Before hermetically sealing the pack with a plastic film by thermal welding, a vacuum is created to extract the air and, immediately thereafter, a mixture of CO_2 and N_2 is injected (50:50–70:30). Nitrogen is an inert gas used as a packaging filler, whereas carbon dioxide is the major antimicrobial factor of MAP (Phillips 1996). The success of MAP is closely related to the permeability of the film used for packaging (Costa et al. 2010), which must maintain the initial condition of the modified headspace for the entire storage period.

10.4　PASTA QUALITY

A food, such as pasta, is an assembly of a variety of "building blocks" that are held together by multiple types of interaction resulting in a large variety of microstructures that will define product quality and stability. Identification and characterization of the most important elements and forces present in a food system are necessary to comprehend its physicochemical, sensory, and nutritional properties. A thorough understanding of foods is sometimes extremely difficult because they are complex systems both from compositional and structural standpoints and are also subjected to very different environmental conditions. However, the relationship between the structure and the final properties is a key factor in food science, food engineering, and plant design and must be considered to accelerate the prototyping stages, to decrease production times, and to reduce costs, so as to deliver a product with the desired properties, functionality, and stability.

Considering the very few raw materials (semolina and water used for basic pasta formulation) and the relatively simple manufacturing process (Dexter 2004), pasta may be thought of as a "simple food." However, the interactions between water and semolina constituents (e.g., starch and gluten proteins) that occur at different time–space scales and the changes they undergo during the entire production process turn pasta into a "complex food." Therefore, the quality of the final product is affected by both raw materials' properties and processing parameters.

But, what is "pasta quality?" Is there a "universal" pasta quality definition? Are there specific requirements for the quality of pasta? No objective and universally recognized parameters are present in the scientific literature to define pasta quality. The consumer judges pasta quality in a very subjective way (Cole 1991; D'Egidio and Nardi 1996; Pagani et al. 2007), and their perception is based on a sensorial analysis of cooked pasta, being largely influenced by the cultural background and the personal experience of the individual. For the Italian consumer, "high-quality pasta" is characterized by a yellow appearance (often provided by the addition of eggs in the formulation), should leave a clear cooking water, and offers "*al dente*" resistance to biting. However, in other countries (such as Asian countries and Argentina), pasta products are expected to be softer, with a more easily breakable texture and "whiter." Moreover, pasta consumption covers a wide variety of products and applications: pasta may be cooked under different conditions [e.g., boiled or baked with tomato sauce (i.e., lasagna)], which actually require different pasta properties. Furthermore, cooked pasta may be consumed in various ways (e.g., drained or in a broth, hot or cold, with or without sauce addition), and the texture of the pasta meal is expected to be different.

On the basis of the consumer's final judgment and the scientific literature, the appearance of pasta and its behavior during cooking are the most important properties that define pasta quality.

10.4.1 Pasta Appearance

The appearance of pasta embraces different characteristics that influence the consumers' perception. Pasta with a good appearance is characterized by a uniform color (which implies the absence of streaks and white spots originating either from uneven semolina hydration, mixing, and shaping conditions or by the presence of tegument residues in the semolina), lack of cracks/fissures, and a homogenous surface texture (due to proper drying or storage conditions).

A yellow color is generally accepted as the most appropriate color, being a desirable characteristic in pasta products; the presence of red and brown components greatly reduces consumer's acceptability of pasta (Feillet and Dexter 1996). The final color of pasta is affected by the properties of the raw materials (semolina and other ingredients, if present) and processing phases, especially some critical "thermal" steps.

10.4.1.1   Effect of Raw Materials

The color of pasta is strictly related to the properties of semolina and, in particular, to its carotenoids [mainly xanthophylls (lutein) and carotene], proteins, and ash content as well as the lipoxygenase activity. The carotenoid content is related to the wheat cultivar and genotype (Borrelli et al. 1999; Clarke 2006), and it is affected by storage, milling, and the oxidating action of lipoxygenase during pasta processing (Borrelli et al. 1999; Carrera et al. 2007; Irvine and Winkler 1950; Irvine and Anderson 1953; McDonald 1979).

A correct milling operation on durum wheat kernels must be carried out to preserve the carotenoid pigments. High extraction rates in the production of semolina result in an increase in the content of lipoxygenases, which are located in the germ, and peroxidases and polyphenol oxidases, which are found in the bran layers (Fraignier et al. 2000; Hatcher and Kruger 1993; Matsuo and Dexter 1980). The pigment content is, therefore, likely to decrease during the subsequent pasta processing operations.

The presence of brown coloration may be considered a negative attribute of pasta (Feillet and Dexter 1996); brown pigments have been attributed to the presence of a water-soluble protein complexated with copper (Matsuo and Irvine 1967) and to the action of oxidizing enzymes during kernel maturation (Kobrehel et al. 1974). A fast milling operation or contamination of external/teguments fractions (bran and aleurone layer) characterized by higher ash contents can also promote brownness in pasta (Cubadda 1988). The brown color component may mask the yellow color (Irvine and Anderson 1953).

10.4.1.2   Effect of Processing

The first phases of pasta processing that allow the transformation of the raw materials (semolina and water) in a dough are mixing, kneading, and shaping. These phases are not commonly considered to have a primary role in determining pasta color; however, the temperature (40°C–50°C) and pH (slightly acidic) conditions established in these steps are optimal for α-amylase activity. The α-amylase activity (determined by the falling number test; Hagberg 1960, 1961; Perten 1964) results in starch depolymerization and in the formation of reducing sugars, which may favor the formation of Maillard reaction–colored compounds during drying. The pressure exerted on the dough during extrusion may partially promote amylase inactivation. To decrease the amount of reducing sugars resulting from amylolytic activity, mixing, kneading, and extrusion times must be minimized (De Noni and Pagani 2010; Seiler 1995).

Very little information was found in the scientific literature on the effect of different mixers and shaping modes on pasta color. The available literature focuses only on fresh pasta. Carini et al. (2010a) measured the L, a^*, and b^* color coordinates in fresh pasta samples (semolina and water) produced with recently designed mixers [Premix (Storci 2005) and Bakmix (Fava and Storci 2006)], which were reported to induce a uniform hydration of the solids and to allow the formation of a dough in 1 to 2 s. Fresh pasta produced with these mixers was only slightly different in the overall color (ΔE parameter) from fresh pasta produced with a traditional mixer.

Different shaping modes were reported to induce altered color in fresh pasta (Carini et al. 2009) and fresh egg pasta (Zardetto and Dalla Rosa 2006). In particular, Carini et al. (2009) observed significantly lower color coordinates (L^*, a^*, and b^*) in extruded pasta samples with respect to laminated samples. Similarly, Zardetto and Dalla Rosa (2006) reported a decrease in L^* and b^* values but an increase in a^* values in fresh egg extruded samples as compared with the laminated ones. The observed differences might be related to the intrinsic divergences between the two processes (different stress conditions in terms of temperature and pressure) that may have induced an altered oxidation pattern of the colored molecules or a different fresh pasta matrix microstructure.

A recently designed machine that carries out the lamination process with the application of vacuum (Storci 2006) was found to further increase the yellow color of fresh pasta (Carini et al. 2009).

Drying is the most determinant processing phase for pasta color alteration due to the potential development of Maillard reaction products that are related to the formation of red–brownish components. In the last decades, the application of high-temperature drying cycles by pasta manufacturers has reduced the processing time but, among other things, it has promoted the formation of undesirable color components in pasta products (Reineccius 1990). The presence of reducing sugars and free amino groups, together with the effect of temperature during drying, favors the development of brown pigments due to Maillard reactions. Moderate temperature drying conditions (80°C) have been reported to increase pasta's yellowness, probably due to a partial inactivation of lipoxygenase, whereas higher temperatures/low moisture conditions (100°C) promote Maillard reaction and the formation of red–brown melanoidins (Zweifel et al. 2003). Moreover, during the application of high-temperature/low-moisture drying, pasta water activity reaches values around 0.80 to 0.90, which are optimal conditions for Maillard reaction to take place (Labuza 1981) and can lead to a higher concentration of colored products (Resmini et al. 1993). The lowest values of furosine, one of the many Maillard reaction products, were found in pasta produced with the mildest conditions of milling, kneading, shaping, and drying (Pagani et al. 1996). On the contrary, the highest values of furosine were detected in low-moisture samples (<15%), manufactured with a higher number of extraction milling cycles and dried at temperatures higher than 75°C (Pagani et al. 1996).

In the production of fresh pasta, in which no drying occurs, the pasteurization phase may have a role in altering pasta color. The wet heat treatment determines an alteration of protein–starch macromolecular interactions, which may induce a change in color (Cencic et al. 1995). It has been reported that pasteurization treatment (99°C ± 1°C for 3 min) on extruded and laminated fresh egg pasta samples can result in a modification of color, with a significant decrease in a^* and b^* values (Zardetto and Dalla Rosa 2006).

10.4.2 Cooking Properties

Cooked pasta properties are fundamental to defining pasta quality. Although the evaluation of pasta quality is very subjective among world consumers, there are some properties that have been selected to define quality in cooked pasta in the scientific literature and gastronomic communities. "Good quality" cooked pasta should be "firm" to the bite (*al dente*) and not sticky, and with minimal loss of solids in the cooking water (D'Egidio et al. 1983; Dalbon et al. 1985; Dexter and Matsuo 1979a,b; Dexter et al. 1981; Pagani et al. 1986; Wyland and D'Appolonia 1982).

Pasta is a low moisture system in which starch and proteins compete for water during cooking. Proteins need water to coagulate, whereas starch requires water to gelatinize. However, the key role determining the proper texture of pasta is the protein network (the continuous phase): a gluten network with proper strength and elasticity allows for the starch entrapment into the matrix and, consequently, for a nonsticky cooked pasta with good texture. In pasta with a weak gluten network, starch can gelatinize before proteins coagulate, thus releasing a larger amount of solids in the cooking water (D'Egidio et al. 1983).

The cooking properties of pasta are, therefore, dependent on both raw materials' characteristics and processing conditions (Scanlon et al. 2005) because they affect the gluten's matrix development.

10.4.2.1 Effect of Raw Materials

Wheat proteins and starch properties are known to affect pasta quality (Cubadda et al. 2007; D'Egidio et al. 1990; Delcour et al. 2000a,b; Dexter et al. 1983; Grant et al. 1993; Oak et al. 2006). Pasta cooking quality is generally related to both protein content and gluten strength of durum wheat (Ames et al. 2003; D'Egidio et al. 1990; Malcolmson et al. 1993; Rao et al. 2001). A higher protein content (higher vitreousness in the wheat kernel) has been associated with an improved pasta cooking quality in terms of textural parameters (firmness and stickiness; Dexter and Matsuo 1977). However, a higher protein content or strength does not always result in better cooking quality as it also depends on the drying process applied (Ames et al. 2003; D'Egidio et al. 1990).

The two gluten protein fractions, glutenins and gliadins, should be considered in the evaluation of pasta's cooking properties; a proper formation of the gluten network in terms of strength and elasticity is ascribable to protein/protein and subunit/subunit aggregations. In particular, low molecular weight gluten subunits (Carrillo et al. 1990; Edwards et al. 2003) and the number of sulfhydryl groups (Alary and Kobrehel 1987) are related to the quality of cooked pasta. Moreover, the content of low molecular weight glutenins (Feillet and Dexter 1996) seems to be related to the quality of cooked pasta.

The rheological performance of the gluten network is also affected by the other main component of semolina, that is, starch. Starch's contribution to pasta quality (dough properties) can be attributed to its interactions with the gluten network (Edwards et al. 2002). Starch distribution in the granule, granule size, and amylose/amylopectin ratio and milling damage have also been related to the appropriate development of the gluten network (Grant et al. 2004; Seiler 2000; Soh et al. 2006).

Polar lipids (monoglycerides) were reported to improve pasta quality in terms of higher cooking stability, being able to form water-insoluble complexes with amylose that reduce stickiness (Matsuo et al. 1986) by decreasing the released amount of amylose in the cooking water. With regard to fresh pasta products containing both hard wheat semolina and common wheat flour, the presence of flour (characterized by lower content and lower quality gluten) strongly affects gluten characteristics and, hence, the cooking quality of pasta.

Changes in the semolina/flour ratio can be adopted to improve dough strength. Eggs are used as an additional ingredient in some fresh pasta formulations and are functional to improve the mechanical and cooking behavior of the product (Alamprese et al. 2005, 2009). Egg properties depend on many factors, such as freshness, composition, and processing treatments that can affect the final product quality. A higher albumen/yolk ratio, dependent on the hen's age but not on the housing systems, was reported to improve pasta cooking quality in the product (Alamprese et al. 2011).

10.4.2.2   Effect of Processing

The effect of the mixing (Carini et al. 2010a) and shaping phases (Carini et al. 2009; Zardetto and Dalla Rosa 2006) on pasta cooking quality has recently been reported. Carini et al. (2010a) studied the effect of two different mixers (that allow for a very short mixing time and was characterized by different mixing and kneading actions) on fresh pasta cooking performance and found larger solid loss during cooking in these samples as compared with those produced with a traditional mixer.

In extrusion shaping, the gluten matrix developed during mixing and kneading undergoes polymerization while the surface of starch granules hydrates. The extrusion temperature is important because it affects the physical state of proteins and starch phases. In particular, if the temperature exceeds 50°C, abnormal protein coagulation and starch swelling may occur, which would damage the protein network, thus resulting (at worst) in pasta disruption during cooking. This phenomenon is prevalent if low quality raw materials are used in the production of pasta (Feillet and Dexter 1996); therefore, starches with high gelatinization temperatures are desirable.

Recent work by Lucisano and coworkers (2008) demonstrated that traditional bronze dies present at the extruder head resulted in pasta with rougher surface and higher porosity than the Teflon-covered dies that gave a more even surface and a smoother texture. These differences may also have an effect on the cooking properties of these pasta samples because the pasta structure is strongly linked to its behavior during cooking.

Pasta shaping by lamination in fresh (Carini et al. 2009) and fresh egg pasta (Zardetto and Dalla Rosa 2009) has been reported to lose a significant lower amount of solids during cooking, as compared with extrusion.

Drying is certainly the most important processing phase that may positively or negatively alter the cooking quality of pasta because the heat may affect the physical state of gluten and starch domains. The traditional low temperature (<60°C) drying process leads to a slight alteration (reorganization) of the gluten network, and no modifications are observed in the starch phase (De Noni and Pagani 2010). High-temperature drying of low moisture products (<15%) induces a nearly complete protein coagulation (inducing the formation of the gluten network) and some changes in the starch granule structure. These changes do not alter the protein–starch interactions during cooking and, therefore, these drying conditions can be considered favorable for pasta quality (De Noni and Pagani 2010). Finally, high drying temperature and high moisture conditions (not specified) may cause a breakage of protein fibrils and a dramatic decrease in pasta quality before cooking, especially when the gluten network is weak (De Noni and Pagani 2010).

In conclusion, pasta cooking quality is dependent on gluten content and strength if low-temperature drying is applied in pasta processing (Ames et al. 2003; D'Egidio et al. 1990). In high-temperature drying (temperature ≥ 60°C), the gluten network in pasta undergoes coagulation, and only the amount of protein dictates pasta cooking quality (D'Egidio et al. 1990).

10.5 PASTA QUALITY MEASUREMENT

10.5.1 COLOR

The color of pasta (uncooked) is usually measured by a reflectance colorimeter in the L^*, a^*, and b^* system or by the image analysis of acquired scans of the product. L^* values (brightness) range from 0 (black) to 100 (white), a^* values (redness) vary from negative (green) to positive (red), and b^* values (yellowness) range from negative (blue) to positive (yellow).

10.5.2 COOKING QUALITY

The evaluation of quality in cooked pasta is related to cooked weight, cooking loss, and textural properties (firmness, stickiness, and bulkiness).

Cooking weight is measured as the water-absorbing capacity of pasta products during cooking, whereas cooking loss is determined as the percentage of solids lost in the cooking water (evaluated as the solid residue in cooking water; AACC 1999).

Regarding textural properties, good quality cooked pasta should be firm to the bite (*al dente*) and not sticky. Selected instrumental methods are used to measure texture properties in cooked pasta. Texture properties are investigated with tests that are based on the application of compression, shear, cutting, or tension stresses, and the material response is measured. The firmness of pasta samples is usually determined as the maximum force recorded in a compression/cut test to fracture or to reach a certain deformation of the sample. Stickiness/adhesiveness is generally measured by touching/deforming the pasta sample in a controlled manner and then assessing the force needed to separate the probe from the sample. Elasticity is usually determined as the tensile force to break the sample (AACC 1999; D'Egidio et al. 1993; Martinez et al. 2007; Tang et al. 1999).

Sensory evaluation of cooked pasta can be also carried out by both consumers or trained judges and can provide information ranging from product acceptance and likeability to specific descriptive analysis of product characteristics. In particular, descriptive analysis investigates a wide range of parameters, such as yellow color, firmness (force required to cut through the noodle using front teeth), chewiness (time needed to masticate the noodle to a state ready to be swallowed), surface stickiness (the product amount adhering on teeth after mastication, or the jamming degree among the spaghetti strands in the plate), and elasticity (the extent to which one piece of spaghetti returns to its original length when stretched; D'Egidio et al. 1993; ISO 1985).

Some authors have proposed a "pasta value" to predict the quality of cooked pasta (Landi and Guarneri 1992). The pasta value tries to correlate some semolina and durum wheat properties (i.e., protein and gluten contents, alveographic semolina properties) and a constant factor (depending on the temperature of the drying conditions used for pasta production) to predict the quality of cooked pasta according to the following formula:

$$\text{Pasta value} = K + 2GLU + 0.04W + 8.5P/L - 2(P/L)^2$$

where GLU is the amount of dry gluten per 100 parts (or the amount of dry protein in dry matter, expressed as nitrogen ×5.70), W is the area of the alveograph curve (cm^2), P is the maximum pressure required for the deformation of the sample (measured in millimeters), L is the length of the alveograph curve (mm), and K is a factor that depends on the drying cycle (42 ± 2 for low-temperature drying and 47 ± 2 for high-temperature drying). The reference scale is from 0 to 100 points, and pasta values equivalent to 80 or higher indicate good-quality pasta.

10.6 NONCONVENTIONAL PASTA PRODUCTS

Diseases related to incorrect food habits, both in western populations (excessive food consumption) and in underdeveloped countries (low energy and macronutrient and micronutrient intakes), as well as the increased incidence of celiac disease, induced researchers and pasta manufacturers to study and produce a wide range of nonconventional pasta products, including nutritionally enriched and gluten-free pasta products.

A wide range of ingredients with well documented nutritional values have been added to standard pasta formulations, and their effect on the quality of pasta (mainly on appearance and cooking quality) and on processing has been investigated. In particular, resistant starch of different types (Bustos et al. 2011; Gelencsér et al. 2008; Sozer et al. 2007), buckwheat (Alamprese et al. 2007) and buckwheat bran flour (Yalla and Mantey 2006), inulin (Bustos et al. 2011; Manno et al. 2009), wheat bran (Sudha et al. 2011), soluble and insoluble nonstarch polysaccharides (Brennan and Todurica 2008), oat bran (Bustos et al. 2011), and carrot and kamut flour (Carini et al. 2010b) are some of the ingredients added to fresh or dry pasta formulation in an attempt to obtain high-fiber pasta. Moreover, protein-enriched pasta has been produced with the inclusion of pea protein concentrate (Mercier et al. 2011), split pea and fava bean flours (Petitot et al. 2010), soy flour (Carini et al. 2010b; Nasehi et al. 2009), and mustard protein isolate (Sadeghi and Bhagya 2008) in the pasta formulation. Generally, the addition of these ingredients has been associated with nonideal gluten development and thus alterations of cooking and textural properties, especially when a significant amount of these nontraditional ingredients was added. The effect of formulation modification on processing has been addressed by only a few researchers. Substitution of 35% of semolina with split pea and fava bean flours (Petitot et al. 2010) required a lower hydration level and a faster mixing speed (from 60 to 120 rpm) to limit the agglomeration of particles during mixing. The specific mechanical energy transferred to the semolina/buckwheat bran flour dough for spaghetti production was found to be lower than the energy transmitted to semolina dough (Manthey et al. 2004) because of the formation of a weaker buckwheat bran flour dough. Similarly, nontraditional ingredients (buckwheat bran flour, flaxseed flour, and wheat bran) used by Yalla and Manthey (2006) reduced specific mechanical energy, mechanical energy during extrusion, as well as the rate of extrusion in spaghetti production.

Drying may also play a crucial role in nonconventional pasta production. It has been reported that improved gluten strength may be obtained if pasta is dried at high or ultrahigh temperatures thanks to protein denaturation, which limits starch granules from rupturing during cooking. In particular, a lower cooking loss in split pea and faba beans flour (up to 35% dry basis) fortified pasta was obtained by applying high (70°C) and very high temperatures (90°C) during the drying cycle (Petitot et al. 2010). The sensory evaluation of the cooking quality in spelt pasta dried at high temperature (90°C) showed good results as compared with pasta dried at low temperature (60°C; Marconi et al. 2002).

Of particular interest are gluten-free pasta products that are intended for population suffering from celiac disease. Celiac disease is a lifelong intolerance to the gliadin fraction of wheat and the prolamins of rye (secalins), barley (hordeins), and possibly oats (avidins; Murray 1999). The only effective treatment for celiac disease is a strict adherence to a gluten-free diet during the patient's lifetime. Consumer demand for gluten-free cereal products is rising steadily with the increase in celiac disease and other allergic reactions to gluten from wheat-based, rye-based, or barley-based foods, and the consumer desire for good-tasting and good-textured products has opened great market opportunities for food manufacturers. The removal of gluten, the proteinaceous complex primarily responsible for pasta products' structure, is difficult when a good quality product is desired.

In gluten-free pasta, starch is responsible for the products' structure, and its gelatinization plays a key role in the manufacturing process. Starch sources can be rice, corn, tapioca, buckwheat, and quinoa flours, among others; modified starches and deglutinated wheat flour can also be used.

Proteins and hydrocolloids are also included in the formulation to improve pasta structure through their water binding ability, gel forming, and viscosity control properties. The most commonly used proteins are obtained from eggs, milk, and leguminous isolates (pea, lupine, and soy), whereas hydrocolloids include gums (guar, carob, and arabic), agar, alginates, cellulose, and methylcellulose, among others. Monoglycerides and diglycerides are used as additives as they allow for an easier processing of pasta.

The gluten-free pasta manufacturing process closely reflects the standard pasta manufacturing steps when pregelatinized raw materials are utilized. If, on the contrary, nonpregelatinized starch sources are used, starch must be gelatinized generally in a mixer basin, in an extruder, or in a steam chamber before further processing. A schematic representation of these technologies is given in Figure 10.8.

Raw materials and water are dosed and hydrated in a premixing step and then subjected to processing. If gelatinization is carried out in the mixing basins (Figure 10.8a), a steam injection is provided; the gelatinized matter is then cooled while it is

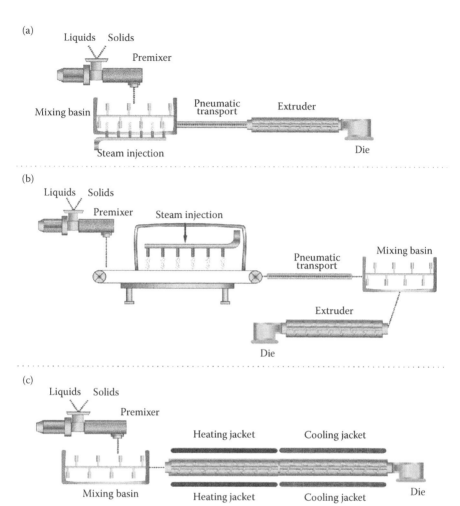

FIGURE 10.8 Gluten-free pasta manufacturing process technologies. Gelatinization technology in a mixer basin (a), in a steam chamber (b), and in an extruder (c).

pneumatically transported toward the extruder, and it is formed in the desired shape through a die.

Another technology (Figure 10.8b) operates using a steam chamber in which a belt moves (from a premixing phase to a cooling pneumatic transport) while steam is injected for gelatinization. The cooled mixture is discharged into a mixing basin and then extruded to the desired shape. Alternatively, after premixing and mixing in basins, the mixture can be gelatinized in an extruder equipped with a heating jacket in the first sector, inducing starch gelatinization, and then with a cooling jacket in the second sector (Figure 10.8c). A die at the end of the extruder defines the proper shape of the final product.

The fundamental parameters of the gelatinization process are time/temperature and the water added in the premixing step, which are strongly dependent on the type and botanic origin of the raw material used. Finally, gluten-free pasta is subjected to the drying process that is usually carried out at very high temperatures to further improve the final structure of pasta.

REFERENCES

Abecassis, J., Abbou, R., Chaurand, M., Morel, M.H., and Vernoux, P. 1994. Influence of extrusion conditions on extrusion speed, temperature, and pressure in the extruder and on pasta quality. *Cereal Chemistry* 71:247–253.

Acquistucci, R. 2000. Influence of Maillard reaction on protein modification and colour development in pasta. Comparison of different drying conditions. *Lebensmittel-Wissenschaft und Technologie* 33:48–52.

Alamprese, C., Iametti, S., Rossi, M., and Bergonzi, D. 2005. Role of pasteurisation heat treatments on rheological and protein structural characteristics of fresh egg pasta. *European Food Research and Technology* 221:759–767.

Alamprese, C., Casiraghi E., and Pagani, M.A. 2007. Development of gluten-free fresh egg pasta analogues containing buckwheat. *European Food Research and Technology* 225:205–213.

Alamprese, C., Casiraghi, E., and Rossi, M. 2009. Modelling of fresh egg pasta characteristics for egg content and albumen to yolk ratio. *Journal of Food Engineering* 93:302–307.

Alamprese, C., Casiraghi, E., and Rossi, M. 2011. Effects of housing system and age of laying hens on egg performance in fresh pasta production: pasta cooking behaviour. *Journal of the Science of Food and Agriculture* 91:910–914.

Alary, R., and Kobrehel, K. 1987. The sulphydryl plus disulphide content in the proteins of durum wheat and its relationship with the cooking quality of pasta. *Journal of the Science of Food and Agriculture* 39:123–136.

Ames, N.P., Clarke, J.M., Marchylo, B.A., Dexter, J.E., Schlichting, L.M., and Woods, S.M. 2003. The effect of extra-strong gluten on durum wheat (*Triticum durum* L.) quality. *Canadian Journal of Plant Science* 83:525–532.

Anese, M., Nicoli, M.C., Massini, R., and Lerici, C.R. 1999. Effects of drying processing on the Maillard reaction in pasta. *Food Research International* 32(3):193–199.

Baiano, A., Conte, A., and Del Nobile, M.A. 2006. Influence of drying temperature on the spaghetti cooking quality. *Journal of Food Engineering* 76:341–347.

Batisti, L., Pusterla, S., and Pollini, C.M. 1995. Trattamento termico della pasta fresca. *Tecnica Molitoria* 46:1041–1055.

Borrelli, G.M., Troccoli, A., Di Fonzo, N., and Fares, C. 1999a. Durum wheat lipoxygenase activity and other quality parameters that affect pasta color. *Cereal Chemistry* 76:335–340.

Borrelli, G.M., Troccoli, A., De Leonardis, A.M., Fares, C., and Di Fonzo, N. 1999b. Qualitative parameters that affect the color expression in durum wheat: genotype and environment involvement. *Tecnica Molitoria* 50:841–845.

Brennan, C.S., and Tudorica, C.M. 2008. Evaluation of potential mechanisms by which dietary fibre additions reduce the predicted glycaemic index of fresh pastas. *International Journal of Food Science and Technology* 43:2151–2162.

Bustos, M.C., Perez, G.T., and León, A.E. 2011. Sensory and nutritional attributes of fibre-enriched pasta. *LWT—Food Science and Technology* 44:1429–1434.

Carini, E., Vittadini, E., Curti, E., and Antoniazzi, F. 2009. Effects of different shaping modes on physico-chemical properties and water status of fresh pasta. *Journal of Food Engineering* 93:400–406.

Carini, E., Vittadini, E., Curti, E., Antoniazzi, A., and Viazzani, P. 2010a. Effect of different mixers on physicochemical properties and water status of extruded and laminated fresh pasta. *Food Chemistry* 122:462–469.

Carini, E., Curti E., Spotti, E., and Vittadini, E. 2010b. Effect of formulation on physico-chemical properties and water status of nutritionally enriched fresh pasta. *Food and Bioprocess Technology* 5:1642–1652.

Carrera, A., Echenique, V., Zhang, W., Helguera, M., Manthey, F., Schrager, A., Picca, A., Cervigni, G., and Dubcovsky, J. 2007. A deletion at the *Lpx-B1* locus is associated with low lipoxygenase activity and improved pasta color in durum wheat (*Triticum turgidum* ssp. *Durum*). *Journal of Cereal Science* 45:67–77.

Carrillo, J.M., Vazquez, J.F., and Orellana, J. 1990. Relationship between gluten strength and glutenin proteins in durum wheat cultivars. *Plant Breeding* 104:325–333.

Cencic, L., Franca, C., and Dalla Rosa, M. 1995. Studio su pasta fresca farcita (ravioli) a diverse attività dell'acqua. *Tecnica Molitoria* 5:449–464.

Clarke, F.R., Clarke, J.M., McCaig, T.N., Knox, R.E., and De Pauw, R.M. 2006. Inheritance of yellow pigment concentration in seven durum wheat crosses. *Canadian Journal of Plant Science* 86:133–141.

Cole, M.E. 1991. Review: prediction and measurement of pasta quality. *International Journal of Food Science and Technology* 26:133–151.

Costa, C. Lucera, A., Mastromatteo, M., Conte, A., and Del Nobile, M.A. 2010. Shelf life extension of durum semolina-based fresh pasta. *International Journal of Food Science and Technology* 45:1545–1551.

Cubadda, R. 1988. Evaluation of durum wheat, semolina and pasta in Europe. In *Durum Wheat: Chemistry and Technology*, eds. Fabriani, G., and Lintas, C., 217–228. AACC, St. Paul, Minnesota.

Cubadda, R., Carcea, M., Marconi, E., and Trivisonno, M. 2007. Influence of gluten proteins and drying temperature on the cooking quality of durum wheat pasta. *Cereal Chemistry* 84:48–55.

D'Egidio, M.G., and Nardi, S. 1996. Textural measurements of cooked spaghetti. In: *Pasta and Noodle Technology*. eds. Kruger, J.E., Matsuo, R.B., and Dick, J.W., 133–156. American Association of Cereal Chemists, St. Paul, Minnesota.

D'Egidio, M.G., De Stefanis, E., Fortini, S., Galtiero, G., Mariani, B.M., and Nardi, S. 1983. Analisi delle caratteristiche legate alla qualità delle paste alimentari: cambiamenti nella composizione dell'amido durante la preparazione e cottura delle paste. *Tecnica Molitoria* 34:564–574.

D'Egidio, M.G., Mariani, B.M., Nardi, S., Novaro, P., and Cubadda, R. 1990. Chemical and technological variables and their relationships: a predictive equation for pasta cooking quality. *Cereal Chemistry* 67:275–281.

D'Egidio, M.G., Mariani, B.M., Nardi, S., and Novarò, P. 1993. Viscoelastograph measures and total organic matter test: suitability in evaluating textural characteristics of cooked pasta. *Cereal Chemistry* 70:67–72.

Dalbon, C., Grivon, D., and Pagani, M. 1996. Continuous manufacturing process. In: *Pasta and Noodle Technology*. eds. Kruger, J.E., Matsuo, R.B., Dick, J.W., 13–58. American Association of Cereal Chemists, St. Paul, Minnesota.

Dalbon, G., Pagani, M.A., Resmini, P., and Lucisano, M. 1985. Einflüsse einer Hitzebehandlung der Weizenstärke ährend des Trocknungsprozesses. *Getreide Mehl und Bro* 39:183–189.

De Noni, I., and Pagani, M.A. 2010. Cooking properties and heat damage of dried pasta as influenced by raw material characteristics and processing conditions. *Critical Reviews of Food Science and Nutrition* 50:465–472.

Del Nobile, M.A., Di Benedetto, N., Suriano, N., Conte, A., Corbo, M.R., and Sinigaglia, M. 2009. Combined effects of chitosan and MAP to improve the microbial quality of amaranth homemade fresh pasta. *Food Microbiology* 26:587–591.

Delcour, J.A., Vansteelandt, J., Hythier, M.C., and Abecassis, J. 2000a. Fractionation and reconstitution experiments provide insight into the role of starch gelatinization and pasting properties in pasta quality. *Journal of Agricultural and Food Chemistry* 48:3774–3778.

Delcour, J.A., Vansteelandt, J., Hythier, M.C., Abecassis, J., Sindic, M., and Deroanne, C. 2000b. Fractionation and reconstitution experiments provide insight into the role of gluten and starch interaction in pasta quality. *Journal of Agricultural and Food Chemistry* 48:3767–3773.

Dexter, J.E. 2004. Grain, paste products: pasta and Asian noodles. In *Food Processing Principles and Applications*, eds. Smith, J.S., and Hui, Y.H., 249–271. Blackwell Publishing, Ames, Iowa, USA.

Dexter, J.E., and Matsuo, R.R. 1977. Influence of protein content on some durum wheat quality parameters. *Canadian Journal of Plant* Science 57:717–727.

Dexter, J.E., and Matsuo, R.R. 1979a. Changes in spaghetti proteins solubility during cooking. *Cereal Chemistry* 56:190–195.

Dexter, J.E., and Matsuo, R.R. 1979b. Effect of water content on changes in semolina proteins during dough-mixing. *Cereal Chemistry* 56:15–19.

Dexter, J.E., Matsuo, R.R., and Morgan, B.C.J. 1981. High temperature drying effect on spaghetti properties. *Journal of Food Science* 46:1741–1746.

Dexter, J.E., Matsuo, R.R., and Morgan, B.C. 1983. Spaghetti stickiness—Some factors influencing stickiness and relationship to other cooking quality characteristics. *Journal of Food Science* 48:1545–1551.

DL, 1993. Attuazione della direttiva 89/437/CEE concernente i problemi igienici e sanitari relativi alla produzione ed immissione sul mercato degli ovoprodotti. Gazzetta Ufficiale. n. 64, 18/03/1993, Italy.

DPR, 2001. Decreto del Presidente della Repubblica 9 Febbraio 2001, n.187. Regolamento per la revisione della normativa sulla produzione e commercializzazione di sfarinati e paste alimentari, a norma dell'Articolo 50 della Legge 22 Febbraio 1994, n. 146, Gazzetta Ufficiale n. 117, 22/05/2001, Italy.

Edwards, N.M., Dexter, J.E., and Scanlon, M.G. 2002. Starch participation in durum dough linear viscoelastic properties. *Cereal Chemistry* 79:850–856.

Edwards, N.M., Mulvaney, S.J., Scanlon, M.G., and Dexter, J.E. 2003. Role of gluten and its components in determining durum semolina dough viscoelastic properties. *Cereal Chemistry* 80:755–763.

European Chilled Food Federation. 1996. Guidelines for the hygienic manufacture of chilled products, http://www.ecff.net/images/ECFF_Recommendations_2nd_ed_18_12_06.pdf (accessed date March 15, 2013).

Fava, E., and Storci, A. 2006. Homogenising mixer. European Patent Application EP1693103A2. Date of publication August 23, 2006.

FDA. 2011. Food and Drug Administration. Department of Health and Human Services. 21CFR139.

Feillet, P., and Dexter, J.E. 1996. Quality requirements of durum wheat for semolina milling and pasta production. In *Pasta and Noodle Technology*, eds. Kruger, J.E., Matsuo, R.B., and Dick, J.W., 95–131. AACC, St. Paul, Minnesota.

Fellows, P. 2001. *Food Processing Technology*, Woodhead Publishing Ltd., Cambridge, United Kingdom.

Fraignier, M.P., Michelle-Ferrière, N., and Kobrehel, K. 2000. Distribution of peroxidases in durum wheat (*Triticum durum*). *Cereal Chemistry* 77:11–17.

Francis, F.J. 1999. *Encyclopedia of Food Science and Technology*, 2nd Edition, Wiley & Sons, NY, USA.

Fu, X., Schlichting, L., Pozniak, C.J., and Singh, A.K. 2011. A fast, simple, and reliable method to predict pasta yellowness. *Cereal Chemistry* 88:264–270.

Gelencsér, T., Gál, V., Hódsági, M., and Salgó, A. 2008. Evaluation of quality and digestibility characteristics of resistant starch-enriched pasta. *Food Bioprocess and Technology* 1:171–179.

Giavedoni, P., Aureli, P., and Lerici, C.R. 1993. Effetto di trattamenti termici su microrganismi inoculati in paste fresche farcite. In *Ricerche e Innovazioni Nell'industria Alimentare*, ed. S. Porretta, 335–364. Chiriotti Editori, Pinerolo, TO.

Grant, L.A., Dick, J.W., and Shelton, D.R. 1993. Effects of drying temperature, starch damage, sprouting, and additives on spaghetti quality characteristics. *Cereal Chemistry* 70:676–684.

Grant, L.A., Doehlert, D.C., McMullen, M.S., and Vignaux, N. 2004. Spaghetti cooking quality of waxy and non-waxy durum wheats and blends. *Journal of the Science of Food and Agriculture* 84:90–196.

Güler, S., Köksel, H., and Ng, P.K.W. 2002. Effects of industrial pasta drying temperatures on starch properties and pasta quality. *Food Research International* 35:421–427.

Hagberg, S. 1960. A rapid method for determining alpha-amylase activity. *Cereal Chemistry* 37:218–222.

Hagberg, S. 1961. Simplified method for determining alpha-amylase activity. *Cereal Chemistry* 38:202–203.

Hatcher, D.W., and Kruger, J.E. 1993. Distribution of polyphenol oxidase in flour millstreams of Canadian common wheat classes milled to three extraction rates. *Cereal Chemistry* 70:51–55.

Irvine, G.N., and Winkler, C.A. 1950. Factors affecting the color of macaroni. II. Kinetic studies of pigment destruction during making. *Cereal Chemistry* 27:205–218.

Irvine, G.N., and Anderson, J.A. 1953. Variation in principal quality factors of durum wheat with a quality prediction test for wheat or semolina. *Cereal Chemistry* 30:334–342.

ISO—International Standard 7304. 1985. Durum Wheat Semolina Alimentary pasta produced from durum wheat semolina—Estimation of cooking quality by sensory analysis. Geneva, Switzerland.

Kobrehel, K., Laignelet, B., and Feillet, P. 1974. Study of some factors of macaroni brownness. *Cereal Chemistry* 51:675–683.

Labuza, T.P. 1981. *Water Activity: Influences on Food Quality*. Academic Press, Inc., London.

Lamacchia, C., Di Luccia, A., Baiano, A., Gambacorta, G., La Gatta, B., Pati, S., and La Notte, E. 2007. Changes in pasta proteins induced by drying cycles and their relationship to cooking behaviour. *Journal of Cereal Science* 46:58–63.

Landi, A., and Guarneri, R. 1992. Durum wheat and pasta industries: twenty years of achievement in science and technology. *Cereal Chemistry and Technology: a Long Past and a Bright Future*. 9th International Cereal and Bread Congress, Paris, France.

Lopez, C.C., Vannini, L., Lanciotti, R., and Guerzoni, M.E. 1998. Microbiological quality of filled pasta in relation to the nature of heat treatment. *Journal of Food Protection* 61:994–999.

Lucisano, M., Pagani, M.A, Mariotti, M., and Locatelli, D.P. 2008. Influence of die material on pasta characteristics. *Food Research International* 41:646–652.

Malcolmson, L.J., Matsuo, R.R., and Balshaw, R. 1993. Textural optimization of spaghetti using response surface methodology: effects of drying temperature and durum protein level. *Cereal Chemistry* 70:417–423.

Manno, D., Filippo, E., Serra, A., Negro, C., De Bellis, L., and Miceli, A. 2009. The influence of inulin addition on the morphological and structural properties of durum wheat pasta. *International Journal of Food Science and Technology* 44:2218–2224.

Manthey, F.A., Yalla, S.R., Dick, T.J., and Badaruddin, M. 2004. Extrusion properties and cooking quality of spaghetti containing buckwheat bran flour. *Cereal Chemistry* 81:232–236.

Marchylo, B.A., and Dexter, J.E. 2001. Pasta production. In: *Cereals Processing*, ed. Owens, G., 109–130. Woodhead Publishing Ltd., Cambridge, United Kingdom.

Marchylo, B.A., Dexter, J.E., and Malcolmson, L.M. 2004. Improving the texture of pasta. In: *Texture in Solid Foods*, Vol. 2, ed. Kilcast, D., 475–500. Woodhead Publishing Ltd., Cambridge, United Kingdom.

Marconi, E., Carcea, M., Schiavone, M., and Cubadda, R. 2002. Spelt (*Triticum spelta* L.) pasta quality: combined effect of flour properties and drying conditions. *Cereal Chemistry* 79:634–639.

Martinez, C.S., Ribotta, P.D., León, A.M., and Añón, C. 2007. Physical, sensory and chemical evaluation of cooked spaghetti. *Journal of Texture Studies* 38:666–683.

Matsuo, R.R. 1993. Durum wheat: production and processing. In: *Grains and Oilseeds: Handling, Marketing, Processing*, 4th edition, vol. 2, 779–807. Canadian International Grains Institute, Winnipeg, Canada.

Matsuo, R.R., and Dexter, J.E. 1980. Comparison of experimentally milled durum wheat semolina to semolina produced by some Canadian mills. *Cereal Chemistry* 57: 117–122.

Matsuo, R.R., and Irvine, G.N. 1967. Macaroni brownness, *Cereal Chemistry* 44:78–85.

Matsuo, R.R., Dexter, J.E., Boudreau, A., and Daun, J.E. 1986. The role of lipids in determining spaghetti cooking quality. *Cereal Chemistry* 63:484–489.

McDonald, C.E. 1979. Lipoxygenase and lutein bleaching activity of durum wheat semolina. *Cereal Chemistry* 56:84–89.

Mercier, S., Villeneuve, S., Mondor, M., and Des Marchais, L.P. 2011. Evolution of porosity, shrinkage and density of pasta fortified with pea protein concentrate during drying. *LWT—Food Science and Technology* 44:883–890.

Milatovic, L., and Mondelli, G. 1991. *Pasta Technology Today*. Chirotti Editori, Pinerolo, Italy.

Murray, J.A. 1999. The widening spectrum of celiac disease. *American Journal of Clinical Nutrition* 69:354–365.

Nasehi, B., Mortazavi, S.A., Razavi, S.M., Tehrani, M.M., and Roselina, K. 2009. Effects of processing variables and full fat soy flour on nutritional and sensory properties of spaghetti using a mixture design approach. *International Journal of Food Sciences and Nutrition* 60:112–125.

Oak, M.D., Sissons, M., Egan, N., Tamhankar, S.A., Rao, V.S., and Bhosale, S.B. 2006. Relationship between gluten strength and pasta firmness in Indian durum wheats. *International Journal of Food Science and Technology* 41:538–544.

OJEC, Official Journal of the European Community, 1222/94 (Annex C) L136, p. 5.

Olivo, S.N. 2005. El Mercado De Pastas En America 2004–2005, III Pasta World Congress 2005 www.internationalpasta.org/resources/extra/file/documentos/Pastamarket2004-2005. ppt (accessed date March 15, 2013).

Pagani, M.A., Gallant, D.J., Bouchet, B., and Resmini, P. 1986. Ultrastructure of cooked spaghetti. *Food Microstructure* 5:111–129.

Pagani, M.A., De Noni, I., Resmini, P., and Pellegrino, L. 1996. Filiera produttiva e danno termico della pasta alimentare secca. *Tecnica Molitoria* 47:345–361.

Pagani, M.A., Lucisano, M., and Mariotti, M. 2007. Traditional Italian products from wheat and other starchy flours. In *Handbook of Food Products Manufacturing*, eds. Hui, Y.H., Chandan, R.C., Clark, S., Cross, N.A., Dobbs, J.C., Hurst, W.J., Nollet, L.M.L., Shimoni, E., Sinha, N., Smith, E.B., Surapat, S., Toldrá, F., and Tichenal, A., 327–388. John Wiley & Sons, Inc., Hoboken, NJ, USA.

Perten, H. 1964. Application of the falling number method for evaluating a-amylase activity. *Cereal Chemistry* 41:127–140.

Petitot, M., Brossard, C., Barron, C., Larré, C., Morel, M.H., and Micard, V. 2009. Modification of pasta structure induced by high drying temperatures. Effects on the in vitro digestibility of protein and starch fractions and the potential allergenicity of protein hydrolysates. *Food Chemistry* 116:401–412.

Petitot, M., Boyer, L., Minier, C., and Micard, V. 2010. Fortification of pasta with split pea and faba bean flours: pasta processing and quality evaluation. *Food Research International* 43:634–641.

Phillips, C.A. 1996. Review: Modified Atmosphere Packaging and its effects on the microbiological quality and safety of produce. *International Journal of Food Science and Technology* 31:463–479.

Rao, V.K., Mulvaney, S.J., Dexter, J.E., Edwards, N.M., and Peressini, D. 2001. Stress–relaxation properties of mixograph semolina–water doughs from durum wheat cultivars of variable strength in relation to mixing characteristics, bread- and pasta-making performance. *Journal of Cereal Science* 34:215–232.

Reineccius, G.A. 1990. The influence of Maillard reactions on the sensory properties of food. In *The Maillard Reaction in Food Processing, Human Nutrition and Physiology*, eds. Finot, P.A., Aeschbacher, H.U., Hurrell, R.F., and Liardon, R., 157–170. Birkhauser Verlag, Basel, Switzerland.

Resmini, P., Pagani, M.A., Pellegrino, L., and De Noni, I. 1993. Formation of 2-acetyl-3-D-glucopyranosylfuran (glucosylisomaltol) from nonenzymatic browning in pasta drying. *Italian Journal of Food Science* 5:341–353.

Sadeghi, M.A., and Bhagya, S. 2008. Quality characterization of pasta enriched with mustard protein isolate. *Journal of Food Science* 73:229–237.

Scanlon, M.G., Edwards, N.M., and Dexter, J.E. 2005. Pasta: strength and structure. *New Food* 8:10–15.

Seiler, W. 1995. Neue Teigaufbereitungstechnologie und Technik bei der Herstellung von Teigwaren. *Getreide Mehl und Brot* 49:182–190.

Seiler, W. 2000. New method for the manufacture of pasta products which takes into account damage to the starch component. *Industries Cereales* 117:31–45.

Soh, H.N., Sissons, M.J., and Turner, M.A. 2006. Effect of starch granule size distribution and elevated amylose content on durum dough rheology and spaghetti cooking quality. *Cereal Chemistry* 83:513–519.

Sozer, N., Dalgic, A.C., and Kaya, A. 2007. Thermal, textural and cooking properties of spaghetti enriched with resistant starch. *Journal of Food Engineering* 81:476–484.

Storci, A., 2005. High speed mixing and homogenization of solid and liquid in e.g., food-, pharmaceutical- and paint manufacture, atomizes liquid and mixes rapidly with powder dispersion in air. Patent DE102005025016, 2005-12-29/IT 2004 MI0111020040601.

Storci, A., 2006. Vacuum dough rolling apparatus. European Patent Application, EP1642503A2.

Sudha, M.L., Ramasarma, P.R., and Rao, V.G. 2011. Wheat bran stabilization and its use in the preparation of high fiber pasta. *Food Science and Technology International* 17:47–53.

Tang, C., Hsieh, F., Heymann, H., and Huff, H.E. 1999. Analyzing and correlating instrumental and sensory data: a multivariate study of physical properties of cooked wheat noodles. *Journal of Food Quality* 22:193–211.

UNIPI. 2009. http://www.unipi-pasta.it/dati/c-mond.html (accessed December 12, 2011).

Villeneuve, S., and Gélinas, P. 2007. Drying kinetics of whole durum wheat pasta according to temperature and relative humidity. *LWT—Food Science and Technology* 40:465–471.

Walsh, D.E., and Gilles, K.A. 1977. Pasta technology: macaroni products. In *Elements of Food Technology*, ed. Desrosier, N.W. AVI Publishing Company, Inc, Connecticut, USA.

Wyland, A.R., and D'Appolonia, B.L. 1982. Influence of drying temperature and farina blending on spaghetti quality. *Cereal Chemistry* 59:199–201.

Yalla, S.R., and Manthey, F.A. 2006. Effect of semolina and absorption level on extrusion of spaghetti containing non-traditional ingredients. *Journal of the Science of Food and Agriculture* 86:841–848.

Yue, P., Rayas-Duarte, P., and Elias, M.E. 1999. Effect of drying temperature on physico-chemical properties of starch isolated from pasta. *Cereal Chemistry* 76:541–547.

Zardetto, S. 2005. Potential application of near infrared spectroscopy for evaluating thermal treatments of fresh egg pasta. *Food Control* 16:249–256.

Zardetto, S., and Dalla Rosa, M. 2006. Study of the effect of lamination process on pasta by physical chemical determination and near infrared spectroscopy analysis. *Journal of Food Engineering* 74:402–409.

Zardetto, S., and Dalla Rosa, M. 2009. Effect of extrusion process on properties of cooked, fresh egg pasta. *Journal of Food Engineering* 92:70–77.

Zardetto, S., Di Fresco, S., and Pasqualetto, K. 1999. La pastorizzazione nella pasta fresca farcita. Parte 2: Valutazione dell'effetto del trattamento termico impostato sui microrganismi presenti e inoculati. *Tecnica Molitoria* 6:643–650.

Zweifel, C., Conde-Petit, B., and Escher, F. 2000. Thermal modifications of starch during high-temperature drying of pasta. *Cereal Chemistry* 77:645–651.

Zweifel, C., Handschin, S., Escher, F., and Conde-Petit, B. 2003. Influence of high temperature drying on structural and textural properties of durum wheat pasta. *Cereal Chemistry* 80:159–167.

11 Wet Milling and Starch Extraction

Sergio O. Serna Saldivar

CONTENTS

11.1 INTRODUCTION

Starch is the most abundant chemical component associated with cereal grains and, historically, man has extracted this constituent for at least 20 centuries due to its functionality, economic value, and nutritional attributes. Nowadays, one of the most significant industrial uses of cereal grains, especially maize (*Zea mays* L.), is as feedstock for the manufacture of refined starches. Starch is a renewable resource and is a mainstay of food- and nonfood-related industries. The food industry employs native starch to produce an array of syrups and sweeteners, as well as modified starches that serve as important food additives. The nonfood uses are mainly for the production of adhesives, paper, textiles, bioplastics, and ethanol fuel. The industry has learned to optimally mill the different kinds of cereals, which differ in pericarp and endosperm structures, and the association between starch granules and the protein matrix. The various types of wet milling industries aim toward the extraction of the maximum possible amount of native or undamaged starch granules. The processes are designed to efficiently take cereals apart and purify their components, which are then used as key ingredients in the food, feed, and nonfood industries. By far, maize is the main source of refined starch worldwide because of its availability, relatively low cost, high starch content (>65%), and the value of its coproducts (steep solids, germ, bran, and gluten). In contrast with the various dry milling industries which separate the grain anatomical parts wet millers separate the kernels' chemical compounds (starch, gluten, oil, and fiber).

Globally, more than 70 million tons of refined starches are produced. Of this output, approximately 33 million tons are sold as starches (25 million tons and 8 million tons of native and modified starches, respectively). Of this production, approximately 77% to 80% is obtained from maize. In 2010, approximately 43.2 out of the 316 million tons of maize produced in the United States was wet milled (Corn Refiners Association 2012; FAO 2012). The value of the maize purchased by refiners' plants was 6.9 billion US dollars. The maize starch produced by the US refining industry produced 11.5 million tons (dry basis) of sweeteners, 3 million tons of starch, 6430 million liters of ethanol, and 13.57 million tons of coproducts. The value added due to manufacturing was estimated at 9.9 billion dollars (Corn Refiners Association 2012).

The aim of the various types of wet milling industries is to obtain the highest possible yield of undamaged starch that meets specifications for color, chemical, and functional properties. In the particular case of wheat (*Triticum aestivum* L.), the wet milling industry aims toward the production of both vital gluten and refined starch. Vital glutens are in high demand from the baking and meat industries to improve bread quality and to use as a relatively inexpensive meat extender, respectively. Starches from regular, waxy, and high-amylose maizes; tapioca (*Manhiot esculenta* Crantz); wheat; and potato (*Solanum tuberosum* L.) are the preferred and most frequently used industrially. These starches have different amylograph properties, pasting behaviors, and functionalities; therefore, they are frequently used to impart different characteristics to processed foods. In this chapter, the extraction of different types of starches associated with cereal grains will be described. In addition, the quality control measurements applied by wet millers will also be discussed.

11.2 MAIZE WET MILLING

Most of the industrially produced starch is obtained from yellow dent maize (*Z. mays*). The optimum physical properties of grain for wet milling are undamaged soft-textured dent kernels with a test weight of 67.5 kg/hL, and thousand kernel weights of more than 300 g. These kernels contain a favorable ratio of endosperm to germ and pericarp, and higher amounts of starch, and their starch granules are easier to extract. Furthermore, the softer endosperm structure hydrates faster during the critical step of steeping—considered the bottleneck of the process. Yellow maize is generally processed because of its relatively lower price and because it favors the production of pigmented gluten that is demanded by the poultry and feed industries (Eckhoff and Watson 2009; Johnson and May 2003; Serna Saldivar 2008, 2010).

11.2.1 CLEANING

Shelled maize is received in bulk and commonly contains pieces of cob, chaff, foreign seeds, and undesirable dockage. Cleaning the maize is an important initial step in the process because the presence of broken kernels can alter the normal flow of steepwater through the maize mass, resulting in nonuniform steeping. In addition, starch granules, eroded from the exposed endosperm surfaces, lixiviate into the steepwater and are gelatinized during the evaporation of the steep liquor, resulting in viscous steepwater (Eckhoff and Watson 2009).

The cleaning process of the shelled maize includes the use of air aspirators to remove dust and chaff and other light particles followed by screening and a destoner or gravity separator. Aspiration is used to remove dust and light chaff; the reciprocating screening cleaner is commonly equipped with a sieve of 4.8 mm orifices to remove broken and smaller pieces of foreign material. The gravity separators are designed and calibrated to remove ferrous contaminants and stones of approximately the same size as the maize kernels. These pieces of equipment usually contain magnets used to trap metal trash (bolts, tools, scrap iron, etc.) from the maize to protect mills from damage (Eckhoff and Watson 2009; Serna Saldivar 2010).

11.2.2 STEEPING WITH SULFUR DIOXIDE

The most critical wet milling step is the steeping of the kernels in a warm solution containing sulfur dioxide (Figure 11.1). The steeping operation involves maintaining the correct balance of water flow, temperature, sulfur dioxide concentration, and pH. The whole maize kernels are normally steeped for 24 to 48 h at a temperature of 48°C to 52°C. By the end of the steeping period, the kernels absorb the sulfur dioxide solution to approximately 45% moisture, release approximately 6.0% to 6.5% of their dry matter into the steep solution, and absorb approximately 0.2 to 0.4 kg of sulfur dioxide per ton of grain. At this point, the treated kernel can be easily squeezed with the fingers and the germ is liberated intact (Eckhoff and Watson 2009; Serna Saldivar 2010).

The typical countercurrent steeping can be divided into three sequential stages known as fermentation, sulfur dioxide diffusion, and sulfur dioxide domination. The

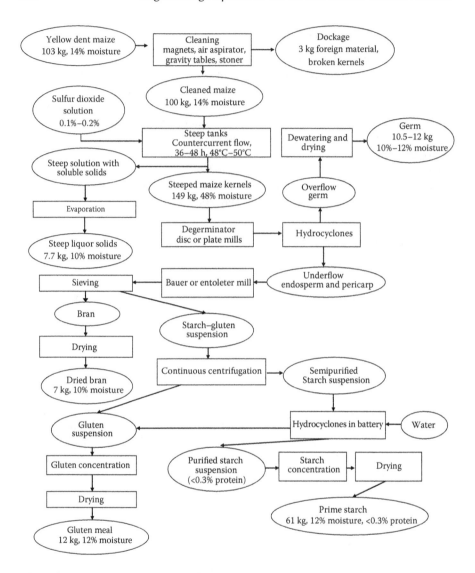

FIGURE 11.1   Flowchart of the sulfur dioxide wet milling process for the production of maize starch.

first stage, which lasts from 8 to 12 h, is when kernels absorb the steep solution containing approximately 2% lactic acid rich in active *Lactobacillus* bacteria. The length of the fermentation phase is controlled by the sulfur dioxide concentration in the steep liquor. The growth of the lactic acid-producing bacteria stops when the sulfur dioxide concentration exceeds 100 to 300 ppm. The second stage of steeping also lasts from 8 to 12 h and is characterized by the sulfur dioxide's diffusion into the kernels. The reducing action of the sulfur dioxide enhances the reactions that lead to the release of starch granules and the weakening of the protein matrix. The sulfur dioxide acts synergistically with the proteolytic action of the *Lactobacillus*, breaking the endosperm cell walls and hydrolyzing the disulfide bonds of the protein matrix that engulfs or traps the starch granules. The final or sulfur dioxide-dominated stage is characterized by the addition of high concentrations of sulfur dioxide (up to 2500 ppm) to the oldest steeped maize. Aside from softening the endosperm structure, the sulfur dioxide and lactic acid prevent germination and the growth of undesirable microorganisms (Eckhoff and Watson 2009; Johnson and May 2003; Serna Saldivar 2010).

The steeping operation is carried countercurrently in large stainless steel tanks positioned in a battery that holds 250 to 600 tons each. The 6 to 12 tanks are built to resist the corrosive action of the sulfur dioxide and lactic acid, designed to be filled with incoming raw maize kernels from an overhead conveyor, and emptied through an orifice at the conical bottom of the container. The inside surface of the cone bottom is covered with a strainer for drainage. In addition, the battery of tanks is equipped with heat exchangers to maintain the desired temperature and pumps to move the steep solution from one tank to another or to withdraw it from the system. The oldest sulfur dioxide solution rich in *Lactobacillus* treats the new incoming kernels, while the new and stronger solution containing 0.1% to 0.2% sulfur dioxide treats the exiting wet kernels (Eckhoff and Watson 2009).

As indicated previously, steeping usually lasts from 30 to 48 h. During this time, the maize kernels with original moisture contents of 12% to 14% gradually absorb the SO_2 solution, increasing the moisture content to 48%. The steep solution enters the kernel typically at the tip cap and moves quickly to the germ and the pericarp's tube cells where it is gradually distributed to the endosperm. The steep solution moves quickly up the sides of the kernel and enters through the dent region into the porous floury endosperm region. According to Eckhoff and Watson (2009), the germ and endosperm tissues become wet in approximately 4 and 8 h, respectively. The water uptake causes a kernel volume expansion of 55% to 65%.

Approximately 5% to 7% of the maize solids solubilize during steeping. Most of these solids lixiviate from the germ tissue. As a result, the original oil content increases from 30% to 38% in raw germ to 55% to 60% in steeped germ. In fact, the germ provides approximately 95% of the solubles that lixiviate to the sulfur dioxide steep solution. These solids are mainly constituted of albumins and globulins, lactic acid, minerals, phytic acid, and B-vitamins (Eckhoff and Watson 2009; Johnson and May 2003; Serna Saldivar 2008, 2010).

11.2.3 Grinding, Degermination, and Fractionation

After steeping, the resulting kernels are wet-milled in plate or disc attrition mills in preparation for degermination. These mills consist of plates covered with pyramidal knobs and the impact ring is absent. The bulk of the germ is freed on the first pass, but a second pass is usually provided after free germ has been removed. The mill gap is adjusted to give the maximum free germ with minimum tissue breakage because the mechanical damage disrupts scutellum cells that will liberate oil, which is mostly absorbed by the gluten. Interestingly, approximately half of the total starch and gluten, especially those associated with the hydrated chalky endosperm, is also liberated during this milling step. The ground particles diluted in water are transported into a series of hydrocyclones to separate the less dense rubbery germ from the rest of the kernel components. Multiple passes may be used to attain clean separations. The germ has a lower density because it contains most of the kernel's oil ($d = 0.9$) and does not contain any starch ($d = 1.5$ g/cm^3). The hydrocyclones used for degermination are 1-m-long conical tubes that are 15.2 to 22.5 cm in diameter at the top. The ground maize with 12% to 14% solids content is forced into the tube under pressure. The orifice angle, aperture, and pump pressure are chosen to produce a rotational velocity sufficient to cause a separation of the particles of differing densities. The heavier endosperm and fiber particles pass out of the bottom of the tube at a dry solids concentration of 20% to 24%, whereas the lighter germ is drawn off the top of the hydrocyclone unit (Eckhoff and Watson 2009; Johnson and May 2003; Serna Saldivar 2008, 2010).

The germ stream is dewatered to 45% to 50% solids concentration by passing it through mechanical squeezers before drying in preparation for oil recovery. The free starch and protein, which comprise about half of the dry substance in the germ cyclone underflow, are separated from the unmilled endosperm and fiber by screening over a screen bend device to reduce the load of solids and water in subsequent milling operations. The slurry of starch and gluten called "prime mill starch" is released mainly from the floury or chalky endosperm (Eckhoff and Watson 2009).

The second milling stage starts when the underflow from the hydrocyclones, which is rich in pericarp and pieces of vitreous endosperm, is further and thoroughly milled to release the starch. The operation is commonly performed with Bauer or Entoleter mills. The latter is most commonly employed in wet milling operations. The Bauer mill employs a combination of attrition and impact milling and further grinds the material between counterrotating grooved plates and by impact on an outer ring. On the other hand, the Entoleter mill further grinds the endosperm chunks by a rotating horizontal disk that impacts with great force against both rotating and stationary pins. The main advantage of this mill is the rapid and complete starch release and the yield of clean pericarp in the form of flakes. Independently of the type of mill, the aim of the grinding operations is to release the starch granules from the protein matrix while causing minimal mechanical damage (Eckhoff and Watson 2009; Serna Saldivar 2010).

The subsequent step consists of the separation of the starch and gluten released by milling from the larger pieces of pericarp. This is best achieved by pumping the slurry with considerable force in a fixed concave wedge bar screen surface arranged in a vertical position. The screens have slit widths of 50, 75, 100, and 150 μm. Generally, a 120° pressure-fed screen with a bar screen slit width of 50 μm is employed by the industry. The separation screen is designed to minimize clogging and for continuous processes. The pericarp mat discharging from the screen is reslurried and reprocessed over other units to recuperate as much endosperm as possible. Typically, the finished pericarp contains 15% to 20% starch. This coproduct is first dewatered in a centrifugal screen, mechanical squeezers, or a continuous discharge centrifuge to a moisture content of 65% to 75%. The fiber is blended with concentrated steep liquor, maize cleanings, and spent or defatted germ and other coproducts, and dried to produce corn gluten feed that contains 18% to 22% protein (Eckhoff and Watson 2009; Johnson and May 2003).

The next steps are aimed at separating the less dense gluten from the denser starch granules. This is accomplished by first passing the starch–gluten slurry through a series of continuous centrifuges followed by a secondary purification using a series of hydrocyclones (10–14 units). Initially, the mill starch streams from both the degermination and fiber washing steps are blended and centrifugally concentrated to reduce solubles and to adjust the concentration of solids in preparation for the final step of starch separation. The mill starch containing 5% to 8% insoluble protein is pumped to a bowl type or nozzle discharge centrifuge equipped with a clarifier assembly as a decanter to permit discharge of water in the overflow and all solids through the nozzles. The low-density stream rich in gluten particles (d = 1.1 g/cm³) is effectively separated from the starch (d = 1.5 g/cm³) by a primary centrifugation step. These continuous pieces of equipment are equipped with a stack of conical disks separated by a narrow space to accentuate the separation of discrete particles with different densities. The heavier starch granules are thrown to the periphery of the centrifuge bowl and ejected through nozzles, whereas the lighter gluten particles are carried up between the disks by a stream of water and are ejected at a low solid concentration. The mill starch is preferably supplied to the centrifuge at 18% to 21% solid content, whereas the starch slurry discharges at 35% to 42% solid content. The semirefined starch contains 1% to 2% protein. On the other hand, the light gluten discharges from the centrifuge at only 1% to 2% solids content and therefore must be concentrated in centrifuges to 12% to 15% solids content for efficient recovery. The protein content of the gluten was 68% to 75% (expressed on a dry matter basis). Gluten recovery is usually achieved with rotary vacuum filters arranged for continuous discharge of cake from the filter cloth belt. An alternative system dewaters the concentrated gluten in a horizontal decanter solids discharge centrifuge in preparation for drying (Eckhoff and Watson 2009).

The next step in the starch process is the removal of the remaining protein by first diluting the starch slurry to 18% to 21% solids content. The starch is further purified in a second centrifugal separator to a final insoluble protein level of 0.27% to 0.32% dry basis. The most common way to do this is through the use of 8 to 14 stages of liquid hydrocyclones designed to simultaneously remove residual gluten and wash the starch. These units are molded plastic devices with a port for entry through which the slurry is pumped into the top of a cone-shaped chamber that is furnished with a couple of outlets to discharge the overflow and underflow. The overflow port positioned on top of the hydrocyclone discharges the low-density fraction, whereas the underflow port located at the bottom of the separator discharges the high-density stream. The small hydrocyclones have an inside diameter of 10 mm at the top of the cone and a length of 16 mm. These small hydrocyclones are assembled into manifolds holding numerous tubes. The internal design is a clamshell in which the manifold containing several hundred tubes is covered top and bottom, forming a three-partitioned unit or a radial configuration in which the individual tubes are radially oriented. Independently of the design, the starch–gluten mix enters the central chamber and is forced by pump pressure (normally 5.4–6.8 atm) simultaneously into each of the individual tubes. The pressure energy creates rotational motion to the liquid entering the conical chamber, producing a centrifugal force throwing the heavier starch granules out the underflow port into a common collection chamber.

On the other hand, the gluten particles discharge through the overflow collection chamber. The concentrated and purer starch is rediluted with water before passing to the next hydroclone unit. Regularly, after six to 10 sequential passes, the starch is completely refined and contains less than 0.33% protein on dry matter basis. The resulting refined starch is continuously dehydrated in drying tunnels to decrease its moisture content to approximately 6%, whereas the gluten stream is first dewatered in basket centrifuges, concentrated with a vacuum filter, and dehydrated to 12% moisture in rotary or drum driers (Eckhoff and Watson 2009; Serna Saldivar 2010). The yields of prime starch and gluten meal from clean maize are from 61% to 65% and 12% to 13%, respectively (Figure 11.1).

11.2.4   WET MILLING OF WAXY AND HIGH-AMYLOSE MAIZES

Some wet milling industries obtain identity-preserved waxy and high-amylose starches for niche markets. By far, the mutant maize most widely processed is the waxy type, which contains more than 95% amylopectin. More than 1 million tons of this particular genotype is produced yearly in the United States, and this is mainly channeled to the wet milling industry. This represents about 1.5% of the maize processed by wet millers in the United States. On the other hand, the high-amylose genotypes, containing 60% to 70% amylose, are processed in lesser amounts (0.2% of the wet-milled maize) to obtain a unique type of starch. These are considered specialty starches commanding premium prices due to their different amylograph properties and functionalities. The commercial processes to obtain waxy and high-amylose starches are practically identical to the one described for regular dent maize with slight but important modifications. The recommended steeping temperatures of the waxy and high-amylose kernels is 3°C lower and higher compared with the temperature commonly used for regular maize (Johnson and May 2003). Also, the high-amylose kernels are steeped for longer periods of time. However, the waxy and high-amylose genotypes commonly harvested in the United States yield about 8.3% (65% vs. 60%) and 34% (65% vs. 43.4%) less refined starch compared with regular dent maize, respectively. The separation of the waxy and high-amylose starches from the gluten is easier and more difficult, respectively. As a result, the refined waxy and high-amylose starches contain lower (0.2%) and higher (0.6%) amounts of residual proteins compared with the regular starch (0.3%) (Johnson and May 2003; Eckhoff and Watson 2009). Anderson et al. (1960) indicated that high-amylose maizes containing 49% and 57% amylose yielded less prime starch, whereas the starch contained higher amounts of protein compared with ordinary maize. The yield of the 57% amylose maize was 43.5% starch containing 0.7% protein. The same authors observed an unusually large swelling of the kernels during steeping (128% vs. 63% increase in volume).

11.2.5   IMPROVEMENT OR MODIFICATION OF THE CONVENTIONAL PROCEDURE

Recently, several investigators have proposed the use of fiber-degrading or proteolytic enzymes (or both) on preground sulfur dioxide-steeped kernels to decrease steeping times without sacrificing yields of prime starch. The cell wall-degrading enzymes disrupt cell walls, aiding in the penetration of the steep solution, whereas the proteases hydrolyze the endosperm proteins that engulf starch granules. Thus, the supplementation of these enzymes can decrease the steeping times to approximately half and reduce the amounts of residual protein associated with prime starch (Eckhoff and Tso 1991; Johnston and Singh 2001; Ling and Jackson 1991; Mezo Villanueva and Serna Saldivar 2004; Moheno-Perez et al. 1999; Serna Saldivar and Mezo Villanueva 2003).

11.2.6 Coproducts

The wet milling of maize yields an array of coproducts that contribute to the economy of the process. The solids of the steep liquor, which are rich in nutrients, are generally used by pharmaceutical industries as a growth medium for molds and other microorganism used to produce antibiotics and related products. The germ is generally channeled to the oil-crushing industry where the oil is mechanically or solvent extracted, refined, and dewaxed, and the protein meal sold to the feed industry or mixed with the gluten feed. The pericarp or bran is generally mixed with the other coproducts for the production of gluten feed, although an increasing amount of this fiber is finely ground for the production of human foods. The gluten rich in protein and yellow pigments (carotenoids and xantophylls) is in high demand by the poultry industry to natural pigment egg yolk and the skin of broilers (Serna Saldivar 2010).

11.3 WHEAT WET MILLING

The starch from wheat (*T. aestivum*) has been extracted by several countries. Historically, it has been documented that wheat starch was used by the ancient Egyptians as glue for papyrus and as an aid to solidify the cloth that covered their mummies. A procedure for starch production was given in some detail in a Roman treatise by Cato in about 184 BC. Briefly, the wheat kernels were steeped in water for 10 days, pressed and mixed with fresh water, and then the resulting slurry filtered on a linen cloth. The filtrate containing the starch was allowed to settle, washed with water, and solar dehydrated (Schwarts and Whistler 2009). The commercial manufacture of wheat starch began in the United Kingdom in the seventeenth century and rapidly reached the United States. The first American wheat starch processing plant was established in New York by around 1807. Forty years later, the same wet milling plant started to extract starch from maize.

Nowadays, about 4.7 million tons of wheat starch are annually produced in the world. This represents about 7.8% of the total starch production worldwide (Maningat et al. 2009; Schwartz and Whistler 2009). More than half and 17% of the wheat starch are produced in the USA and Europe, respectively. The modern wet milling plants process refined flours with two major goals: the production of vital gluten and native starch. The vital gluten is a food additive used in the baking and meat processing industries whereas the starch has analogous uses as maize starch. The wheat endosperm possesses two distinctive types of starch granules which differ in size and shape. These starches are classified as A or B types. The A starch granules are large and lenticular with a mean diameter of 14.2 μm whereas the B starch granules are small and spherical with a mean diameter of only 4.1 µm. Quantitative separations of wheat starch granules indicated that wheat starch contains approximately 70% by weight of A granules. A range of 23 to 50 by weight of B granules was found in 130 lines of Australian wheat, with the lowest levels observed in soft wheats. The physical, chemical characteristics and susceptibility to amylase hydrolysis of these types of starches are described by Maningat et al. (2009). Both types of wheat starches contain sufficient amounts of gluten to trigger the allergic reactions associated to gluten intolerance or celiac disease.

Kempf and Rohrmann (1989) described 15 different processes for industrial production of wheat starch and wheat gluten using as raw materials wheat kernels or refined flours. According to Schofield and Booth (1983) there are six general manufacturing processes of wheat starch and gluten separation known as dough, batter, aqueous dispersion, chemical dispersion, wet-milling of whole kernels and non-aqueous separation. However, the last three are currently in disuse due to high operation costs, reduced process efficiency, poor product quality and effluent problems. Thus, only three of the processes are actually commercially practiced. All these manufacturing processes employ refined wheat flour as feedstock and follow the same basic principles of starch vital gluten separation. These processes are known as dough-Martin, dough-batter and batter (Maningat et al. 2009; Serna Saldivar 2010).

11.3.1 CONVENTIONAL DOUGH OR MARTIN PROCESS

The Martin or dough ball process was proposed in Paris in approximately 1835 and is still considered the most practiced method (Figure 11.2). This method uses refined wheat flour as the raw material and starts when the flour is mixed with water in a 2:1 ratio to form a uniform, consistent, properly developed, and smooth dough. Hard wheat flours require more water compared with soft counterparts. The starch is simply extracted by washing the dough with water. The washing step gradually releases the starch granules without breaking the gluten into small particles. The industry uses several washing devices such as a rotary drum, mixers with twin screws, and blenders equipped with a couple of sigma blades that counterrotate at different speeds. Regardless of the equipment, the water extracts most of the starch granules and some soluble compounds (proteins and soluble carbohydrates). The insoluble part stays on the screens and consists mainly of wet gluten (prolamins or gliadins and glutelins). Excess gluten water is removed by compression through a pair of rolls, and the resulting gluten carefully dehydrated to 8% to 10% moisture in vacuum or drum driers. The starch suspension containing A and B starch granules with approximately 10% solids content is first passed through a set of vibrating screens to remove small pieces of gluten and other contaminants. The denser A starch is easily separated from contaminants such as soluble proteins with continuous centrifuges or hydrocyclones (or both). The resulting starch is dewatered and dried to decrease its moisture content to approximately 10%. The industrial specification for prime wheat starch does not allow more than 0.3% protein. The Martin process yields approximately 65% A and B starch and 14% vital gluten (Figure 11.2). The rest is inseparable or B starch and other solids lost during the washing and centrifugation steps (Maningat et al. 2009; Serna Saldivar 2010).

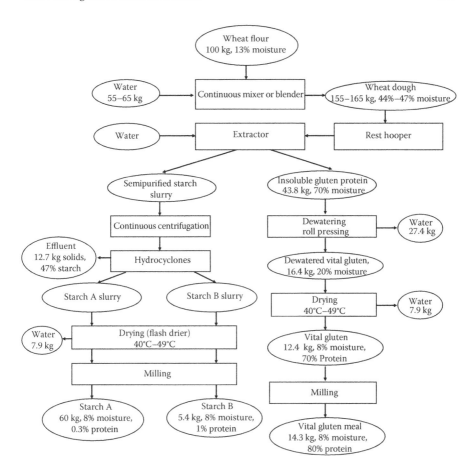

FIGURE 11.2 Flowchart of the dough–Martin process for the production of wheat starch and vital gluten. (Adapted from Knight, J.W., and Olson, R.M., Wheat starch: production, modification and uses. Chapter 15 In: *Starch: Chemistry and Technology*, R.L. Whistler, J.N. Bemiller and E.F. Paschall (eds). Second Edition, Academic Press, Orlando, FL, 1984.)

11.3.2 Dough–Batter Process

The dough–batter process utilizes two basic and critical steps: agglomeration of gluten with high-shear homogenizers and centrifugal separation of a gluten-rich fraction from starch using hydrocyclones or decanters. The process is industrially named "hydrocyclone" because this equipment is key for the clean separation of the starch from the gluten and other contaminants. The process is still widely used because the hydrocyclone needed is compact and relatively inexpensive and requires less water. The dough–batter or hydrocyclone process detailed in Figure 11.3 starts when refined wheat flour is kneaded with warm water to form a slack dough. After 10 to 12 min of resting, the dough is strongly agitated with water to form gluten threads and a slurry. The slurry is filtered through a strainer and then pressure-pumped to a set of hydrocyclones where the gluten migrates in the light stream leaving the unit.

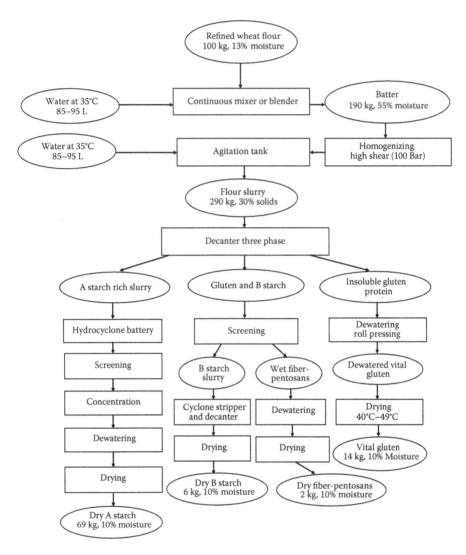

FIGURE 11.3 Flowchart of the batter high-pressure disintegration process for the production of wheat starch and vital gluten. (Adapted from Maningat, C.C. et al., Wheat starch: production, properties, modification and uses. Ch. 10 In: *Starch: Chemistry and Technology*, J.N. Bemiller and R. Whistler (eds). Third Edition, Academic Press, New York, 2009.)

Agglomeration of the gluten strands occurs in the exit port because of the increased protein concentration and intense mixing. The agglomerated gluten in the overflow fraction is isolated by screening, washed to remove B starch, bran, and cell wall debris, and then dewatered before drying. A set of hydrocyclones is used to separate, wash, and concentrate the A starch in the underflow fraction in a countercurrent fashion. Further purification of the A starch from the fiber involves static and rotary screening. The remaining starch slurry separated in the overflow fraction of the set of hydrocyclones undergoes screening to remove fine fibers and is then separated into B starch and a second A starch fraction. Wet coarse and fine fibers are removed as a separate stream. The remaining solubles constitute the effluent stream. The advantages of the dough–batter process compared with the dough–Martin counterpart include better yield of vital gluten, lower water and energy requirements, and lower volume of effluents (Maningat et al. 2009).

11.3.3 BATTER PROCESS

The first batter process, named Fesca, was invented in the 1920s to obtain both wheat starch and vital gluten. The original process was gradually modified and improved to obtain better yields of high protein gluten and refined starch. The principles of the old Fesca process were used to develop more efficient processes currently known as Alfa-Laval/Raisio and high-pressure disintegration processes. These technologies process refined wheat flour with water to yield a thin batter that is washed and later on subjected to centrifugal forces to separate vital gluten, A and B starches, and contaminants.

11.3.3.1 Fesca

The Fesca process processes refined wheat flour to extract starch and vital gluten. In this process, the wheat flour is mixed with water to form a thin batter, which is dispersed by shearing to prevent gluten development. The batter is centrifuged to separate the starch from the protein concentrate. The starch is reslurried, refined, and dried, whereas the protein concentrate is usually spray-dried. The main advantage of this process is that water usage is minimal and solids recovery is nearly 100%. The use of ammonium hydroxide in the Fesca process was proposed in 1969. The modified process allows for better separation of the starch but has a detrimental effect on the amino acid residues of the protein concentrate (Knight and Olson 1984; Maningat et al. 2009; Serna Saldivar 2010).

11.3.3.2 Alfa-Laval/Raisio Process

This extraction method was developed from the modified Fesca process and has the advantage of significant savings in the consumption of water (three parts water per part flour). A wheat flour dispersion with approximately 30% solids content is mixed at 30°C using a high-shear mixer. The resulting batter is separated by a decanter into a starch fraction containing approximately 1% protein and a gluten fraction with 40% protein. The starch is further refined with conical screens and decanters, whereas the gluten phase is screened, dewatered, and dehydrated (Maningat et al. 2009).

11.3.3.3 High-Pressure Disintegration Process

The latest batter technology for wheat starch production is known as high-pressure disintegration. The process, mainly used in Europe, Australia, and Canada, is based on starch extraction and separation by centrifugal forces from a highly sheared batter. The procedure (Figure 11.4) starts when refined wheat flour is mixed with 0.85 to 0.95 parts water at 35°C in a continuous dough mixer to achieve a smooth batter.

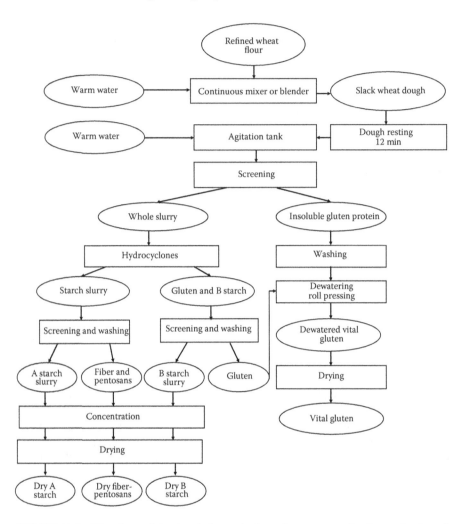

FIGURE 11.4 Flowchart of the dough–batter hydrocyclone process for the production of wheat starch and vital gluten. (Adapted from Maningat, C.C. et al., Wheat starch: production, properties, modification and uses. Ch. 10 In: *Starch: Chemistry and Technology*, J.N. Bemiller and R. Whistler (eds). Third Edition, Academic Press, New York, 2009.)

The batter is pumped at high pressure (100 bar) through a homogenizing valve to release starch granules from the hydrated gluten, which is disrupted into tiny particles during shearing. The main advantage of this technology is the comparatively lower water requirement. Subsequently, the homogenized batter is diluted with water to a 30% solid concentration and fed into a three-phase decanter centrifuge designed to separate and concentrate the denser A starch from the intermediate density stream rich in B starch granules and gluten, and the least dense viscous stream containing pentosans (arabinoxylans) and other contaminating solubles. The medium dense stream is pumped into a sieve drum where B starch granules are washed from the fine gluten particles. The A starch slurry from the centrifugal decanter is screened and then refined either with multistage hydrocyclones or with separators and decanters. Likewise, the B starch stream is first passed through vibrating screens and concentrated in a decanter. The major advantages of the process are the reduction in water consumption (to as low as three parts of water per part of wheat flour) and a 10% increase in A starch yield.

11.3.4 SULFUR DIOXIDE PROCESS

Although it is not commercially used today, whole wheat kernels can be wet-milled following the same principles described for maize. The main disadvantage of this process is that the sulfur dioxide steeping denatures the gluten, yielding a nonvital or nonfunctional product. The steeping time requirements are significantly reduced compared with maize. Steeped wheat kernels are wet-milled, followed by screening to remove bran and germ. The slurry containing starch granules and gluten proteins is next sent to a series of centrifuges or hydrocyclones. The resulting starch and gluten slurries are concentrated before drying. According to Knight and Olson (1984), prime starch yields, based on unclean wheat, vary from 53% to 59%. According to Maningat et al. (2009), this wet milling process is not commercially used today due to the bran contamination in the starch and gluten, and because the process generates a high volume of effluents with low pH.

11.3.5 PRODUCTS AND COPRODUCTS

Wheat starch may be considered a by-product in the manufacture of vital gluten. Most of the starch is sold in unmodified form mainly to the food and paper industries. The properties of A and B starch granules are listed in Table 11.1. The thickening power of wheat starch is less than that of maize starch, but paste texture, clarity, and strength are about the same. The lower gelatinization temperature of wheat starch gives an advantage over maize starch for use as adhesives. The wet milling of wheat yields either vital or nonvital glutens and other by-products such as germ and bran when whole kernels are used as raw materials. Vital gluten commands a high market value because it is extensively used to increase the gluten content of flours. Most formulations of whole wheat baked products contain vital gluten to compensate for the loss of functionality. Vital gluten is also used to produce more functional flours especially in Europe and Japan. Nonvital gluten, bran, and germ are generally channeled to the feed industry (Serna Saldivar 2010).

11.4 RICE WET MILLING

The commercial production of rice (*Oryza sativa* L.) starch is limited due to the relatively high cost of the rice relative to other cereals and the difficulty of extracting the starch due to the endosperm's structure. The cost of mill run rice is usually two to three times that of maize. Rice starch can be obtained using two major industrial methods: traditional and mechanical. The first is the most widely

TABLE 11.1

Physical and Viscoamylograph Properties of Cereal Starches Compared with Potato and Tapioca Starches

	Maize			Wheat		Rice		Potato	Tapioca
	Regular	Waxy	High Amylose	A Granules	B Granules	Regular	Waxy	Potato	Tapioca
Type of Granule	Single	Single	Single	Single	Single	Compound	Compound	Single	Single
Granule Shape	Round, angular	Round, angular	Polygonal, elongated and spherical	Lenticular	Spherical	Polyhedral, pentagonal	Polyhedral, pentagonal	Lenticular	Spherical, egg shaped
Granule Size (μm)	5–30	5–25		15–19	5–7	1.6–8.7	1.9–8.1	10–100	3–28
Granule Diameter (μm)	13.8–14.5	13.9–14.2		14.1	4.12	2–10	5.4–5.6	34–36	13.8–14.2
AMY/AMP	25:75	1:99	52–68:32–48	30:70	26:74	19:81	1:99	20:80	17:83
Paste Temperature (°C)	72	69	75	64	65	79	78.5	61	59
Peak Viscosity (BU)	470	1000	Not detected	298		420	—	2500	1400
Viscosity at 95°C (BU)	470	400	Not detected	280		400	—	850	520
Viscosity Hold 95°C/30 min (BU)	350	250	86	230		350	—	340	280
Viscosity at 50°C (BU)	830	390	355	543		840	—	600	500
Viscosity Hold at 50°C/30 min (BU)	760	370	265	530		760		630	510
Retrogradation Rate	High	Low	High	High		Medium	Low	Medium	Medium

Sources: Huang, D.P., and Rooney, L.W., Starches for snack foods. Ch. 5 In: *Snacks Foods Processing*. E. Lusas and L.W. Rooney (eds). Technomic Publishing Co., Lancaster, PA, 2001; Jacobson, M.R. et al., *Cereal Chem.* 74(5):511–518, 1997; Shuey, W.C. and Tipples, K.H., *The Amylograph Handbook*. American Association of Cereal Chemists, St. Paul, MN, 1982; Serna Saldivar, S.O., Manufacturing of cereal-based dry milled fractions, potato flour, dry masa flour and starches. Ch. 2 In: *Industrial Manufacture of Snack Foods*. Kennedys Publications Ltd., London, 2008; Snyder, E.M., Industrial microscopy of starches. Ch. XXII In: *Starch: Chemistry and Technology*, R.L. Whistler, J.N. Bemiller, and E.F Paschall (eds). Second Edition, Academic Press, Orlando, FL, 1984; Zobel, H.F., Gelatinization of starch and mechanical properties of starch pastes. Ch. IX In: *Starch: Chemistry and Technology*. R.L. Whistler, J.N. Bemiller, and E.F. Paschall (eds) Second Edition. Academic Press, San Diego, 1984.

Note: Tests were conducted with 35 g starch/500 mL, 60 min holds at 95°C and 50°C.

accepted and used to manufacture most of the 25,000 metric tons manufactured annually worldwide (Mitchell 2009). However, rice starch is unique in terms of properties and niche functionalities because it synthesizes compound starch granules that, upon milling, release tiny starch angular-shaped subunits that only measure 3 to 10 µm (Table 11.1).

11.4.1 RAW MATERIAL

Most of the rice starch is obtained from milled broken rice, which is composed of second heads and brewer's rice, which represents approximately 18% of the milled rice. Needless to say, the broken rice has a lower market value compared with head rice (Mitchell 2009).

11.4.2 ALKALI WET MILLING PROCESS

Rice starch is commonly isolated by alkali extraction because the process provides high yield and high purity and is low in capital costs. The main disadvantage of this process is that it yields an alkaline effluent that contributes significantly to the general costs of wastewater treatment. To obtain prime starch, the white grain or broken kernels have to be treated with sodium hydroxide. The alkali enhances the softening of the vitreous endosperm via the solubilization of the glutelins that comprise most of the endosperm proteins. The resulting starch usually contains less than 1% protein. The broken milled rice is steeped in a 0.3% to 0.5% NaOH solution adjusted to 50°C for up to 24 h (Figure 11.5; Mitchell 2009).

The steeped stock is fed to disc or attrition mills and the starch slurry is allowed to rest for 10 to 24 h to further dissolve the proteins. The fiber is removed by filtration and the starch recovered by centrifugation. The starch is washed with water to remove excess alkali and then dehydrated to approximately 10% moisture. Protein could be recovered from the steeping and centrifugation effluents after first adjusting the pH to 6.4. The protein is recovered using filter presses or centrifuges (Mitchell 2009).

11.4.3 MECHANICAL WET MILLING PROCESS

The main advantage of this particular process is the assortment of starches with different proteins and functionalities that may be produced. Furthermore, the mechanical wet milling of rice requires less water and has the least detrimental or negative effect on the environment.

The mechanical process of rice starch production is based on the physical separation of the starch granules and endosperm protein from the white polished rice. The released starch granules exist in small aggregates of 10 to 20 µm in diameter. Depending on the process, the starch can be completely refined (<0.25% protein) or may contain up to 7% protein. The refined starch is similar in appearance to its alkali-processed counterpart but possesses different amylograph properties (Mitchell 2009).

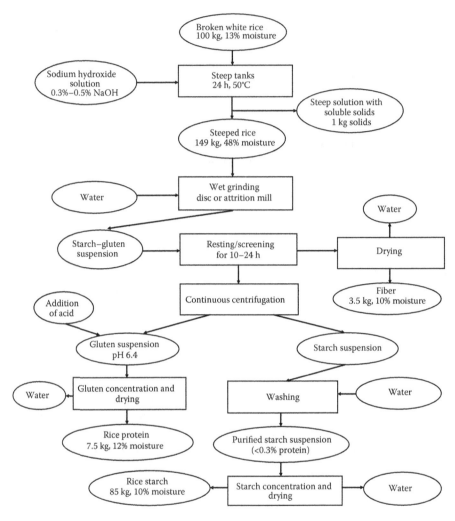

FIGURE 11.5 Flowchart of the alkaline wet milling process for the production of rice starch.

11.4.4 ENZYMATIC WET MILLING PROCESS

Puchongkavarin et al. (2005) developed an experimental rice starch extraction procedure that uses a couple of enzymes applied to liberate starch granules. The production process of starch from polished white rice was affected by the application of cellulase under slightly acidic conditions to degrade cell walls followed by protease under neutral conditions to loosen the protein bodies that are associated with starch granules. In comparison with the conventional alkaline process, the proposed enzyme technology yielded starch with slightly higher residual protein (0.48% vs. 0.6%) but damaged granules to a lesser extent (4.14% vs. 2.65%). An additional washing of the enzyme-isolated starch with 0.2% sodium hydroxide or 0.5% sodium dodecylsulfate solutions further improved the purity of rice starch. The physicochemical properties (viscoamylograph, differential scanning calorimetry [DSC], and swelling power) of the enzymatically isolated starches were comparable with starch from the alkaline process. The authors claim that the process they have developed allows the replacement of the alkaline process and thus eliminates the levels of harmful effluents from rice refiner plants.

11.4.5 PRODUCTS AND COPRODUCTS

The physical and chemical properties of rice starches are unique especially due to their relatively smaller size and wide range of amylose/amylopectin ratios. The small starch granules, upon cooking and cooling, have a high freeze–thaw stability and better acid resistance compared with other cereal starches. Starch is mainly used as a cosmetic dusting powder and as a pudding, especially for the production of baby foods (Mitchell 2009; Serna Saldivar 2010). Rice starch is also ideally suited for the production of gluten-free foods. Depending on the amylopection/amylose ratio, the starch can exhibit a range of gelatinized textures and strengths. Regular starches have a higher peak, cooked and cooled viscosities, and paste textures classified as short and pasty (Table 11.1). On the other hand, the waxy rice starch, upon cooking and cooling, yields long and stringy pastes. The wet milling of rice yields protein-rich coproducts that upon drying can be used as animal feed. Unlike the alkali-solubilized protein, the protein resulting from the mechanical process is a valuable coproduct suitable for the food industry.

11.5 SORGHUM WET MILLING

Sorghum (*Sorghum bicolor* L. Moench) can be used as a raw material for the production of starch. The wet milling process of sorghum is practically identical to the process previously described for maize (Eckhoff and Watson 2009; Munck 1995; Rooney and Serna Saldivar 2000; Serna Saldivar 2010). The major difference between maize and sorghum is the difficulty of separating sorghum starch and gluten. According to Eckhoff and Watson (2009), the sorghum pericarp also contributes to the difficulty because it is more fragile compared with the maize counterpart. Small particles of sorghum pericarp interfere with the starch and gluten separation. In addition, the phenolics associated to the sorghum pericarp may impart a pink coloration to the starch (Moheno Perez et al. 1999; Mezo Villanueva and Serna Saldivar 2003; Rooney and Serna Saldivar 2000; Serna Saldivar and Mezo Villanueva 2003).

Decortication techniques can be used to selectively remove most of the bran and concentrate the starch in the stock. In fact, steeping of decorticated kernels improves the color of the starch and lowers its protein content. The processing of pearled sorghum gluten also yields gluten with a higher protein content and allows a better separation of the germ, which is more attractive for oil extraction. Alternatively, decorticated kernels can be degerminated before steeping by roller milling and germ separation by flotation. Steeping of the resulting endosperm pieces can be significantly reduced to only 8 to 10 h instead of 30 to 48 h. Other documented disadvantages of sorghum are that it contains a smaller germ and produces a cream-colored gluten instead of the characteristic highly pigmented yellow coloration of maize. However, the properties of regular and waxy sorghum starches are similar to those extracted from their respective maize counterparts. Research clearly indicated that regular and waxy sorghum kernels subjected to sulfur dioxide steeping and treated with a cocktail of cell wall-degrading enzymes or proteases yielded more prime starch compared with their untreated counterparts and produced similar amounts of starch compared with conventionally treated maize (Eckhoff and Watson 2009; Moheno Perez et al. 1999; Mezo Villanueva and Serna Saldivar 2004; Serna Saldivar and Mezo-Villanueva 2003).

11.6   WET MILLING OF RYE, TRITICALE, BARLEY, OR OATS

The industrial production of rye (*Secale cereale*), triticale (*Triticum secale*), barley (*Hordeum vulgare*), and oat (*Avena sativa*) starches is very limited or practically nonexistent today. However, rye and oat starches were produced during the Second World War when maize, wheat, and potato starches became scarce. The extraction of rye starch is more difficult compared with wheat because of its higher pentosan content and the poor gluten-forming ability of its endosperm proteins. The industrial production of rye starch uses refined flour as feedstock. The flour is treated with sulfur dioxide (0.05%–0.15%) and *Lactobacillus* culture media tempered to 30°C to 48°C for 5 to 24 h. After the incubation step, starch is separated by centrifugation and sieving. The average yield of starch using this particular process is approximately 50% (Autio and Eliasson 2009a). The starch of triticale can be extracted following the same procedure or by using any of the wheat wet milling technologies described previously. The industrial production of oat starch is very limited mainly due to difficulties in the clean separation of the starch. The starch content of dry milled oat flour is 67.0% to 73.5%. Unlike wheat starch, oat starch cannot be separated from the grain by selective hydration and centrifugation because of the hydrated bran and protein layers. Autio and Eliasson (2009b) described a commercial fractionation process in which dehulled oats or groats are first dry milled and the resulting meal soaked in a solution of cellulases and hemicellulases. The oat starch can be purified with repeated washings to obtain a refined starch with protein and lipid contents of 0.44% to 0.6% and 0.67% to 1.11% (dry basis), respectively. In addition to starch, the process also yields fiber and protein fractions. There are several methods to isolate barley starch, even though it is not commercially produced. In contrast with maize starch, the separation of barley starch presents difficulties mainly attributed to its high β-glucan content, which produces high viscosities in aqueous solutions and impairs the separation of starch by screening and subsequent centrifugation. The methods are classified as acid, mercuric oxide, or enzymatic. The major steps of the acid method are steeping with acid-cracked kernels followed by neutralization, grinding, and screening. The key to the mercuric chloride method is the steeping of the barley kernels in an aqueous solution (0.01 M), screening, and repeated toluene washings to remove residual proteins. On the other hand, the enzymatic method hydrolyzes the grain components with proteases, cellulases, xylanase, and glucanases to release starch granules or produce less viscous solutions rich in starch that is separated by screening and centrifugal force. The enzymatic process yields several products: starch, protein, β-glucans, bran, and tailings resulting in starch yields in the range of 44% to 54% of the total dry matter. The resulting barley starch exists in two distinct populations: large and lenticular (A type), and small and irregularly shaped (B type) granules (Vasanthan and Hoover 2009).

11.7 PROPERTIES OF STARCHES

Cereal grains store most of its energy in starch granules. These granules contain the linear and branched starches named amylose and amylopectin, respectively. Amylose is essentially a linear polymer composed of glucose units linked with α,1-4 glycosidic bonds, whereas amylopectin also has several α,1-6 glycosidic bonds. In amylopectin, only 4% to 5% of the total glycosidic bonds are α,1-6. Amylose is usually composed of approximately 1500 glucose units, whereas amylopectin is a larger molecule containing up to 600,000 glucose units. The ratio of amylose/amylopectin in regular starches is generally 25:75, whereas in waxy starches, it is more than 95% amylopectin. Although amylose is called a linear starch, the molecule forms helixes. The helicoidal conformation allows amylose to complex with iodine, free fatty acids, alcohols, and emulsifiers. One important characteristic of amylose is that it stains purple with iodine and is the most important molecule related to retrogradation and the formation of cohesive gels (Serna Saldivar 2008, 2010).

Native starch granules are water-insoluble and swell reversibly when placed in water. The most important feature of native starches is that they show birefringence when exposed to the polarized light plane. This phenomenon occurs due to the high level of organization within starch granules, which makes these structures behave like pseudocrystals. Due to its larger size and more complex structure, amylopectin stains brownish-red with iodine. Amylopectin has a low retrogradation rate and forms weak and sticky gels (Shelton and Lee 2000; Shuey and Tipples 1982; Snyder 1984; Zobel 1984). The gelatinization temperatures of high-amylose starches are higher than those of normal and waxy counterparts (Case et al. 1998; Boltz and Thompson 1999). High-amylose starches are commonly used in the confectionery industry because starches have excellent gelling and film-forming properties.

The comparative physical and viscoamylograph properties of various types of starches are summarized in Table 11.1. The properties of wheat starches are similar to those of maize starch; however, maize starch is a better thickener. Among the cereal starches, rice has one of the smallest starch granules, varying in size from 3 to 10 μm. The granules are formed as polyhedral or pentagonal dodecahedrons. The main industrial use of rice starch is in the cosmetics industry. However, rice starch has unique viscoamylograph properties that make it ideal for use in extrusion processes (Shuey and Tipples 1982; Snyder 1984; Zobel 1984; Serna Saldivar 2008, 2010).

11.8 QUALITY CONTROL AND ASSURANCE

The quality control for wet milling industries is divided into two major types: the value of cereals or dry-milled fractions used as raw materials and the quality of refined starches. Table 11.2 summarizes the quality control measurements recommended for the different segments of the wet milling industries.

TABLE 11.2
Quality Control Parameters Most Commonly Used to Assess Quality and Functionality of Starches

Quality Control	Equipment Instrument	Importance
Chemical Tests		
Moisture content	NIRA, moisture meters, moisture balances, air oven method (AACC 44-15A)	Moisture is one of the most important parameters because it affects stability during storage and affects water required for optimum processing of starches into syrups, sweeteners, and other products
Ash content	Muffle furnace (AACC Method 08-01)	This is one of the most important parameters associated with milling efficiency
Protein content	NIRA or Kjeldhal methods (AACC 46-10, 46-12, or 46-13)	Protein content is one of the most important indicators of degree of starch refining. Prime starches usually contain less than 0.3% protein
Color score	Color meters (Hunter Lab, Agtron. and others)	Starch color is greatly affected by the effectiveness of the wet milling process especially in yellow maize rich in carotenoids
Sulfur dioxide/sulfite content	Monier Williams method (AOAC 1990)	Sulfur dioxide is distilled in a Monier Williams apparatus and the distillate titrated with NaOH. These sulfurous compounds are used in the wet milling process of corn. Residual sulfites are regulated because they cause allergic reactions
Starch content	NIRA, enzymatic analysis of total starch (AACC 76-11 and 13)	Determination of the most important product of wet milling processes. Most analytical methods involve total hydrolysis of starch with α-amylase and amyloglucosidase followed by glucose determination by oxidase–peroxidase reagent
Amylose and amylopectin contents	Colorimetric assay (Martinez and Proboliet 1996; Serna Saldivar 2012)	Amylose is usually quantified by the iodine colorimetric assay in which iodine binds with amylose to produce a blue complex. The starch is gelatinized with NaOH, neutralized and reacted with iodine. The amount of amylose is related to absorbance at 620 nm. Amylopectin is usually determined by difference
Iodine dye test	Iodine test (0.3% iodine–potassium iodide solution)	Iodine staining is routine practice to distinguish between the red staining waxy starch and their normal blue staining counterparts

(*continued*)

TABLE 11.2 (Continued)
Quality Control Parameters Most Commonly Used to Assess Quality and Functionality of Starches

Quality Control	Equipment Instrument	Importance
Congo red dye test	Congo red solution	Dye test used to differentiate between native and gelatinized and damaged starch. Undamaged native starch granules do not stain whereas damaged or gelatinized granules stain red
Birefringence endpoint temperature	Light microscope equipped with a Kofler heating platform	Starch gelatinization is measured by loss of granule birefringence in dilute granule suspensions subjected to an increased temperature profile, because not all the granules in any one starch species gelatinizes at the same temperature but rather over a range of about 10°C to 12°C. The birefringence endpoint temperature is the temperature at which 95% of the starch granules lose birefringence
Starch gelatinization	Birefringence with microscope equipped with polarized filters	Microscopic technique used to determine the relative amount of gelatinized versus native starch granules (starch granules show Maltese cross)
Starch damage	Enzyme methods AACC 76-30A and 76-31 (Williams and LeSeelleur 1970)	Damaged starch is the portion of starch that is mechanically disrupted. Starch damage is based on susceptibility to α- and β-amylases or amyloglucosidases. Undamaged starch granules are resistant to β-amylase whereas damaged counterparts are attacked at a measurable rate
Resistant starch	Enzyme digesting method (Saura Calixto et al. 1993)	The method uses the principle of determining residual starch from the insoluble dietary fiber residue. It consists of the enzymatic hydrolysis of starch with heat-stable α-amylase followed by a proteolytic degradation with a protease, and finally the hydrolysis of the starch with amyloglucosidase to yield glucose. Glucose is determined in the supernatant with glucose oxidase and peroxidase

(continued)

TABLE 11.2 (Continued)
Quality Control Parameters Most Commonly Used to Assess Quality and Functionality of Starches

Quality Control	Equipment Instrument	Importance
	Functionality Tests	
Viscosity	Brabender and rapid viscoamylograph (Deffenbaugh and Walker 1989; Shuey and Tipples 1982; Thiewes and Steeneken 1997; Walker and Hazelton 1996)	One of the most important functional tests, because after analyzing the viscoamylograph curve, the temperature at the start of gelatinization, peak viscosity, shear thinning and set back viscosity (retrogradation) could be determined. These properties are considered the fingerprint of the starch. The traditional and standard method to determine viscoamylograph properties is the Brabender and, more recently, the rapid viscoamylograph or improved version of the Brabender. The Rapid Visco Analyzer greatly reduced analysis time and correlates with the Brabender
	Viscosimeters	The determination of paste viscosity is one of the major factors affecting the functionality of starches. Viscosity tests should be carefully controlled in terms of the addition of water and temperature
Thermal analysis	DSC (Zobel 1984; Serna Saldivar 2010, 2012)	The DSC is a thermoanalytical assay in which the difference in the amount of heat required to increase the temperature of a sample are measured as a function of temperature (temperature increases linearly as a function of time). DSC analysis of starch–water slurries have been used to determine quantitatively starch gelatinization as an enthalpy ($-\Delta Hg$) of gelatinization. Enthalpy measurements can also be used to measure the return to crystallinity in aged starch gels. The endotherm excursions show a broad temperature range over which crystal structure is being melted. The temperature at which gelatinization is initiated is in general agreement with values reported from loss of birefringence measurements. In addition, the temperature needed to provide maximum starch disruption and high viscosity are in close agreement with the maximum or peak viscosity development in the viscoamylograph

(continued)

TABLE 11.2 (Continued)
Quality Control Parameters Most Commonly Used to Assess Quality and Functionality of Starches

Quality Control	Equipment Instrument	Importance
X-ray diffraction	X-ray crystallography (Thomas and Atwell 1999)	Starch granules behave like crystals and, when irradiated with x-rays, form distinctive patterns according to their crystal structure. Three general x-ray patterns have been identified in native starch. Most cereal starches produce an A pattern. The characteristic pattern is disturbed when starch is subjected to cooking and other thermal treatments
Gel rigidity and strength	Texture analyzer, penetrometer	Starch gelation occurs as the hydrated and dispersed molecules reassociate. Amylose containing starches set up quickly because linear amylose associates more readily than branched amylopectin molecules. Gelation is related to gel strength. Both parameters are measured with texture analyzers or with penetrometers

Sources: AACC Approved Methods of Analysis, Tenth Edition, American Association of Cereal Chemists, St. Paul, MN, 2000; AOAC, *Official Methods of Analysis of the Association of Official Analytical Chemists*. K. Helrich (ed). Fifteenth Edition, Arlington, VA, 1990; Deffenbaugh, L.B., and Walker, C.E., *Cereal Chem.* 66:493–499, 1989; Martinez, C., and J. Prodoliet, *Starch/Starke* 48(3):81–85, 1996; Saura Calixto, F. et al., *J. Food Sci.* 58:642–643, 1993; Serna Saldivar, S.O., Manufacturing of cereal-based dry milled fractions, potato flour, dry masa flour and starches. Ch. 2 In: *Industrial Manufacture of Snack Foods*. Kennedys Publications Ltd., London, 2008; Shuey, W.C. and Tipples, K.H., *The Amylograph Handbook*. American Association of Cereal Chemists, St. Paul, MN, 1982; Thiewes, H.J., and Steeneken, P.A.M., *Starch/Starke* 49(3):85–92, 1997; Thomas, D.J., and Atwell, W.A., Starch analysis methods. Ch. 2 In: *Starches*. Eagan Press, St. Paul, MN, 1999; Walker, C.E., and Hazelton, J.L., *Application of the Rapid Visco Analyser*. New Port Scientific Pty. Ltd., Warriewood, NSW, Australia, 1996; Williams, P.C., and LeSeelleur, G.C., *Cereal Sci. Today* 15:4–19, 1970; Zobel, H.F., Gelatinization of starch and mechanical properties of starch pastes. Ch. IX In: *Starch: Chemistry and Technology*. R.L. Whistler, J.N. Bemiller, and E.F. Paschall (eds). Second Edition. Academic Press, San Diego, 1984.

11.8.1 RAW MATERIALS

The determinations of grain or flour qualities are relevant for refiners because it relates to yields and quality of the starches. Grains are usually inspected in terms of moisture, dockage or foreign material, test weight, density, kernel weight, grain hardness, and starch content. Moisture and dockage are crucial because the cost of any lot of grain is adjusted according to these parameters and negatively influences starch and other chemical components as well as storage stability. The reference moisture content of cereal grains is, in most instances, 14%. This parameter is usually assessed by gravimetric assays (Methods 44-15, 44-19, and 44-40; AACC 2000). The most practical way to determine moisture in grain elevators is by the use of electronic probers (Method 44-11, AACC 2000). The nondestructive assay is performed on whole grains in a matter of seconds and is based on the principles that bound and free water conduct electricity differently. Another popular and widely used method is the determination of moisture with the near-infrared analyzer. Dockage is an important factor in the commercial value of any lot of grain. The foreign material includes other grains, stones, rodent feces, vegetative material such as sticks, cob pieces, and metal and glass contaminants. There are several methods to estimate foreign material (Methods 28-00, 28-01, and 28-03; AACC 2000). The amount of foreign material is important because this is closely related to wet milling yields and quality of the starch (color, aroma, and flavor). In addition, dockage is a source of insect and mold contamination and negatively affects grain stability during storage. The average kernel weight usually assessed by the 1000 kernel weight is frequently employed because it is an excellent indicator of grain size, which is related to wet milling yields. The industry prefers uniform and large kernels because they contain a higher proportion of endosperm or starch. The test is simple, practical, and fast and is usually performed using an automatic seed counter. Apparent and true grain densities are closely related to grain quality and endosperm texture or hardness. The most widely used is the bushel or volumetric weight (Method 55-10, AACC 2000), which is performed with a Winchester Bushel Meter. On the other hand, true grain density, generally expressed in grams per cubic centimeter, is generally determined by measuring the weight of a given volume that is displaced by a known weight of test material. True density can be determined by ethanol displacement or by air, nitrogen, or helium displacement using a pycnometer. The wet milling industry typically uses less dense or softer kernels because they require shorter steeping and yield more starch. There are an ample number of subjective tests to estimate grain hardness. Hardness is mainly affected by the ratio of corneous to floury endosperm and apparent and true density values. The assays most practiced consist of subjecting a lot of grain to the abrasive action of a mechanical decorticator, such as the tangential abrasive dehulling device, for a given amount of time, and the flotation index. Germ viability is important because it is closely related to grain soundness. Grains can die or lose their physiological activity due to heat damage and insect or mold attack during storage. Germ viability is usually determined by the tetrazolium test, which gives an indication of enzyme activity. The advantage is that it is fast and relatively inexpensive and can be performed rapidly (Serna Saldivar 2012).

11.8.2 LABORATORY WET MILLING

The classic SO_2 laboratory wet milling procedure was developed by Watson in the 1950s (Watson et al. 1951). Since then, the laboratory procedures have been perfected (Eckhoff et al. 1993, Johnson and May 2003). The basic steps of the milling assay are steeping in SO_2 and lactic acid solution, milling of the steeped kernels, and then separation and drying of the milled fractions. The procedures are useful because they are relatively fast and require small amounts of grain, and the recovery of the starch is very reproducible. Today, the Eckhoff et al. (1996) procedure is recognized as the standard method. Continuous countercurrent steeping systems have been developed for use in the laboratory (Steinke et al. 1991; Johnson and May 2003) to emulate commercial operations. These methods allow the recovery of prime and tail starches and its coproducts (fiber, germ, and gluten), which are then further characterized (i.e., color, protein content, and viscoamylograph properties). Laboratory methods for the production of wheat and rice starches are described by Serna Saldivar (2012).

11.8.3 STARCH QUALITY AND FUNCTIONALITY

Many different analytical and functional methods are extensively used to determine the chemical, physical, and functional properties of starches obtained from the various wet milling processes described previously. There are various methods used to quantify the total starch content in foods (Table 11.2). In most of these assays, the starch is first gelatinized and then enzymatically hydrolyzed into glucose that is colorimetrically assayed after reaction with glucose oxidase, peroxidase, and dihydrochloride *o*-dianasidine. Starch can also be quantified by the polarimetric AACC Method 76-20 (AACC 2000). Basically, the starch is first solubilized in an alkaline solution (mercury chloride and ethanol) that is treated with calcium chloride acidified with acetic acid and further treated with stanium chloride. The filtrated sample is polarimetrically analyzed with a sodium light beam (Serna Saldivar 2010, 2012). The ratio between amylose and amylopectin is critically important because it greatly affects the functionality of starch. Amylose is usually quantified by iodine colorimetric assays in which iodine binds with amylose to produce a blue complex that is read in a spectrophotometer, whereas amylopectin is calculated by difference. Native starch is usually quantified by birefringence by observing the amount of starch granules showing the typical Maltese cross under a microscope equipped with polarized filters. The birefringence endpoint temperature test (BEPT) is measured using the same basic principle, but the microscope is equipped with a starch slurry heating device that heats at a controlled rate. The BEPT is considered the temperature at which 95% of the granules lose birefringence or crystallinity. Starch damage is based on the susceptibility of the starch to α-amylase or β-amylases and amyloglucosidases. Undamaged starch granules are resistant to β-amylase, whereas damaged counterparts are attacked at a measurable rate (Serna Saldivar 2010, 2012).

The amylograph measures the viscosity changes of starch slurries with a certain amount of solids that are subjected to a standardized and programmed temperature regime. The classic instrument is the Brabender apparatus. The complete viscoamylograph curve is obtained after four distinctive and sequential stages: temperature increase, hot temperature hold, temperature decrease or cooling, and cold temperature hold. The instrument records the viscosity of the starch slurry expressed in viscoamylograph units. Generally, the assay starts at 50°C with a temperature increase of 1.5°C/min until the maximum temperature of 95°C is achieved. Next, the temperature is held for 15 to 30 min followed by gradual cooling (–1.5°C/min) until reaching 50°C. The final stage is the temperature hold at 50°C. The entire test or run can last for 120 min (Thomas and Atwell 1999). Table 11.1 summarizes the amylograph properties of different kinds of starches. The Rapid Visco Analyzer (RVA) is an instrument that operates under the same principles as the Brabender amylograph. The advantage of the RVA is that it considerably shortens the run time while maintaining a high correlation with the Brabender (Crosbie and Ross 2007; Deffenbaugh and Walker 1989; Thiewes and Steeneken 1997). Regardless of the type of instrument, the amylograph curve is considered as the fingerprint of starches because it determines the initial change in viscosity related to gelatinization temperature, peak viscosity during heating, the viscosity fall after the peak (shear thinning), and viscosity changes through the cooling cycles related to retrogradation. These instruments are the most frequently used to determine native and modified starch properties related to gelatinization, viscosity, and retrogradation (Serna Saldivar 2010, 2012).

DSC is a thermoanalytical assay in which the difference in the amount of heat required to increase the temperature of a sample is measured as a function of temperature (temperature increases linearly as a function of time). In cereals, DSC analysis of starch-rich and starch–water slurries has been used to quantitatively determine starch gelatinization as an enthalpy ($-\Delta Hg$) of gelatinization. Enthalpy measurements can also be used to measure the return to crystallinity in aged starch gels. The endotherm excursions show a broad temperature range over which crystal structure is being melted. The temperature at which gelatinization starts generally agrees with the values reported from loss of birefringence measurements. In addition, the temperature needed to provide maximum starch disruption and high viscosity is in accordance with the maximum or peak viscosity development in the viscoamylograph (Zobel 1984). The instrument is used to determine the degree of gelatinization in bakery products and measure their rate of retrogradation. The parameters usually evaluated are enthalpy of crystal fusion, onset temperature, and transition temperature (Serna Saldivar 2010, 2012).

REFERENCES

AACC Approved Methods of Analysis. 2000. Tenth Edition, American Association of Cereal Chemists, St. Paul, MN.

AOAC. 1990. *Official Methods of Analysis of the Association of Official Analytical Chemists.* K. Helrich (ed). Fifteenth Edition, Arlington, VA.

Anderson, R.A., Vojnovich, C., and Griffin, E.L. 1960. Wet milling high-amylose corn containing 49- and 57-percent amylose starch. *Cereal Chem.* 37:334–342.

Autio, K., and Eliasson, A.C. 2009a. Rye starch. Ch. 14 In: *Starch: Chemistry and Technology,* J.N. Bemiller and R. Whistler (eds). Third Edition, Academic Press, New York.

Autio, K., and Eliasson, A.C. 2009b. Oat starch. Ch. 15 In: *Starch: Chemistry and Technology,* J.N. Bemiller and R. Whistler (eds). Third Edition, Academic Press, New York.

Boltz, K.W., and Thompson, D.B. 1999. Initial heating temperature and native lipid affects ordering of amylose during cooling of high amylose starches. *Cereal Chem.* 76:204–212.

Case, S.E., Capitani, T., Whaley, J.K., Shi, Y.C., Trzasko, P., Jeffcoat, R., and Goldfarb, H.B. 1998. Physical properties and gelation behavior of a low-amylopectin maize starch and other high-amylose maize starches. *J. Cereal Sci.* 27:301–314.

Corn Refiners Association. 2012. The positive economic impact of corn wet milling. *Corn Refiners Association 2011 Annual Report.* Washington, DC.

Crosbie, G.B., and Ross, A.S. 2007. *The RVA Handbook*. American Association of Cereal Chemists, St. Paul, MN.

Deffenbaugh, L.B., and Walker, C.E. 1989. Comparison of starch pasting properties in the Brabender viscoamylograph and the rapid visco-analyzer. *Cereal Chem.* 66:493–499.

Eckhoff, S.R., and Tso, C.C. 1991. Wet milling of corn using gaseous SO_2 addition before steeping and the effect of lactic acid on steeping. *Cereal Chem.* 68:248–251.

Eckhoff, S.R., and Watson, S.A. 2009. Corn and sorghum starches: production. Ch. 9 In: *Starch: Chemistry and Technology*, J.N. Bemiller and R. Whistler (eds). Third Edition, Academic Press, New York.

Eckhoff, S.R., Rausch, K.D., Fox, E.J., Tso, C.C., Wu, X., Pan, Z., and Buriak, P. 1993. A laboratory wet-milling procedure to increase reproducibility and accuracy of products yields. *Cereal Chem.* 70:723–727.

Eckhoff, S.R., Singh, S.K., Zher, B.E., Rausch, K.D., Fox, E.J., Mistry, A.K., Haken, A.E. et al. A laboratory wet corn-milling procedure. *Cereal Chem.* 73:54–57.

FAO (Food Agriculture Organization). 2012. Statistical Database. Rome, Italy. http://faostat. fao.org.

Huang, D.P., and Rooney, L.W. 2001. Starches for snack foods. Ch. 5 In: *Snacks Foods Processing*. E. Lusas and L.W. Rooney (eds). Technomic Publishing Co., Lancaster, PA.

Jacobson, M.R., Obanni, M., and Bemiller, J.N. 1997. Retrogradation of starches from different botanical sources. *Cereal Chem.* 74(5):511–518.

Johnson, L.A., and May, J.B. 2003. Wet milling: the basis for corn biorefineries. Ch. 12 In: *Corn: Chemistry and Technology*. P.J. White and L.A. Johnson (eds). Second Edition. American Association of Cereal Chemists, St. Paul, MN.

Johnston, D.B., and Singh, V. 2001. Use of proteases to reduce steep time and SO_2 requirements in a corn wet milling process. *Cereal Chem.* 78(4):405–411.

Kempf, W., and Rohrmann, C. 1989. Process for the industrial production of *wheat* starch from whole *wheat*. Ch. 31 In: *Wheat is Unique*. Y. Pomeranz (ed). American Association of Cereal Chemists, St. Paul, MN.

Knight, J.W., and Olson, R.M. 1984. Wheat starch: production, modification and uses. Chapter 15 In: *Starch: Chemistry and Technology*, R.L. Whistler, J.N. Bemiller and E.F. Paschall (eds). Second Edition, Academic Press, Orlando, FL.

Ling, D., and Jackson, D.S. 1991. Corn wet milling with a commercial enzyme preparation. *Cereal Chem.* 68:205–206.

Maningat, C.C., Seib, P.A., Bassi, S.D., Woo, K.S., and Lasater, G.D. 2009. Wheat starch: production, properties, modification and uses. Ch. 10 In: *Starch: Chemistry and Technology*, J.N. Bemiller and R. Whistler (eds). Third Edition, Academic Press, New York.

Martinez, C., and Prodoliet, J. 1996. Determination of amylose in cereal and noncereal starches by a colorimetric assay: collaborative study. *Starch/Starke* 48(3):81–85.

Mezo Villanueva, M., and Serna Saldivar, S.O. 2004. Effect of protease addition on starch recovery from steeped sorghum and maize. *Starch/Starke* 56(8):371–378.

Mitchell, C.R. 2009. Rice starches: production and properties. Ch. 13 In: *Starch: Chemistry and Technology*, J.N. Bemiller and R. Whistler (eds). Third Edition, Academic Press, New York.

Moheno Perez, J.A., Almeida Dominguez, H.D., and Serna Saldivar, S.O. 1999. Effect of fiber degrading enzymes on wet milling and starch properties of different types of sorghums and maize. *Starch/Starke* 51:16–20.

Munck, L. 1995. New milling technologies and products: whole plant utilization by milling and separation of the botanical and chemical components. Ch. 8 In: *Sorghum and Millets: Chemistry and Technology*. D.A.V. Dendy (ed). American Association of Cereal Chemists, St. Paul, MN.

Puchongkavarin, H., Varavinit, S., and Bergthaller, W. 2005. Comparative study of pilot scale rice starch production by an alkaline and an enzymatic process. *Starch/Stärke* 57:134–144.

Rooney, L.W., and Serna Saldivar, S.O. 2000. Sorghum. Ch. 5 In: *Handbook of Cereal Science and Technology*, K. Kulp and J.G. Ponte (eds). Second Edition. Marcel Dekker, Inc., New York.

Saura Calixto, F., Goñi, I., Bravo, L., and Mañas, E. 1993. Resistant starch in foods: modified method for dietary fiber residues. *J. Food Sci.* 58:642–643.

Schofield, J.D., and Booth, M.R. 1983. Wheat *proteins* and their technological significance. pp. 1–65 In: *Developments in Food Proteins-2*. B.J.F. Hudson (ed). Applied Science Publishers, New York.

Schwartz, D., and Whistler, R.L. 2009. History and future of starch. Ch. 1 In: *Starch: Chemistry and Technology*, J.N. Bemiller and R. Whistler (eds). Third Edition, Academic Press, New York.

Serna Saldivar, S.O. 2008. Manufacturing of cereal-based dry milled fractions, potato flour, dry masa flour and starches. Ch. 2 In: *Industrial Manufacture of Snack Foods*. Kennedys Publications Ltd., London.

Serna Saldivar, S.O. 2010. *Cereal Grains: Properties, Processing and Nutritional Attributes*. CRC Press (Taylor & Francis Group), Boca Raton, FL. 705 p.

Serna Saldivar, S.O. 2012. *Cereal Grains: Laboratory Reference and Procedures Manual*. CRC Press (Taylor & Francis Group), Boca Raton, FL. 368 p.

Serna Saldivar, S.O., and MezoVillanueva, M. 2003. Effect of cell wall degrading enzymes on starch recovery and steeping requirements of sorghum and maize. *Cereal Chem.* 80(2):148–153.

Shelton, D.R., and Lee, W.J. 2000. Cereal carbohydrates. Ch. 13 In: *Handbook of Cereal Science and Technology*, K. Kulp and J.G. Ponte (eds). Second Edition. Marcel Dekker, Inc., New York.

Shuey, W.C. and Tipples, K.H. 1982. *The Amylograph Handbook*. American Association of Cereal Chemists, St. Paul, MN.

Snyder, E.M. 1984. Industrial microscopy of starches. Ch. XXII In: *Starch: Chemistry and Technology*, R.L. Whistler, J.N. Bemiller, and E.F. Paschall (eds). Second Edition, Academic Press, Orlando, FL.

Steinke, J.D., Johnson, L.A., and Wang, C. 1991. Steeping maize in the presence of multiple enzymes. II. Continuous countercurrent steeping. *Cereal Chem.* 68:12–15.

Thiewes, H.J., and Steeneken, P.A.M. 1997. Comparison of the Brabender viscograph and the Rapid Visco Analyser. *Starch/Starke* 49(3):85–92.

Thomas, D.J., and Atwell, W.A. 1999. Starch analysis methods. Ch. 2 In: *Starches*. Eagan Press, St. Paul, MN.

Vasanthan, T., and Hoover, R. 2009. Barley starch: production, properties, modification and uses. Ch. 16 In: *Starch: Chemistry and Technology*, J.N. Bemiller and R. Whistler (eds). Third Edition, Academic Press, New York.

Walker, C.E., and Hazelton, J.L. 1996. *Application of the Rapid Visco Analyser*. New Port Scientific Pty. Ltd., Warriewood, NSW, Australia.

Watson, S.A., Williams, C.B., and Wakely, R.D. 1951. Laboratory steeping procedure used in a wet-milling research program. *Cereal Chem.* 28:105.

Williams, P.C., and LeSeelleur, G.C. 1970. Determination of damaged starch in flour. *Cereal Sci. Today* 15:4–19.

Zobel, H.F. 1984. Gelatinization of starch and mechanical properties of starch pastes. Ch. IX In: *Starch: Chemistry and Technology*. R.L. Whistler, J.N. Bemiller, and E.F. Paschall (eds). Second Edition. Academic Press, San Diego.

12 Extrusion Cooking of Cereal-Based Products

Rolando J. González, Silvina R. Drago,
Roberto L. Torres, and Dardo M. De Greef

CONTENTS

12.1 INTRODUCTION

Cereals and their products have been revalued as food because the consumption of starch as a carbohydrate source has been recommended for a healthy diet (Lusas 2001). The 2005 US Dietary Guidelines Advisory Committee has recommended that consumers make "half their grain whole" (Miller Jones and Engleson 2010).

Besides that, changes in the food consumption model verified in modern society have induced consumers to demand for new foods. These changes have converted the food product concept into a more complex one. Today, a food product is composed of two components: the first one is related to food value itself (alimentary utilities), which includes food's nutritional components that satisfy physiological needs; the second one is related to nonalimentary utilities, which includes variety, novelty, convenience, healthiness, label information, etc., and satisfies psychological and sociological needs.

Modern consumers are willing to buy healthy, novel, and convenient foods. They also like having a variety of foods that is as wide as possible so that they can choose freely according to their own wishes.

The wide variety of cereal food products available in the market nowadays is directly related to the above concept, and extrusion technology allows cereal processors to produce such variety in response to the consumer's demands.

Food extrusion has been practiced for more than 60 years. It was initially used for forming macaroni and ready-to-eat cereal products. In these operations, relatively little heating or cooking was accomplished within the extruder. The extruder was used to mix and pressurize high-moisture food dough and force it through a shaping die at temperatures below the boiling point of water, where little or no puffing occurs. Screw cooking extruders were developed in the 1940s for making puffed snacks from cereal flours or grits.

Nowadays, food extrusion includes a variety of extruder designs, which allow the manufacture of a large number of products (having different shapes, colors, and appearances). Moreover, cooking extrusion is considered a high-temperature, short-time process, and the future of this technology seems to lie in its ability to produce novel foods (Guy and Horne 1988; Harper 1989; Fast 1991; Bouzaza et al. 1996; González et al. 2000).

The most prevalent use of food-type extruders is for the manufacture of dry and semimoist pet foods. Another important extruded product is textured soy protein, which is used as a meat-like ingredient or as a meat extender.

12.2 MATERIALS AND PRODUCTS

12.2.1 CEREALS

Cereals are members of the grass family (Gramineae). They produce one-seeded fruit, which is the caryopsis, commonly called the kernel or grain. It consists of a fruit coat or pericarp that surrounds the seed and adheres tightly to a seed coat. The seed consists of an embryo or germ and an endosperm enclosed by a nucellar epidermis and a seed coat. In general, all cereal grains have these same parts in approximately the same relationship with each other (Hoseney 1994).

The chemical constituents of cereals are often compartmentalized, which plays an important role in grain storage stability by preventing enzymes and their corresponding substrates from being in contact. In general, the chemical composition of cereals is characterized by its high starch content (>60%), which is located in the endosperm. Their protein content (from 6% to 14%) is lower than that of legumes, but is higher than that of tubers. Ash content varies from 1% to 3%, although that from bran and germ fractions is much higher (from 6% to 10%).

Wheat (*Triticum vulgaris*, *Triticum aestivum*, and *Triticum durum*), corn (*Zea mays* L.), also called maize and Indian corn, and rice (*Oryza sativa* L.) are the three crops of greatest worldwide production. Wheat and rice are mainly food crops, whereas maize is primarily a feed crop. The cereal most commonly used in the extrusion cooking process is maize; however, others such as rice or sorghum can also be used. Barley and rye have not been extensively used in extrusion.

Precooked cereal products, such as "snacks," breakfast cereals, flakes, instant flours, "arepas," and "nixtamalized flour," are good examples of convenience food, the demand for which has greatly increased (González et al. 1998).

12.2.1.1 Maize

The largest proportion of maize production is used as feed. Other uses include wet milling for starch production and dry milling for special types of grits and flours, used in the elaboration of different products such as beer, flakes, snacks, and "polenta." There are commercial quality grades to classify maize, but these quality criteria are not enough for food applications. Thus, the food industry is becoming aware about the need to establish quality parameters directly related to each use.

For some popular products, color is a determinant factor. For example, yellow maize is preferred for polenta and white maize is preferred for arepas. In Mexico, white maize has traditionally been preferred for making "tortillas," but because of the insufficient supply of this type of maize, the use of yellow maize has been accepted.

Figure 12.1 shows the three types of maize: dent, flint, and pop. Dent maize is characterized by its low mechanical resistance, mainly due to its high proportion of

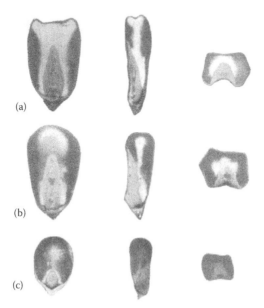

FIGURE 12.1 Maize types: (a) dent, (b) flinty, and (c) pop.

floury endosperm. On the contrary, flinty maize (hard and corneous) is characterized by its high proportion of horny endosperm, and finally, pop maize is constituted by almost 100% of horny endosperm and its pericarp is much thicker than that of the other two. These characteristics make popcorn unique for popping (Torres 2005).

Among the different commercial types of corn, those having hard or flinty endosperms are preferred by food processors for making products such as snacks and corn flakes (Robutti et al. 2002).

Several works have demonstrated the importance of the endosperm's characteristics in different applications. In dry milling, the proportion of "break flour" is related to the proportion of floury endosperm. In extrusion, snacks made from "waxy" maize are dense and have low expansion. Similarly, yellow dent maize gives less expanded products than those obtained from maize having harder endosperms (González et al. 2004). In the case of "puffing," hybrids having a higher proportion of flinty endosperm give higher product volume (Matz 1993). According to Whalen (1998), some physicochemical properties such as water distribution in the grain and pasting temperature are determinants of "corn flakes" quality.

Some authors have focused their studies on the physicochemical properties of the starch. They have found that percentage of crystallinity, amylose content, gelatinization temperature, and amount of free and bound lipids vary with the maize genotype (Guzmán et al. 1992; Chi-Tang and Izzo 1992; Bhatnagar and Hanna 1994a,b; Yamin et al. 1997). Others have studied the proteins and their relation with grain characteristics, such as grain density and breaking susceptibility (Wu and Bergquist 1991; Wasserman et al. 1992; Dombrink-Kurtzman and Knutson 1997; Pratt et al. 1995).

The physical properties of the maize endosperm affect its performance during the cooking process. For example, when maize grits are cooked in boiling water, grits from hard endosperm take more time to cook than grits from soft endosperm, whereas the contrary happens in extrusion cooking. In the first case, the differences are attributed to the restriction for starch swelling caused by the protein matrix (González et al. 2005), and in the second case, the difference in particle hardness is the key factor. During passage through the initial part of the extruder channel, more energy would be applied to harder particles, causing a temperature increase and speeding up the cooking process. Thus, harder particles would cook and become fluid more rapidly than softer ones. This is supported by die pressure values, which are lowest for the hardest particles (Robutti et al. 2002).

12.2.1.2 Wheat

Wheat is cultivated all over the world, and approximately 30,000 wheat varieties from 14 species are processed. However, only about 1000 varieties are of commercial significance (Posner and Hibbs 2005). Wheat is classified into white or red and durum varieties. The standard of wheat (*T. vulgaris*) is made according to the kernel hardness (from soft to hard). Durum wheat (*T. durum*), much harder than bread type one, is used for the production of speck-free semolina for pasta and is processed differently from common bread wheat, also being graded differently to assure pasta quality. In the case of extrusion cooking process, soft wheat should be used because no special property of gluten is needed for such a process. Moreover, because of its gluten content, higher temperatures and moisture are needed to obtain expanded products.

12.2.1.3 Rice

Rice is consumed principally as a grain (whole or polished) and, therefore, the texture of the cooked grain is a matter of primary importance. Different waxy and nonwaxy cultivars are usually classified according to their grain dimensions, amylose content, gelatinization properties of the extracted starches, and texture of cooked rice. Grain quality denotes different properties for different rice market sectors, ranging from that of breeders to consumers (Juliano 1982, 1985; Mackenzie 1994).

When rice is selected to elaborate extruded products, the amylose content has to be taken into account because it affects the cooking degree (CD) attained during extrusion and, consequently, extrudate texture will depend not only on extrusion process variables, but also on amylose content (González et al. 2000). Low-amylose rice is cooked easier than high-amylose varieties, either in boiling water or when rice grits are cooked through extrusion.

12.2.2 PRECOOKED CEREAL PRODUCTS

Precooked cereals can be found as flour, flake, puffed grains, expanded products, or even "pellets" with different shapes and degrees of expansion. Processes to obtain these products differ, according to the way of cooking; popcorn being a special case.

Extruded flakes differ from those made using traditional processes in that, instead of cooking the individual endosperm pieces (~4–5 mm) in a batch steam cooker and then flaking them, previously moistened grits (approximately 0.6 to 1.0 mm in size) are continuously cooked in the extruder and then the "extrudate pieces" coming out from the extruder are flaked. A good description of these processes was done by Fast (1991).

Regarding precooked flour, extrusion cooking has the advantage of producing a variety of CDs, which determines the final use. For example, to prepare an instant nixtamalized maize flour or even precooked flour for arepas or polenta, extrusion conditions needed to obtain a relatively low degree of cooking have to be used. Such conditions could be high moisture and low shear. To prepare instant flour for cream soup formulations, a high degree of cooking is necessary (low moisture and high shear).

Expanded products can be obtained by direct extrusion using conditions that allow a high degree of cooking (low grit moisture content and high temperature). "Pellets" are another type of product with varied shapes, colors, and flavors. Cooked dough obtained from a low shear cooking extruder is fed to a forming extruder to produce pellets, which can be expanded in hot air or in frying oil.

12.3 MANUFACTURING: TECHNOLOGICAL AND ENGINEERING ASPECTS

12.3.1 CEREAL COOKING PROCESS

The cereal cooking process can be classified into two main types: hydrothermal cooking and thermomechanical cooking. The first process is characterized by relatively high moisture content (>30%). This process can be done at atmospheric pressure, as in the case of nixtamalized maize, or at higher pressure (steam cooker), as in the case of the production of cereal flakes and parboiled rice. During this treatment, starch granules partially or totally lose their crystalline structures, but little to no amount of granule destruction is attained. This characteristic is very important for the texture perceived when the product is eaten.

The second type of cooking process is characterized by two important cooking factors: the relatively low moisture content (<30%) and the high shear stress applied to the starchy material being processed. As a consequence, a considerable proportion of starch granules are destroyed. Extrusion cooking belongs to this type of process.

An important point to consider when cereals are cooked is the relationship between the gelatinization temperature (GT) and the amount of water surrounding the starch granules (Blanshard 1987). When a dispersion of starch granules is heated in excess of water, GT is independent of the water content, but from approximately 50% water, GT starts to increase as water content decreases.

12.3.2 EXTRUSION PROCESS

Cooking–extrusion of starchy material has become a highly used technique for obtaining a wide range of products such as snacks, breakfast cereals, and special flours (for instant soup mixes, porridge, etc.; Guy and Horne 1988; Harper 1989; Bouzaza et al. 1996; González et al. 2000). It is a versatile technological process that allows for a variety of CDs and, consequently, a variety of textures by controlling extrusion process variables such as moisture, die diameter, and degree of external heating transferred from the barrel.

The process converts a noncohesive granular material (grits and flour), composed of biopolymers (starch and proteins), into a restructured solid in one operation. This thermomechanical cooking involves the conversion of solid material in a viscoelastic fluid or "melt." That is, the transport mechanism through the extruder changes along the screw from solid to fluid flow. As a consequence of the pressure built-up during fluid flow, high shear stresses are developed, which cause structural transformations in the material. These transformations include the loss of starch's crystalline structure, destruction of granular structure, rupture of glycoside bonds, and new molecular interactions (Davidson 1992; Mitchell and Areas 1992; González et al. 2002). Finally, as the melt comes out from the die, it expands as a consequence of the water flashing from the liquid state to the vapor state. Such expansion will depend on the melt's viscoelastic properties (Arhaliass et al. 2003). In summary, the rupture of crystalline and granular structures determines the CD of the starchy material.

Viscous energy dissipation together with shear stress fields are the main factors responsible for the material's transformation intensity. These factors depend on extruder design and process variables, such as screw compression ratio (CR), die diameter, screw speed, material moisture, residence time, type of material, etc. (González et al. 2002; Meuser and Van Lengerich 1992; Llo et al. 1996).

Figure 12.2 shows die viscosity (in pascal seconds) versus die pressure (in pascals) corresponding to red and dent maize, high amylose and waxy rice, and amaranth samples extruded at 15% and 30% moisture, for three rpm (77, 120, and 173)

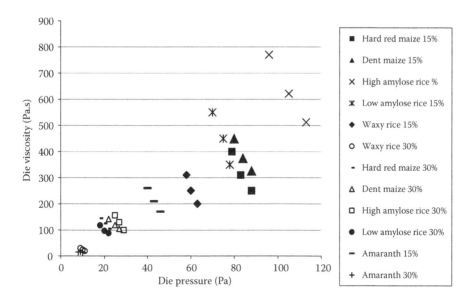

FIGURE 12.2 Die viscosity (η) versus pressure (Pa) corresponding to red maize, high amylose, and waxy rice and amaranth samples extruded at 15% and 30% moisture, for three rpm (77, 120, and 173), using a 10 DN Brabender extruder, a 3:1 CR, and 3 × 20 ($d × l$) mm die.

using a 10 DN Brabender extruder, a 3:1 CR, and 3 × 20 ($d × l$) mm die. Die viscosity decreases as P increases due to the pseudoplastic behavior of the melt because P is directly related to shear stress. An additional factor that reduces die viscosity is the degree of material structure destruction caused by an increasing amount of shear, an effect which is more evident at low moisture contents. At 30% moisture level, die viscosity and pressure are much lower than the corresponding values at 15% because the effect of the type of material is also lower.

12.3.3 EXTRUDER DESIGNS

Despite the extensive capability and broad range of applications developed for single-screw extruders, the food industry is evaluating and using a variety of twin-screw machines in food processing applications. Interest in twin-screw extruders occurred because of their operational capabilities, which have extended their range of applications. In addition to their use in manufacturing goods produced on single-screw extrusion equipment, twin-screw extruders are being applied for the production of confectionery products, including caramels and toffees, clear hard candies and gums, expanded cereal bases for candy production, and chocolate (Harper 1992).

Single-screw extruders can be a single piece, but a splinted shaft that accepts screw sections of varying configurations increases the versatility and reduces the cost of replacing worn sections (Harper 1992). In twin-screw extruders, two parallel screws are placed in a figure-eight sectional barrel. The primary objective of this geometry is to overcome problems of slipping at the wall (Janssen 1989).

Twin-screw extruders can be classified into closely intermeshing counterrotating or corotating, self-wiping, and nonintermeshing extruders. Figure 12.3 (single-screw designs) and Figure 12.4 (double-screw designs) show some of the designs (single and double screws) used for food extrusion.

Corotating machines have screws turning in the same direction. Such machines have been the most popular for food processing because of their higher capacity resulting from the relatively large void volume of the screws and higher screw speeds (Harper 1986). Single-screw extruders and nonintermeshing twin-screw extruders can best be compared with a drag pump. Closely intermeshing twin-screw extruders, whether corotating or counterrotating, can be modeled as positive displacement pumps. Corotating, self-wiping, twin-screw extruders belong to an intermediate class (Janssen 1989).

Both types of extruder designs, single and double screw, have similar functions. Compression is achieved in several ways, with a gradual decrease in flight in the direction of discharge being the most common (usually in a single-screw extruder). Another is a decrease in the pitch (usually in a double-screw extruder). Regarding mixing, kneading discs, kneading paddles, mixing discs, reverse pitch screws, and cut flight screws are employed (Harper 1992; Chao-Chi Chuang and Yeh 2004).

The major difference between a single-screw extruder and a twin-screw extruder lies in the mechanism of transportation. A single-screw extruder has a screw rotating in a closely fitting barrel. It is easy to see that if the material adheres to the screw and slips at the barrel wall, there will be no output and the material will rotate freely with the screw. To achieve maximal output, the material must slip as freely as possible at the screw surface and adhere as much as possible to the barrel wall. The

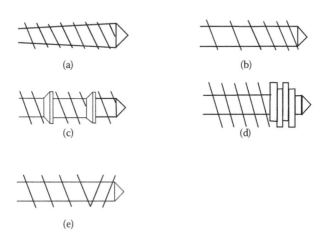

FIGURE 12.3 Single-screw designs: (a) increasing root diameter; (b) decreasing pitch, constant root diameter; (c) constant root diameter, constant pitch screw with restrictions in constant diameter barrel; (d) constant root diameter, constant pitch screw with mixing disc, or kneading paddles in constant diameter barrel; (e) constant root diameter and constant pitch screw with reverse pitch screws.

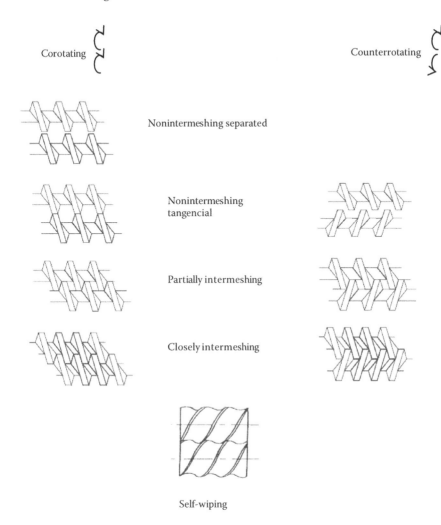

Corotating   Counterrotating

Nonintermeshing separated

Nonintermeshing
tangencial

Partially intermeshing

Closely intermeshing

Self-wiping

FIGURE 12.4   Corotating and counterrotating double screw designs.

latter restriction can be met by introducing extra friction at the barrel using grooves. Moreover, the screw must be as smooth as possible (Janssen 1989).

In single-screw extruders, the conveying action is a result of two friction effects: first, the friction between the screw and the product, and second, the friction between the barrel and the product. In twin-screw extruders with closely intermeshing screws, the product is enclosed between the screws and the barrel in C-shaped chambers. Therefore, it is prevented from rotating with the screws, and so is conveyed positively toward the die. Here, depending on the process conditions, friction at the barrel wall is less important (van Zuilichem et al. 1984).

Table 12.1 shows the main differences between single-screw and twin-screw extruders.

TABLE 12.1

Main Differences between Single-Screw and Twin-Screw Extruders

	Single-Screw Extruder	Twin-Screw Extruder	Comments
Transport mechanism	Friction between metal and food material	Positive displacement	
Conveying angle screw	~10°	~30°	About three times the effective conveying angle of single-screw extruders allows them to handle sticky and/or other difficult-to-convey feed ingredients
Throughput capacity	Dependent on moisture and fat content and pressure	Independent	
Relative residence time distribution spread	1.5	1.0	
Self-cleaning	No	Self-wiping	
Main energy supply	Viscous dissipation	Heat transfer to barrel	
Heat distribution	Large temperature differences	Small temperature differences	
Approximate specific power consumption per kilogram of product	900–1500 kJ/kg	400–600 kJ/kg	
Heat transfer	Poor-jackets control barrel wall temperature and slip at wall	Good in-filled sections	
Mixing	Poor	Good	
Mechanical power dissipation	Large shear forces	Small shear forces	To increase mechanical energy dissipation and enhance mixing, kneading discs are employed. Reverse pitch elements cause the kneading disc section to be filled and dissipate substantial energy
L/D	4–25	10–25	L/D increase is required if the twin-screw extruder's function is to accomplish a substantial amount of heat transfer. Machines having a L/D of more than 20:1 were used for this purpose

(*continued*)

TABLE 12.1 (Continued)
Main Differences between Single-Screw and Twin-Screw Extruders

	Single-Screw Extruder	Twin-Screw Extruder	Comments
Minimum water content	10%	8%	
Maximum water content	30%	95%	
Ingredients	Flowing granular materials	Wide range	
Capital costs (extruder)	Low	High	Twin-screw extruders are 1.5–2.0 times as expensive for a given throughput

12.4 KEY FACTORS AFFECTING QUALITY OF EXTRUDED COOKING CEREAL PRODUCTS

12.4.1 COOKING DEGREE

The quality of extrusion-cooked products is highly dependent on the CD of the starch fraction. Other factors are physical properties and size distribution of the particles. The term "cooking degree" (CD) is directly related to the proportion of starch structure (either crystalline or granular) that has been lost during the cooking process.

12.4.2 RESIDENCE TIME DISTRIBUTION

The transport of particles inside the extruder is characterized by its residence time distribution, which presents some dispersion and contributes to the heterogeneity of the degree of granular structure destruction. Even though the temperature levels were high enough to accomplish starch gelatinization, at a particular moisture level, it is possible to find native starch granules together with gelatinized granules (and fragments), and dispersed amylose and amylopectin in several degrees of aggregation in the extrudate. The first ones correspond to those particles that have the shortest residence time, and thus could not be cooked because their residence time was not enough for the cooking process to be accomplished. The proportion of each type of starch structures will affect the hydration properties (Haller 2008). Therefore, the extrusion cooking process converts starch granules into different "structural states" (native, gelatinized, fragments of granules, and macromolecular aggregates), which have to be taken into account when the effects of factors such as temperature (T) and moisture (M) are being analyzed. These two factors act in opposite directions on the friction exerted inside the extruder and, consequently, the effect of T and M on CD will depend on the relative magnitude of the effect of each factor. For extrusion conditions corresponding to low CD, native and gelatinized granules predominate, whereas for conditions corresponding to high CD, fragments of granules and macro-molecule dispersion predominate.

12.4.3 EXTRUSION CONDITIONS

The combination of the independent variables defines the cooking process and the final product quality. Figure 12.5 shows the wide range of extrudate characteristics that can be obtained from maize grits by changing grits moisture (M), extrusion temperature (T), and screw CR. Samples obtained at high T, high CR, and low M will show the highest CD in comparison with those obtained at low T, low CR, and high M. All these samples showed the disappearance of native starch granules, except those samples obtained at a 2:1 CR and 15% moisture at the three temperature levels. These three samples showed a very low CD; they contained grit particles and native starch granules, indicating that extrudate came out of the extruder without a complete conversion of the "solid flow" into "fluid flow" transport mechanism. This happened for that particular combination of the three process variables with a 4.25 mm die diameter and 120 rpm. This problem is solved by increasing the CR (4:1) or decreasing the die diameter.

Some of the response variables are directly related with CD. This is the case for specific mechanical energy consumption (SMEC), extrudate-specific volume (SV), and water solubility. Although others such as expansion, water absorption (WA), and extrudate water dispersion viscosity present a maximum at a certain value of CD, beyond this point, these properties are inversely related with CD.

Regarding the hydration properties of extrudates, water dispensability and WA are frequently used to evaluate CD. At low CD, native and gelatinized starch granules predominate, but as CD increases, the proportion of fragmented granules also increase, to the detriment of the proportion of gelatinized (swollen) granules. At a particular level of CD, a further increment of CD is characterized by a decrease in

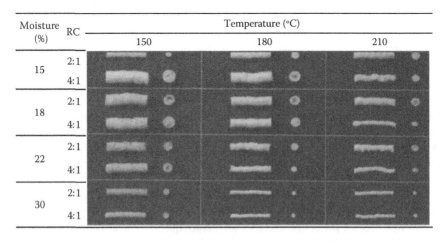

FIGURE 12.5 Corn grits extruded under the following conditions: Brabender 20 DN single-screw extruder, 120 rpm screw speed, and 4.25 mm die diameter.

WA. Figure 12.6 shows this effect. These data correspond to extruded whole rice samples, using a Brabender 20 DN single-screw extruder at the following conditions: moisture content, 14%, 16.5%, and 19%; temperature, 160°C, 175°C, and 190°C; a 4:1 CR; and 3 × 20 ($d × l$) mm die. It has been observed that WA reaches a maximum value of approximately 8.5 at a water solubility ($S\%$) value of approximately 17. Samples with $S\%$ values of less than 17 would contain more native starch granules and lower amounts of fragment than those samples with much higher $S\%$ values. However, when rice grits obtained from dehulled and degermed grains were extruded at the same extrusion conditions, the expected inverse relationship between $S\%$ and WA was observed: $S\% = -15.414$ WA $+ 117.76$; with $R^2 = 0.849$. In this regard, it has been observed that at the same extrusion conditions, CD of extruded whole-grain particles is reduced in comparison with those that are degermed and dehulled.

The cooking process and the degree of structure destruction can be also evaluated through the measurement of viscosity of extrudate water dispersion. Another way of seeing the effect of extrusion conditions on CD can be done by relating extrusion responses, such as SMEC and suspension viscosity of the extruded product, as observed in Figure 12.7. Samples were obtained from red maize grits, using a Brabender 20 DN extruder, at 150 rpm, 3 mm die, 4:1 screw CR, grits moisture level of 15% to 19%, and extrusion temperature from 150°C to 180°C. As CD increases (high T and low M), suspension viscosity decreases.

CD plays an important role in cereal food-based formulation; for example, in cream soup base formulations, it is known that to obtain a rice or maize extrudate dispersion with a high smoothness score, no grittiness, homogeneity, and stability, high CD is necessary and the $S\%$ value of the extrudate should be higher than 30. On the other hand, $S\%$ values lower than 10 are needed if the extruded maize flour is going to be used to prepare an "instant polenta" formulation. Extrusion conditions such as 180°C temperature, 15% moisture level, and a 4:1 CR would be needed for an extrudate to be used in a cream soup base formulation, whereas for "instant polenta," adequate extrusion conditions would be 150°C temperature, 30% moisture level, and a 2:1 CR.

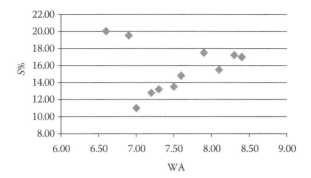

FIGURE 12.6 Relationship between water solubility and WA corresponding to extruded whole rice samples. A Brabender 20 DN single-screw extruder was used under the following conditions: moisture content, 14%, 16.5%, and 19%; temperature, 160°C, 175°C, and 190°C; a 4:1 CR and 3 × 20 ($d × l$) mm die.

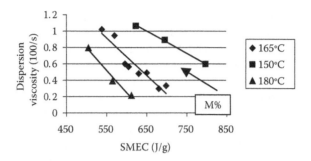

FIGURE 12.7 Relationships between SMEC and suspension viscosity of extruded maize grits measured at 100 s^{-1} shear stress and for 15% solid concentration. A Brabender 20 DN single-screw extruder was used under the following conditions: moisture content, 15%, 17%, and 19%; temperature, 150°C, 165°C, and 180°C; a 4:1 CR and 3 × 20 ($d \times l$) mm die.

12.4.4 RAW MATERIAL CHARACTERISTICS

The physical properties of the particles composing the material being extruded affect transport inside the extruder. Also, they affect the extrusion cooking process and, consequently, the CD reached by extrudates. Hard particles (such as those from horny endosperm maize) are transported faster than soft particles, as a consequence of the higher friction exerted on the barrel surface. A particular case would be the extrusion of material containing particles coming from the germ or pericarp, which reduces friction levels, interferes with the starch cooking process, and could broaden residence time distribution, as previously stated.

Die viscosity could be considered as an indicator of CD reached by the material being extruded and it can be related with other indicators such us $S\%$. Figure 12.8 shows the relationship between die viscosity and water solubility ($S\%$), corresponding to extruded maize grits from different endosperm hardness levels and rice grits from different rice amylose contents. Data were obtained using a Brabender 20 DN under the following conditions: 15% moisture, 3:1 CR, 2 × 20 ($d \times l$) mm die,

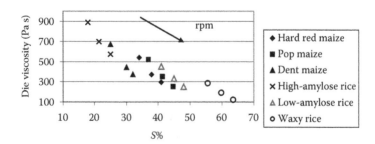

FIGURE 12.8 Die viscosity versus water solubility ($S\%$) corresponding to maize grits from different endosperm hardness levels and rice grits from different rice amylose contents extruded with a Brabender 20 DN under the following conditions: 15%, three rpm (77, 120, and 173), 3:1 CR, and 2 × 20 ($d \times l$) mm die.

and three rpm (77, 120, and 173). As screw speed increases, CD increases (higher *S%*) and, consequently, die viscosity is reduced because much more of the material structure is destroyed. It has also been observed that material characteristics (maize endosperm hardness and rice amylose content) affect CD. Softer maize endosperm (dent) cooked less than harder ones (hard red and pop); these effects were discussed by Robutti et al. (2002). They showed that grits from harder maize expanded more, consumed less energy upon extrusion, and cooked more rapidly compared with grits from softer maize. On the other hand, high-amylose rice expanded more, cooked less, and consumed more energy compared with low-amylose rice. These effects were discussed by González et al. (1998, 2000, 2006).

12.5 QUALITY OF EXTRUDED PRODUCTS

12.5.1 PHYSICOCHEMICAL AND SENSORY PROPERTIES RELATED WITH QUALITY

Some of the quality characteristics that may be relevant for food applications are hydration properties and texture, which affect product acceptability. As previously mentioned, extruded product properties are dependent on the CD reached upon extrusion and the raw material structure characteristics (maize endosperm hardness and amylose content). In this regard, precooked flours having different CDs allow the preparation of suspensions with different viscosities for the same solid concentration (González et al. 1991).

For products that are consumed as cooked suspensions, such as "atoles" and polenta, the energy density of the eaten portion is an important nutritional aspect and it is directly related to the suspension's solid concentration. Because product acceptability will depend on its consistency (or viscosity), the degree of swelling of the hydrated particles reached during cooking will ultimately determine the energy density of the eaten portion. It is obvious that CD will determine the final size of the hydrated particles and consequently the suspension's viscosity. As it was shown in Figure 12.7, extrudate viscosity suspension is inversely related with SMEC, which is directly related with CD.

Another factor to take into account is the presence of bran and germ particles, which reduces CD. It has been observed that extrudates obtained from whole grains, at extrusion conditions normally used for expanded products, contain native starch granules indicating an incomplete conversion of the fluid flow mechanism, which affects the expansion and other properties. Figure 12.9 shows the differences in expansion and specific volume from expanded products made with maize grits (dehulled and degermed) and whole maize grits. This interference in the cooking process of starch caused by the presence of oil and fiber could be overlapped by changing the extrusion conditions to increase DC, for example, by reducing die diameter, increasing screw CR, reducing moisture content, and reducing particle size.

Regarding cereal expanded products, the extrudate structure varies according to CD reached during extrusion. Extrudate texture plays an important role in product acceptability. It has been observed that to have an expanded product with good texture, CD should not be too high; otherwise, crispiness would be reduced and stickiness would be high enough to reduce product acceptability.

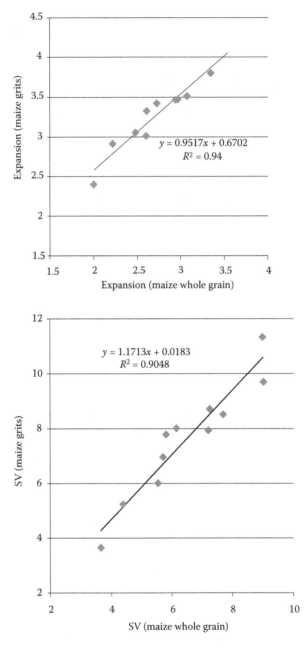

FIGURE 12.9 Differences in expansion and specific volume from expanded products made with maize grits (dehulled and degermed) and whole maize grits. A Brabender 20 DN single-screw extruder was used under the following conditions: moisture content, 14%, 16.5%, and 19%; temperature, 160°C, 175°C, and 190°C; a 4:1 CR and 3 × 20 ($d \times l$) mm die.

Another aspect to take into account is the loss of crispiness due to moisture sorption during the storage period. The increase of moisture content causes plasticizing of the starch–protein matrix and, consequently, an increase in toughness (or decrease in crispiness) of the product (Biliaderis 1992; Katz and Labuza 1981). The product's water activity is the main factor in controlling the shelf life of these products. The loss of crispiness has been detected at more than 0.33 a_w value, which corresponded to a moisture content of 6.5% (wet basis), which is in agreement with the changes observed in the shape of the compression curve during mechanical resistance analysis using an Instron testing machine (González et al. 2004). At low a_w, structures were fragile and corresponded to a crispy texture. At 0.33 a_w, a transition of the way fracture was observed, whereas at high a_w, the structure was no longer fragile and the product failed through flattening.

Both mechanical and sensory methods are normally used to evaluate the structural properties and texture of dry cereal foods, such as crackers and expanded products. Case et al. (1992) studied the effect of gelatinization on the physical properties of several extruded products obtained from maize flour, wheat flour, and wheat starch using the Instron universal testing machine and Kramer press. Drago et al. (2011) showed good correlation between hardness sensory score and extrudate mechanical resistance (r^2, 0.8677) as well as a good inverse correlation between extrudate mechanical resistance and SMEC (r^2, 0.7478) for samples obtained from extruded maize–soybean mixtures. This means that as SMEC increases, CD increases and consequently the extrudate structure is less resistant.

It is clear that the quality requirements of grain processors are becoming increasingly diversified. Thus, breeding efforts must aim for new cultivars that have specific and suitable quality attributes that are closely associated with the biochemical and biophysical properties of kernels.

12.6 USING EXTRUSION TO DEVELOP NEW NUTRITIONALLY IMPROVED FOODS

12.6.1 EXTRUDED LEGUMES AND MAIZE–LEGUME BLENDS

Extrusion has been used to develop expanded products other than maize or blends with maize. In these cases, the trends have been novelty or increase the nutritional value of traditional maize expanded products.

Regarding new product development, the effects of extrusion conditions on potential mineral availability (measured as mineral dialyzability) of extruded bean flour have been studied by Drago et al. (2007a). *Phaseolus vulgaris* beans of the agronomic cultivar "flor de mayo" were ground and dehulled to obtain grits and then extruded at different temperatures (140°C, 160°C, and 180°C) and moisture contents (17%, 20%, and 23% M), according to an experimental design. The extrusion variables slightly affected iron (FeD%) and zinc (ZnD%) dialyzability. Only the extruded sample with the lowest CD (140°C, 23% M) presented DFe% significantly different from that having the highest one (180°C, 17% M): 8.6 ± 0.3 versus 10.7 ± 1.6, respectively. Regarding zinc, the extrusion lightly decreased DZn%, depending on the extrusion conditions: 32.8 ± 0.3, 35.7 ± 0.8, and 37.9 ± 1.8 for 140°C, 23% M and 180°C, 23% M extrusion conditions and raw grit samples, respectively.

Regarding nutritional improvement, when making foods for feeding programs, the content, protein quality, and bioavailability of nutrients should be considered. An alternative is to use bean flours, ensuring the organoleptic quality of the product by lipoxygenase inactivation treatments. Sosa Moguel et al. (2009) studied the influence of lipoxygenase inactivation and extrusion cooking on the physical and nutritional properties of corn/cowpea (*Vigna unguiculata*) blends. Corn was blended in an 80:15 proportion with cowpea flour treated to inactivate lipoxygenase (CI) or noninactivated cowpea flour (CNI). Extrusion variables were temperature (150°C, 165°C, and 180°C) and moisture (M: 15%, 17%, and 19%). Based on their physical properties, the 165°C/15% M corn/CNI, 165°C/15% M corn/CI, and 150°C/15% M corn/CI blends were chosen for nutritional quality analysis. Extrudate chemical composition indicated high crude protein levels compared with standard corn-based products (10.3% vs. 7%). With the exception of lysine, essential amino acid contents in the three treatments met Food and Agriculture Organization requirements. Mixtures of corn and cowpea showed a significant improvement in protein quality (from 2.3 to 4.3 g Lys/100 g of protein) with respect to corn. Drago et al. (2007b) analyzed the effects of extrusion on the availability of minerals using these samples. The results showed that the inactivation did not affect FeD% from cowpea flour (13.0 ± 1.3 vs. 11.4 ± 1.4 for native and inactivated cowpea), whereas the extrusion improved it (20.3 ± 1.4 vs. 20.6 ± 1.2 for CNI and CI, respectively). Although both processes had a slight adverse effect on ZnD%, the value was quite high: 52.5 ± 1.2 versus 46.9 ± 3.0 for native and inactivated cowpea and 40.8 ± 3.0 versus 43.5 ± 1.0 for CNI and CI, respectively.

Extrusion and lipoxygenase inactivation are promising options for developing corn/cowpea extruded snack products with good physical properties and nutritional quality.

12.6.2   WHOLE GRAIN

Nowadays, the use of whole grains in food formulations is highly recommended, and extrusion cooking allows us to obtain precooked whole grain products.

Whole maize and rice grits were extruded using a Brabender 20 DN (Germany) single-screw extruder to obtain expanded products. The following conditions were selected based on a preliminary work/screw CR of 4:1, cylindrical die (diameter/length) of 3/20 mm, screw speed of 150 rpm, extrusion temperature of 175°C (die and extruder barrel), and 14% moisture content. The samples were analyzed regarding mineral availability as dialyzable minerals. DFe% was 7.4 ± 0.0 and 6.4 ± 0.2 and DZn% was 17.7 ± 0.0 and 24.3 ± 0.1 for rice and maize, respectively.

The higher protein, fiber, and mineral contents from whole grains in comparison with refined grains confirmed their nutritional advantages; thus, the use of whole grains to produce snack products is highly recommended.

12.6.3   WHOLE GRAIN BASE BLENDS

The effects of the addition of wild legumes (*Lathyrus* from the south of Spain) on the physical and nutritional properties of extruded products based on whole corn and brown rice were studied (Pastor Cavada et al. 2011). Samples were obtained with a Brabender single-screw extruder. Chemical composition, amino acid content, protein digestibility, and potential availability of iron and zinc from expanded products were determined. With only 15% of rice replacement by the legume, a significant increase of protein content and quality, and fiber and mineral content was obtained. Protein digestibility was in the range of 82% to 84%, whereas mineral availability was in the 7.4% to 16.3% range for iron and in the 10% to 14.6% range for zinc. The performance of each mixture during extrusion and the physical properties of the extrudates were considered to be in the range of those expected for snack-type products. Three advantages were remarkable for these products: they were whole grain food-grade, they had better nutritional quality than a traditional extrudate, and they were also suitable for people with celiac disease.

Another example is an extruded product made from two varieties of amaranth and their mixtures with maize at two levels of replacement (30% and 50%). Amaranth is a native culture of America appreciated for its high nutritional properties including high mineral content. Mineral availability was estimated using the dialyzability method. The mineral content of for *A. cruentus* and *A. caudatus* was listed as the following: Fe, 64.0 and 84.0 mg/kg; Ca, 1977.5 and 2348.8 mg/kg; and Zn, 30.0 and 32.1 mg/kg, respectively. These values were higher in amaranth than in maize products (6.2, 19.1, and 9.7 mg/kg, respectively). Mineral availability was in the range of 2.0% to 3.6%, 3.3% to 11.1%, and 16.0% to 11.4% for Fe, Ca, and Zn, respectively. Extruded amaranth and amaranth/maize products supply a higher intake of Fe and Ca than extruded maize. Amaranth-extruded products and the addition of amaranth to maize may be an interesting way to increase the nutritional value of snacks such as extruded products and a way to incorporate whole grains into the diet.

12.6.4 Fortified Extruded Product

It is possible to develop nutritionally improved maize/soy expanded products with good sensory characteristics and to select the most appropriate iron source and enhancers for obtaining a fortified product with good mineral availability (Drago et al. 2011). A mixture prepared with 88% of commercial maize grits and 12% of dehulled soybean grits made from lipoxygenase-inactivated beans was used. With such a base mixture, 21 different samples containing different mineral sources (ferrous sulfate [FS], ferric bisglicinate [FBG], NaFeEDTA, zinc sulfate, and calcium carbonate), with and without selected absorption enhancers (EDTA, sodium citrate, or ascorbic acid), were prepared. Each mixture was extruded using a 20 DN Brabender extruder, with a 4:1 screw CR, 3 mm die, 150 rpm screw speed, 180°C extrusion temperature, and 14% moisture content, which are considered adequate for a snack-type expanded product.

On each extruded sample, physicochemical properties such as SMEC, expansion, specific volume, water solubility, mechanical resistance, hardness (sensory test), and mineral dialyzability were determined. Results showed that mixture composition significantly affected SMEC. Two sample groups were distinguished: samples containing citrate and samples containing calcium. These latter samples presented the lowest CEEM values, whereas samples containing citrate presented intermediate values of CEEM. Besides that, samples with calcium presented low values of expansion and specific volume, which is in agreement with their lower CEEM values.

The addition of soybean to maize not only allowed significantly improved protein value without impairing the sensorial attributes of the expanded products but also improved the supply of Fe, Zn, and Ca from these products. The use of absorption enhancers, such as citrate or EDTA, could be a suitable strategy to increase the intrinsic mineral (Fe, Zn, and/or Ca) supply of these products without fortification. Of the three evaluated sources of iron, NaFeEDTA displayed higher availability with respect to FS and to FBG. Its use did not affect the availability of endogenous Zn and Ca. Nevertheless, the sample fortified with FS displayed better Zn availability. This fact, together with the high cost of the NaFeEDTA, makes FS more appropriate for these products.

Although the iron availability from iron-fortified products was lower than that from nonfortified samples (except for NaFeEDTA), the iron supply from these fortified products was greater. EDTA was the best enhancer for FS-fortified and FBG-fortified products. The effect of the ascorbic acid and the citrate depended on the iron source used. These results show that there is a relationship between the iron source and the enhancer. To obtain good iron availability, it is necessary to study which is the most appropriate enhancer for each iron source. Interactions between the mineral sources were demonstrated. The use of Ca fortification impaired the availability of Fe and Zn. The negative effect was greater in the presence of Zn and Ca simultaneously. In these multifortified products, the use of EDTA as an enhancer lightly improved mineral availability. When a multifortified product is formulated, it is very important to evaluate the interactions among different minerals and to select the combination of mineral sources and absorption enhancers more appropriate for the elaboration of the product.

12.6.5 FUNCTIONAL FOODS

Four hydrolysates were obtained from bovine hemoglobin concentrate (BHC) and were used to fortify extruded maize products (5 g/kg of BHC hydrolysates). Antioxidant capacity (AOC) measured using [2,29-azinobis-(3-ethylbenzothiazoline-6-sulfonic acid)] radical cation (ABTS) assay and expressed as Trolox equivalent antioxidant capacity (TEAC) inhibition of angiotensin-converting enzyme (ACE I), and iron availability (using dialyzability method) were measured, both on hydrolysates and corn/hydrolysate blends. The physicochemical properties of the extruded products were not affected by fortification. The ACE inhibition and TEAC values from hydrolysates were significantly higher than those from BHC. The ACE inhibition and AOC from extruded products with added hydrolysates were higher than those from maize control; however, the extrusion process modified both ACE inhibition and AOC formerly present in hydrolysates (Cian et al. 2011). Regarding iron availability, the enzymatic hydrolysis increased the iron dialyzability with respect to the substrate. The highest value of iron dialyzability corresponded to the more hydrolysated sample. Extruded products fortified with BHC hydrolysates showed higher iron dialyzability than that fortified with BHC. However, iron dialyzability corresponding to BHC was lower than that expected from a heme–iron. Therefore, heme–iron availability was low when it was determined in the absence of meat proteins, and hydrolysis could increase potential iron availability (Cian et al. 2010).

REFERENCES

Arhaliass, A., Bouvier, J.M., and Legrand, J. 2003. Melt growth and shrinkage at the exit of the die in the extrusion-cooking process. *J. Food Eng.* 60(2): 185–192.

Bhatnagar, S. and Hanna, M.A. 1994a. Amylose-lipid complex formation during single-screw extrusion of various corn starches. *Cereal Chem.* 71: 582–587.

Bhatnagar, S. and Hanna, M.A. 1994b. Extrusion processing conditions for amylose-lipid complexing. *Cereal Chem.* 71: 587–593.

Biliaderis, C.G. 1992. Structures and phase transitions of starch in foods systems. *Food Tech.* 18: 98–109.

Blanshard, J.M.V. 1987. Starch granule. Structure and function: a physicochemical approach. In *Starch: Properties and Potential. Critical Reports and Applied Chemistry*, ed. T. Galliard, Chap. 2, Vol. 13. New York: The Society of Chemical Industry by John Wiley and Sons.

Bouzaza, D., Arhaliass, A., and Bouvier, J.M. 1996. Die design and dough expansion in low moisture extrusion-cooking process. *J. Food Eng.* 29(2): 139–152.

Case, S.E., Hamann, D.D., and Schwartz, S.J. 1992. Effect of starch gelatinization on physical properties of extruded wheat- and corn-based products. *Cereal Chem.* 69: 401–404.

Chao-Chi Chuang, G. and Yeh, A. 2004. Effect of screw profile on residence time distribution and starch gelatinization of rice flour during single screw extrusion cooking. *J. Food Eng.* 63: 21–31.

Chi-Tang, H.O. and Izzo, M.T. 1992. Lipid-protein and lipid-carbohydrate interactions during extrusion. In *Food Extrusion Science and Technology*, eds. J.L. Kokini; C.T. Hao, and M.V. Karwe, 415–425. New York: Marcel Dekker.

Cian, R.E., Drago, S.R., De Greef, D.M., Torres, R.L. and González, R.J. 2010. Iron and zinc availability and some physical characteristics from extruded products with added concentrate and hydrolysates from bovine hemoglobin. *Int. J. Food Sci. Nutr.* 61(6): 573–582.

Cian, R.E., Luggren, P., and Drago, S.R. 2011. Effect of extrusion process on antioxidant and ACE inhibition properties from bovine haemoglobin concentrate hydrolysates incorporated into expanded maize products. *Int. J. Food Sci. Nutr.* 62(7): 774–780.

Davidson, V.J. 1992. The rheology of starch-based material in extrusion process. In *Food Extrusion Science and Technology*, eds. J.L. Kokini, C. Ho, and M.V. Karwe, 263–275. New York: Marcel Dekker.

Dombrink-Kurtzman, M.A. and Knutson, C.A. 1997. A study of maize endosperm hardness in relation to amilose content and susceptibility to damage. *Cereal Chem.* 74: 776–780.

Drago, S.R., Velasco González, O., Torres, R.L., González, R.J. and Valencia, M.E. 2007a. Effect of the extrusion on functional properties and mineral dialyzability from *Phaseolus vulgaris* bean flour. *Plant Foods Human Nutr.* 62(2): 43–48.

Drago, S.R., González, R.J., Chel-Guerrero, L., and Valencia, M.E. 2007b. Evaluación de la disponibilidad de minerales en harinas de frijol y en mezclas de maíz/frijol extrudidas. *Rev. Inform. Tecnol.* 18(1): 41–46.

Drago, S.R., Lassa, M.S., Torres, R.L., De Greef, D.M. and González, R.J. 2011. Use of soybean in cereal based food formulation and development of nutritionally improved foods. In *Soybean and Nutrition/Book 5*, ed. H.A. El-Shemy, Chap. 3, 45–66. Rijeka, Croatia: INTECH open access Publishers.

Fast, R.B. 1991. Manufacturing technology of ready-to-eat cereals. In *Breakfast Cereal and How They Are Made*, eds. R.B. Fast and E.F. Caldwell, 15–42. Minnesota, USA: American Association of Cereal Chemists St. Paul Inc.

González, R.J., Torres, R.L., De Greef, D.M., Gordo, N.A., and Velocci, M.E. 1991. Influencia de las condiciones de extrusión en las características de la harina de maíz para elaborar sopas instantáneas. *Rev. Agroquím Tecnol. Aliment.* 31(1): 87–96.

González, R.J., Torres, R., and De Geef, D.M. 1998. Diferencias entre variedades de arroz y maíz utilizando el amilógrafo y dos diseños de extrusores como métodos de cocción. *Rev. Inform. Tecnol.* 9(5): 35–44.

González, R.J, Torres, R.L., and Añón, M.C. 2000. Comparison of rice and corn cooking characteristics before and after extrusion. *Polish J. Food Nutr. Sci.* 9(50)1: 29–34.

González, R.J., Torres, R.L., and De Greef, D.M. 2002. Extrusión—cocción de cereales. *Bol. Soc. Bras Cienc Tecnol. Aliment.* 36(2): 83–136.

González, R.J., De Greef, D.M., Torres, R.L., Borrás, F. and Robutti, J. 2004. Effects of endosperm hardness and extrusion temperature on properties of products obtained with grits from two commercial maize cultivars. *Lebensm Wiss U Technol.* 37: 193–198.

González, R.J., Torres, R., De Greef, D., Bonaldo, A., Robutti, J., and Borrás, F. 2005. Efecto de la dureza del endospermo del maíz sobre las propiedades de hidratación y cocción. *ALAN* 55(4): 354–360.

González, R.J., Torres, R.L., and De Greef, D.M. 2006. El arroz como alimento: el grano y la harina, parámetros de caracterización y de calidad. In *El Arroz: su Cultivo y Sustentabilidad en Entre Ríos*, ed. R. Benavides, Chap. 1, 19–52. Santa Fe: Univ. Nac. del Litoral.

Guy, R.C.E. and Horne, A.W. 1988. Extrusion and co-extrusion of cereals. In *Food Structure—Its Creation and Evaluation*, eds. J.M. Blanshard and J.R. Mitchell, 331–349. London: Butterworth.

Guzman, L.B., Tung-Ching, L., and Chichester, C.O. 1992. Lipid binding during extrusion cooking. In *Food Extrusion Science and Technology*, eds. J.L. Kokini, C.T. Ho, and M.V. Karwe, 427–436. New York: Marcel Dekker.

Haller, A.D. 2008. Evaluación de las condiciones de extrusión necesarias para elaborar harina de maíz con características similares a las de una harina nixtamalizada. *Magister Thesis in Food Science and Technology.* Santa Fe, Argentina: F.I.Q.-U.N.L.

Harper, J.M. 1986. Processing characteristics of food extruders. In *Food Engineering and Process Applications Vol 2. Unit Operations*, eds. M. Le Maguer and P. Jelen, 101–114. London: Elsevier Applied Science Publishers.

Harper, J.M. 1989. Food extruders and their applications. In *Extrusion Cooking*, eds. C. Mercier, P. Linko, and J.M. Harper, 1–15. St. Paul, MN, USA: American Association of Cereal Chemists, Inc.

Harper, J.M. 1992. A comparative analysis of single and twin-screw extruders. In *Food Extrusion Science and Technology*, eds. J.L. Kokini, C.T. Ho, and M.V. Karwe. Chap. 8. 139–164. New York: Marcel Dekker, Inc.

Hoseney, R.C. 1994. Structure of cereals. In *Principles of Cereals Science and Technology*, Chap. 1, 1–28. St. Paul, MN: American Association of Cereal Chemists, Inc.

Jansen, L.P.B.M. 1989. Engineering aspects of food extrusion. In *Extrusion Cooking*, eds. C. Mercier, P. Linko, and J.M. Harper, Chap. 2, 17–38. St. Paul, MN: American Association of Cereal Chemists, Inc.

Juliano, B. 1982. An international survey of methods used for evaluation of the cooking and eating qualities of milled rice. *International Rice Research Institute. Res. Paper Ser.* 77: 1–27.

Juliano, B. 1985. Criteria and tests for rice grain qualities. In *Rice Chemistry and Technology*, ed. B. Juliano, Chap. 12, 443–514. Minnesota: AACC.

Katz, E.E. and Labuza, T.P. 1981. Effect of water activity on the sensory crispness and mechanical deformation of snack food products. *J. Food Sci.* 46: 403–409.

Llo, S., Tomschik, U., Berghofer, E., and Mundigler, N. 1996. The effect of extrusion operating conditions on the apparent viscosity and the properties of extrudates in twin-screw extrusion cooking of maize grits. *Lebensm Wiss U Technol.* 29: 593–598.

Lusas, W.E. 2001. The snack foods setting. In *Snack Foods Processing*, eds. W.E. Lusas and W.L. Rooney, 3–28. Pennsylvania: Technomic Publishing.

Matz, S.A. 1993. *Snack Food Technology*, 3rd Ed. New York: AVI Publishing Co.

McKenzie, K.S. 1994. Breeding for rice quality. In *Rice Science and Technology*, ed. W.E. Marshal and J.I. Wadsworth, Chap. 5, 83–111. New York: Marcel Dekker, Inc.

Meuser, F. and Van Lengerich, B. 1992. System analytical model for the extrusion of starches. In *Food Extrusion Science and Technology*, eds. J.L. Kokini, C. Ho, and M.V. Karwe, 619–630. New York: Marcel Dekker.

Miller Jones, J. and Engleson, J. 2010. Whole grains: benefits and challenges. *Annu. Rev. Food Sci. Technol.* 1: 19–40.

Mitchell, J. and Areas, J.A.G. 1992. Structural changes in biopolymers during extrusion. In *Food Extrusion Science and Technology*, eds. J.L. Kokini, C. Ho, and M.V. Karwe, 345–360. New York: Marcel Dekker.

Pastor Cavada, E., Drago, S.R., Pastor, Díaz, J., Alaiz, M., Vioque Peña, J., and González, R.J. 2011. Effects of the addition of wild legumes (*Lathyrus annuus and Lathyrus clymenum*) on the physical and nutritional properties of extruded products based on whole corn and brown rice. *Food Chem.* 128(4): 961–967.

Posner, E.S. and Hibbs, A.N. 2005. Wheat: the raw material. In *Wheat flour milling*, 1–46, St. Paul, MN: American Association of Cereal Chemists Inc.

Pratt, R.C., Paulis, J.W., Miller, K., Nelsen, T., and Bietz, J.A. 1995. Association of zein classes with maize kernel hardness. *Cereal Chem.* 72(2): 162–167.

Robutti, J.L., Borrás, F.S., González, R.J., Torres, R.L., and De Greef, D.M. 2002. Endosperm properties and extrusion cooking behavior of maize cultivars. *Food Sci. Technol./LWT.* 35: 663–669.

Sosa Moguel, O., Ruiz-Ruiz, J., Martínez-Ayala, González, R.J., Drago, S.R., Betancur Ancona, D., and Chel Guerrero, L. 2009. Effect of extrusion conditions and lipoxygenase inactivation treatment on the physical and nutritional properties of corn/cowpea (*Vigna unguiculata*) blends. *Int. J. Food Sci. Nutr.* 60(S7): 341–354.

Torres, R.L. 2005. Estudio de las características fisicoquímicas de diferentes genotipos de maíz y su comportamiento durante la extrusión termoplástica. *Magister Thesis in Food Science and Technology.* Santa Fe, Argentina: F.I.Q.-U.N.L.

van Zuilichem, D.J., Stolp, W., and Janssen, L.P.B.M. 1984. Engineering aspects of single- and twin-screw extrusion-cooking of biopolymers. In *Extrusion Cooking Technology*, ed. R. Jowitt, 75–92. London: Elsevier Applied Science Publishers.

Wasserman, B.P., Lu-Fang, W., and Kin-Yu, C. 1992. Molecular transformations of starch and protein during twin-screw extrusion processing of cornmeal. In *Food Extrusion Science and Technology*, eds. J.L. Kokini, C.T. Ho, and M.V. Karwe, 325–333. New York: Marcel Dekker.

Whalen, P.J. 1998. Detection of differences in corn quality for extrusion processes by rapid visco analyzer. *Cereal Food World.* 43: 69–72.

Wu, Y.V. and Bergquist, R.R. 1991. Relation of corn grain quality to yields of dry-milling products. *Cereal Chem.* 68: 542–544.

Yamin, F.F., Svendsen, L., and White, P.J. 1997. Thermal properties of corn starch extraction intermediates by differential scanning calorimetry. *Cereal Chem.* 74: 407–411.

13 New Trends in Cereal-Based Products

Concha Collar

CONTENTS

13.1 INTRODUCTION

As the core of human nutrition, cereals are ubiquitous and multifaceted staple foods addressed to the population worldwide as carriers of either macronutrients and micronutrients or tailored foods for specific targeted groups as healthy, convenient, and indulgent raw materials, providing perfect vectors for diversity and innovation. Approximately half of the world's cropland is devoted to growing cereals. The combination of direct intake with indirect consumption, in the form of foods like meat and milk, account for cereals' presence in approximately two-thirds of all human calorie intake worldwide (Dyson 1999).

According to the Economic Research Service of the US Department of Agriculture, per capita grain availability, adjusted for losses, increased from 43.2 to 61.7 kg/year between 1970 and 2009, which corresponds to an increase in energy availability of 432 to 619 kcal/day during this interval (USDA 2011). Historically, grain intake decreases as meat intake increases, but given the role of grains in grain-based and nongrain-based goods, grain availability has paralleled the increase in meat and cheese availability (Barnard 2010). Although the creation of some entirely new product templates should be expected for this new decade, the evolution of existing product platforms should also be considered. Drivers and major trends currently influencing the bakery and cereals market including health, premiumization, guilt-free indulgence, convenience, and ethical consumerism are analyzed, and new and dynamic reports are launched to provide detailed market insights into global, regional, and category level to forecast trends accurately. To maintain and develop market share, consumers' expectations should be met. Bakery and cereals manufacturers have been forced to address the "staple" image of their products. To accomplish these goals, manufacturers have innovated by (1) reformulating their products to make them more indulgent, using alternative basic ingredients with an extra nutritional value or fortifying their products with functional ingredients to make them healthier (or both), and (2) developing innovative processes and applying classic and novel technological processes and delivery methods.

13.2 NEW TRENDS IN CEREAL-BASED GOODS MATCHING CONSUMERS' DEMANDS AND PERCEPTION

Intense competition within the bakery market, combined with evolving consumer expectations, has continued to drive innovation, particularly in health, indulgence, and convenience. Simplicity heads the top 10 new food/new product development trends in Europe for 2010, according to Innova Market Insights (2009). Although according to a recent article published in *Food Technology*, the magazine of the Institute of Food Technologists (2011), the past few years of economic confusion and destabilization have caused consumers' food attitudes to shift and shape the top 10 food trends for 2011. Getting real and new nutrients constitutes the main identified trends. Consumers are increasingly concerned about the contents of the food in their diets and are shifting away from getting nutrients via fortified foods and turning toward products that are naturally high in vitamins/minerals and those that have been blended with other foods to create even higher nutrient levels (Angioloni and Collar 2011a,b). Whole grain was the most sought after health claim on packages in 2010, followed by high fiber, low sodium, low fat, no trans fat, low sugar, low calorie, no chemical additives, no preservatives, and low/lower cholesterol (Collar 2008; Collar and Angioloni 2010, 2011; Angioloni and Collar 2011a,c).

Hot cereal tends to be marketed on the intrinsic nutritional value of oatmeal, with more focus on novel flavors and textures. The addition of nuts facilitates a new focus on mouthfeel for hot cereal. Adult-oriented flavors have also been developed, and dessert-inspired flavors, such as cinnamon roll, cheesecake, banana cream pie, and chocolate chip cookie dough are expected to debut shortly (Prepared Foods Network 2011). Products pushing the nutritional envelope with probiotics (Altamirano-Fortoul 2012) or Ω-3s are likely to proliferate. Time also figures in the grain-based ingredients used for some of 2010's most noteworthy cereals—time, as in ancient grains (spelt and quinoa). The increasing popularity of gluten-free foods and the seed's versatility have helped quinoa because it is naturally gluten-free and can be used in many grain-based foods: from breads and pastas to cookies, tortilla chips, and frozen entrees. Flax, which contains Ω-3 fatty acids, such as α-linolenic acid, has become a familiar addition to an array of packaged foods. As with cereal bars, probiotics seem to have a promising future for ready-to-eat breakfast cereals. Efforts are being made to move away from formulations perceived as being highly processed. A novel approach includes naturalness and innovative flavors and tastes.

13.3   NOVEL CEREAL PRODUCTS MADE FROM GRAINS ALTERNATIVE TO WHEAT AND RYE

There has recently been a great deal of increased interest in minor cereals, ancient crops, and pseudocereals—aside from wheat—that constitute highly nutritional grains with potential bread making applications. The concept of using traditional South American, African, and Asian raw materials and fermented foods as templates for wheat, wheat-free, and gluten-free-based foods in Europe and North America is in good accordance with the interest in westernized countries for ethnic foods with revisited value addition. Indigenous foods from different cultures and civilizations with ethnic eating habits are making headway in a globalized world with strong immigration movements, encompassing the use of traditional raw materials as ingredients in novel foods (Collar 2011). The growing general demand for novel tasty and healthy foods, together with the increasing number of people with celiac disease, have given birth to a new market consisting of cereal products made from gluten-free grains (Arendt and Dal Bello 2008; Salovaara et al. 2011). In this challenging market, oat (Flander et al. 2007), rice (Rosell and Marco 2008), corn (Brites et al. 2010), sorghum, millet (Taylor et al. 2006), quinoa, amaranth, and buckwheat (Angioloni and Collar 2011a; Alvarez-Jubete et al. 2010; Rosell et al. 2009) have gained a special position as basic ingredients used singly or in associated blends to make gluten-free, wheat-free, and wheat composite breads with variable sensory acceptabilities and nutritional values (Angioloni and Collar 2011a, 2012a). Also, the interest in legumes as ingredients for bread production is growing. High legume–wheat blends seem to be an efficient strategy to obtain sensorially accepted and nutritionally enhanced breads with no dramatic technological impairment when structuring agents (gluten/hydrocolloids) are incorporated (Angioloni and Collar 2012b). The incorporation of chickpea and green pea into bread formulas has decreased and delayed starch hydrolysis and concomitantly reduced the expected glycemic index. The use of defatted soybean flour promoted a boost of bread's antiradical activity. Structuring agents helped restore dough viscoelasticity in highly legume-replaced wheat matrices and apparently obstructed starch-degrading enzyme accessibility, causing a slowdown of starch hydrolysis kinetics.

13.4   APPLICATION OF NEW/EMERGING TECHNOLOGIES IN CEREALS AND CEREAL-BASED GOODS

The development of innovative baking processes (Patel et al. 2005) addresses the use of different thermal and nonthermal technologies as described by Sumnu et al.

(2007). In a recent study, consumers' levels of concern for all technologies were investigated (Cardello 2003). Individuals who demonstrated a willingness to consume foods processed using one novel technology (e.g., irradiation) had lower concern ratings for other technologies. Ratings of concern were negatively correlated with expected liking for products believed to be processed using a new technology. Expected liking ratings were positively influenced by visual exposure to the product and by a safety and benefit statement.

In this section, advances in the application of new/emerging technologies for creating new textures and flavors (high hydrostatic pressure [HHP]), improving technological and nutritional features (ultrasound, vacuum cooling, radiating technologies, dry processes), monitoring online processes (ultrasounds, radiating technologies), process analyzing (near-infrared), and characterizing molecular dynamics (nuclear magnetic resonance [NMR] and magnetic resonance imaging [MRI]) in cereals and cereal-based goods are presented and discussed in detail.

13.4.1 HIGH HYDROSTATIC PRESSURE

The use of HHP technology in cereal grain processing offers many advantages over traditional technologies involving thermal treatment for reducing the microbial population of spoilage microorganisms, reducing allergenicity, inactivating unwanted food enzymes and compounds, and consequently, increasing the shelf life of storage products (Estrada-Girón et al. 2005). In addition, HHP also offers advantages that may be used to develop novel textured foods. To generate novel textures and products, Apichartsrangkoon et al. (1998) subjected hydrated wheat gluten to pressure heat treatment in the range of 200 to 800 MPa at temperatures ranging from 20°C to 60°C with holding times of 20 to 60 min, and found a time-dependent effect when HHP was used to develop textured products. The effect of hydrostatic pressure (0.1–800 MPa) in combination with various temperatures (30°C–80°C) on the chemical and physical properties of wheat gluten, gliadin, and glutenin has been recently studied (Kieffer et al. 2007). Treatment of gluten with low pressure (200 MPa) and temperature (30°C) increased the proportion of the ethanol–soluble fraction and decreased gluten strength. Increased pressure and temperature induced a significant strengthening of gluten, and under extreme conditions (e.g., 800 MPa, 60°C), gluten cohesivity was lost. Several studies have investigated the effects of HHP on pure starch (Buckow et al. 2007; Gomes et al. 1998; Katopo et al. 2002) or protein (Kieffer et al. 2007; Mozhaev et al. 1996) systems. HHP has been shown to cause starch gelatinization depending on the pressure applied, treatment time and temperature, concentration and type of starch, as well as moisture content of the samples (Stolt et al. 2001). The primary benefits of pregelatinized starch are moisture retention, improvement of texture, increase in volume, and enhanced shelf life of baked goods (Thomas and Atwell 1997). HHP can also irreversibly change the structural and functional properties of proteins (Winter 2003) due to disruption of noncovalent interactions within proteins, with subsequent reformation of intramolecular and intermolecular bonds within or between proteins. A study by Gomes et al. (1998) investigated the effect of HHP on amylases in wheat and barley flours and observed a significant increase in the activity of starch-degrading enzymes at 400 to 600 MPa, whereas activity was decreased at 600 MPa. Ahmed et al. (2007) investigated the effect of HP treatment of basmati rice slurries and found complete gelatinization and denaturation of starch and protein components, respectively, as well as increased mechanical strength of the HHP-treated rice slurries. High-pressure processing in corn tortillas has been shown to slightly affect the structural properties whereas significantly affecting the molecular properties of water in terms of decreasing proton mobility as determined by NMR (Vittadini et al. 2004), and thus affecting the quality and microbial stability of corn tortillas. Later studies on the microbial, physical, and structural changes in high-pressure processed wheat dough, as a function of pressure level (50–250 MPa) and holding time (1–4 min), revealed a drastic reduction of the endogenous microbial population of wheat dough and an increased dough hardness and adhesiveness (Bárcenas et al. 2010).

Scanning electron micrographs suggested that proteins were affected when subjected to pressure levels higher than 50 MPa, but starch modification required higher pressure levels. HHP-treated yeasted doughs led to wheat breads with different appearance and technological characteristics: the crumb acquired a brownish color and heterogeneous cell gas distribution with increased hardness due to a new crumb structure. The authors concluded that high hydrostatic processing in the range of 50 to 200 MPa could be an alternative technique for obtaining novel textured cereal-based products.

Because HHP could cause conformational changes in cereal starch and proteins, the technology may represent a valuable alternative to improving the baking performance of oats and gluten-free cereals. A fundamental evaluation of the effect of HHP on oat batters has recently been carried out and showed that the treatment significantly affected oat batter microstructure and both starch and proteins (Hüttner et al. 2009). HHP treatment significantly improved batter viscosity and elasticity. At pressures lower than 300 MPa, the increase in the viscous component was higher than the increase in the elastic component. On the contrary, at pressures higher than 350 MPa, the elastic component was predominant. Differential scanning calorimetry revealed that high HHP induced starch gelatinization, which started at 300 MPa and was almost complete after treatment at 500 MPa. Overall, the extent of starch gelatinization and protein modification was dependent on the applied pressure, but the results collected thus far clearly show that HHP can be used to improve the functionality of oat batters. The effect of HHP treatment on the bread-making performance of oat flours revealed significantly improved bread volume upon the addition of 10% oat batter treated at 200 MPa (Hüttner et al. 2010). The staling rate was reduced in all breads containing oat batter treated at 200 MPa. In contrast, bread quality deteriorated due to the addition of oat batters treated at pressures higher than 350 MPa, regardless of the addition level. Overall, weakening of the protein structure, moisture redistribution, and possibly changed interactions between proteins and starch were responsible for the positive effects of HHP treatment at 200 MPa on the bread making performance of oat flour. Protein network formation and pre-gelatinization of starch did not improve oat bread quality. The application of HHP of sorghum batters was investigated to evaluate the potential of pressure-treated sorghum as a gluten replacement in the production of sorghum breads (Vallons et al. 2010). The results revealed a weakening of the batter structure at pressures lower than 300 MPa. Freeze-dried sorghum batters treated at 200 MPa (weakest batter) and at 600 MPa (strongest batter) were added to a sorghum bread recipe, replacing 2% and 10% of untreated sorghum flour, respectively. The results showed a delayed staling for breads containing 2% of sorghum treated at 600 MPa. However, adding 10% resulted in a low specific volume and poor bread quality. The quality of breads containing different amounts of sorghum treated at 200 MPa was not significantly different from the control bread. The potential of HHP treatment to promote structural formation in gluten-free flours—buckwheat, white rice, and teff batters (40 g/100 g)—was investigated when treated for 10 min at 200, 400, or 600 MPa (Vallons et al. 2011). Pasting profiles revealed HHP-induced starch gelatinization, and Lab-on-a-Chip capillary gel electrophoresis revealed protein polymerization by thiol/disulfide interchange reactions in white rice and teff batters. For buckwheat proteins, no cross-linking mechanism was observed, which was explained by the absence of free sulfhydryl groups. An increase in viscoelastic properties at higher pressures was observed and could be explained by the modifications occurring in the starch and protein structures. Overall, this study has shown that HHP treatment has the potential to improve the functional properties of gluten-free batters. In a very recent article, the effect of HHP on the gelatinization/gelling attributes of hydrated oats, millet, sorghum, and wheat flours, and on the small and large rheological properties of blended wheat doughs, was investigated to evaluate the significance of HHP on dough viscoelastic reinforcement of highly replaced composite cereal matrices (Angioloni and Collar 2012c).

Single hydrated cereal flours, giving slurries of dough yield at 160 and 200, were treated for 10 min at 200, 350, or 500 MPa. Regardless of the nature of the flour, changes in the flour's viscometric features after HHP treatment seem more evident in samples prepared at a dough yield of 200 (Figure 13.1). A strengthening effect associated with the incorporation of pressure-treated flours into the bread dough formulation was achieved at HHP treatments not exceeding 350 MPa. Highly replaced composite dough samples treated at 500 MPa proved to be extremely stiff, resistant to stretch, weakly cohesive, and weakly extensible, and thus was not suitable for bread making.

13.4.2 Ultrasounds

Ultrasonic techniques provide a nondestructive, rapid, and low-cost technique for the measurement of physical food characteristics, including the viscoelastic properties of doughs (Garcia-Alvarez et al. 2006). The capability of ultrasound measurements for discriminating flours for different purposes has been studied, proving the potential of using ultrasound as an alternative measurement method to discriminate types of flours for different purposes. The technique can therefore be used for the creation of control charts to monitor the production and control of processes. Low frequency (50 kHz) ultrasonic velocity and attenuation measurements have been successfully used to investigate the role of gas bubbles and the surrounding matrix in determining the mechanical properties of mechanically developed dough (Elmehdi et al. 2004). These experimental results show that both ultrasonic velocity and attenuation are sensitive to the presence of gas bubbles in dough. The attenuation coefficient

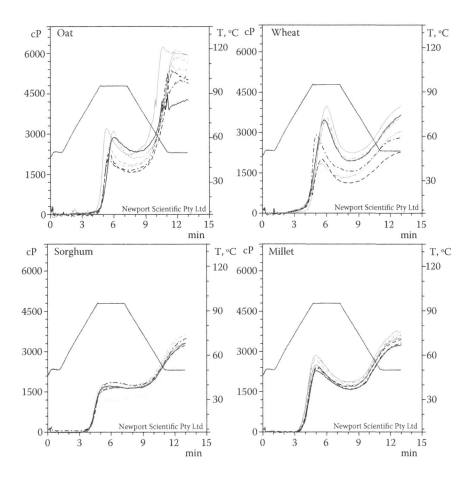

FIGURE 13.1 Viscometric profile of untreated (—, 0 MPa) and HHP-treated (–·–, 200 MPa; – – –, 350 MPa; ·····, 500 MPa) hydrated flours at a flour–water ratio of 1:0.6 -w/w- (black lines), and 1:1 -w/w- (gray lines). (From Angioloni, A. and Collar, C., 2012c, reprinted with permission.)

increased in proportion to the amount of gas occluded within the dough. This observation suggests that the attenuation can be modeled by a background matrix value plus a contribution due to scattering and absorption from the bubbles given by the product. The physical interpretation of these attenuation measurements suggests that the ultrasound technique not only provides information about the contribution of the gas bubbles to the mechanical properties of the dough but also points to another potential application of the ultrasonic velocity measurements, providing a means of probing changes in the intermolecular bonding in the dough matrix as the mixing environment is manipulated. The ultrasound technique has proven to be a promising tool for probing how additives (surfactants) interact with gas bubbles in dough systems during mixing, especially because the opacity of dough precludes using optical techniques for this purpose (Mehta et al. 2009). Because of its ability to probe the effect of mixing times and ingredients on dough properties, the ultrasound technique has the potential to be deployed as an online quality control tool in the baking industry. Characterization of cake batters by ultrasound measurements has evidenced the feasibility of the technique in the detection of deficient batters because significant correlations between the acoustic impedance and the batter consistency, the elastic and the complex moduli, have been obtained (Gómez et al. 2008). A high-power ultrasound bath system was used as a processing aid during sponge cake batter mixing in enhancing the mixing process to produce a better quality of cake texture (Tan et al. 2011). The ultrasound was able to enhance the mixing process by producing a lower batter density and flow behavior index as well as higher overrun and viscosity compared with the nonaided mixing process. With the 2.5 kW ultrasound-assisted mixing for the entire batter mixing duration of 9 min, a better cake quality was produced in terms of lower cake hardness and higher cake springiness, cohesiveness, and resilience. Low-intensity ultrasound was evaluated as a means to detect changes in the mechanical properties of yellow alkaline noodle dough prepared with different formula ingredients (Bellido and Hatcher 2010). Results showed that ultrasonic velocity and attenuation increased and decreased, respectively, with an increase in NaCl concentration (1%–3%), or the inclusion of transglutaminase (2%) to the noodle dough formula. These changes in dough formulation resulted in noodle dough of increased mechanical strength, as confirmed by changes in longitudinal mechanical moduli. The velocity, attenuation, and storage mechanical modulus obtained from ultrasonic experiments were significantly correlated with the maximum stress detected. The ultrasonic technique proved to be a simple and reliable method for ascertaining and discriminating fundamental mechanical properties in yellow alkaline noodles. The ultrasonic treatment of brown rice at different temperatures (with regard to cooking properties and quality) evidenced a loss in natural morphology of the rice bran, allowing water to be easily absorbed by a rice kernel, particularly at high-temperature treatment (Cui et al. 2010). In addition, ultrasonic treatment increased the peak, hold, and final viscosities and significantly decreased the onset and peak temperatures. Thus, ultrasonic treatment can be used for reducing the cooking time of brown rice. The sensitivity of ultrasonic waves to changes in the size and shape of crumb cells demonstrates the potential for using ultrasounds as a tool for characterizing the mechanical and structural properties of bread crumb, and hence for measuring some of the determining factors of bread quality (Elmehdi et al. 2003).

13.4.3 Vacuum Cooling

In the past decade, the application of vacuum cooling has been extended to other sectors of the food industry, including the bakery and particulate foods processing sectors (Wang and Sun 2001). The major characteristic of vacuum cooling is that the product can be cooled at extremely higher speeds compared with conventional cooling methods (Zheng and Sun 2005). Bread rolls, crusty breads, biscotti bread, cakes, and baked biscuits have all been proven to be suitable vacuum-cooled baked products. Vacuum cooling takes place immediately after bread and similar products are removed from the oven and before packaging to avoid vapor condensation in the wrapping, particularly in plastic bags. In terms of product quality, shape, and texture, vacuum-cooled baked products are superior to those that are air-cooled because less contraction and collapse occurs during storage as vacuum-cooled baked products have a more uniform distribution of internal temperature and moisture (Wang and Sun 2001). Furthermore, vacuum cooling helps to extend product shelf life because spore contamination only occurs at the end of the cooling process when air is allowed to enter the vessel.

13.4.4 RADIATING TECHNOLOGIES

13.4.4.1 Infrared

Infrared technology offers several advantages when it is used for heating bakery products, in particular, with the possibility of reaching high heat flux and heating at a distance with better control compared with convective heat transfer. However, the high power density applied to the material can significantly alter the usual drying kinetics and bring about important processes (Salagnac et al. 2004). Baking of bread using infrared technology has been studied by several authors, as reported by Dessev et al. (2011). Keskin et al. (2004) showed that short infrared radiation combined with microwaves can be used to bake bread with quality that is similar to conventional baking processes. Infrared radiation gives better textural qualities, thinner crust, and finer crumb structure, which are characteristics that can be desirable for some consumers. At the dough level, weight loss, surface temperature, and short to medium infrared absorptivity of the dough measured in a series of radiant heating experiments showed a strong positive correlation between surface temperature and total heat flux absorbed by the dough; with the variation in the infrared absorptivity of the dough being explained by the state of the water (vapor or liquid) in the samples (Dessev et al. 2011).

13.4.4.2 Microwave Radiations

Being a measurement technique that is nonhazardous, nondestructive, and noninvasive, this technique is still preferred and developed for instrumentation and sensor purposes in the food industry (Shiinoki et al. 1998). Unlike infrared measurements, microwave techniques can measure the bulk properties of a material and not just the surface. The measured microwave properties are able to illustrate discrepancies in variables such as fat, sugar, salt, and bulk density aside from water within a material (Kent et al. 2000; Kent 2001). Investigations measuring dough properties using microwave measurements began in the last decade and were limited to moisture investigations using frequencies ranging from 0.2 to 20 GHz to measure the dielectric properties of dough (Chin et al. 2005).

The basic mechanisms involved in the microwave technique for dough include microwave power transmission and absorption in which the material's polarity is studied. The study of polarization (redistribution of the charges comprising a material) is contributive to its dielectric properties, which are usually expressed in terms of complex permittivity. Complex permittivity consists of the real part (the permittivity or the dielectric constant) and the imaginary part (the loss factor; Kent 2001).

In practice, other related variables are measured using various methods such as the transmission methods (attenuation measurements), resonator methods, reflectance measurements, or microwave spectrometry. In an article by Chin et al. (2005), some measurements using the microwave transmission method to characterize bread doughs mixed at various densities, salt, and water contents were presented, and the relationship between dough density and microwave transmission variables were stated using a simple linear regression analysis. This nondestructive measurement method is potentially useful and can be developed for online density monitoring in the dough processing industry. In a more recent research, the dielectric properties of microwave-baked cakes and its constituents over a frequency range of 0.915 to 2.450 GHz were investigated (Al-Muhtaseb et al. 2010). The study reported that sugar showed higher dielectric properties than flour; therefore, formulations with higher sugar contents will exhibit higher power absorption, which is more appropriate for microwave baking. However, prediction of the formulation cannot be based on the microwave baking characteristics only; other quality parameters such as taste and color also need to be considered. Prediction of the degree of starch gelatinization in wheat flour dough can be achieved in pregelatinized dough during microwave heating for the production of quick-boiling noodles (Xue et al. 2010). The change in the temperature distribution was described with numerical prediction during intermittent heating, and the progress of starch gelatinization according to the heat transfer in the sample dough was predicted by the Runge–Kutta gel method. Despite the good accordance between the predicted and experimental results, accurate control of the degree of gelatinization needs further examination concerning heat analysis before the method can be applied in the food industry.

13.4.4.3 Ionizing Irradiation

The use of ionizing irradiation in cereals and cereal-based goods is mainly focused on the reduction of antinutrients (Arvanitoyannis and Stratakos 2010). The effects of cooking followed by irradiation (10 kGy) on the antinutritional factors, phytic acid and nitrates, in a ready-to-eat meal of sorghum indicated a decrease in the phytic acid level from 135 mg/100 g of the raw sorghum endosperm meal to 80.5 mg/100 g for the cooked and irradiated sample (Duodu et al. 1999).

13.4.5 Near-Infrared Spectroscopy

Near-infrared (NIR) spectroscopy has become a key element in process analytical technology. NIR spectroscopy technology is an extremely successful tool in the cereal industry, allowing fast screening methods for the prediction of specific analytes. NIR spectroscopy analyses make quality control instantaneous and moves it from the laboratory to the production line. A recent review by Jespersen and Munck (2009) focused on the next step in NIR spectroscopy technology—spectral classification—which promises new, challenging options in identifying functional factors in plant breeding, single-seed sorting, and process control as well as in food production and design.

Today, inline and online NIR spectroscopy technology (Lee 2007) in cereal industrial process control has replaced wet chemical analyses to a large extent, reducing the need for a laboratory. In the grain and feed industry, there are now inline NIR analyzers that can continuously evaluate moisture and protein, as well as NIR spectroscopy equipment for online control of, for example, extract and alcohol during brewing and fermentation. In principle, the development of a spectral fingerprint from a window in a process can be followed as a trajectory by the self-modeling principal component analysis (PCA) algorithm without a standard (Munck et al. 1998). The integration among NIR spectroscopy instruments, chemometrics, and knowledge about processes is now taking place as "process analytical technology" with multiway (Smilde et al. 2004) chemometric models including the time dimension built on *n*-way partial least squares (PLS) (Bro 1996) and parallel factor analysis (PARAFAC). Allosio et al. (1997) demonstrated the use of NIR spectroscopy in the reflectance mode (1100–2500 nm) evaluated by PARAFAC in the transformation from barley to malt.

13.4.6   NMR AND MRI

NMR spectroscopy and MRI are based on the magnetic properties of the atomic nucleus and many elements have isotopes with such properties, above all the omniabundant proton. NMR and MRI have some distinct advantages over other instrumental methods: they are noninvasive and nondestructive, most systems are transparent for the excitation, they measure volumes instead of surfaces, and the methods allow the extraction of both physical and chemical information. By exploiting the magnetic resonance, it is possible to obtain unique knowledge about, for example, flow, diffusion, and water distribution without perturbing the system. Similarly, processes such as heating, freezing, salting, hydration, and dehydration can be monitored absolutely noninvasively. NMR is an important tool to investigate the dynamics of food materials, and water, at a molecular level.

Low-resolution ^{1}H NMR spectroscopy has been previously applied to bread staling studies using multiple experiments to observe different windows of relation times. Water molecular dynamics during bread staling have been investigated by NMR relevant during storage both in gelatinized waxy maize starch (Farhat et al. 2003) and bread (Engelsen et al. 2001). ^{1}H FID rigidity increase was attributed to a reduced mobility of the bread matrix due to both recrystallizing amylopectin and loss of water from the crumb. ^{1}H T_2 relaxation by low-resolution NMR experiments has been reported (Engelsen et al. 2001) to adequately represent changes in the mobility of bakery products during storage at 23 MHz (Sereno et al. 2007) and at 20 MHz (Chen et al. 1997). These studies reported multiple ^{1}H T_2 populations in baked products that underwent major changes during storage, resulting in reduced mobility (shorter ^{1}H T_2 relaxation time) in stored products. ^{1}H T_1 relaxation was reported to have monoexponential behavior (at 20 MHz) and to slightly decrease in mobility (from 90 to 80 ms) during storage (Leung et al. 1983; Vittadini and Vodovotz 2003), providing a less informative insight into the molecular dynamics of bread staling.

Fast field cycling (FFC) is an NMR technique that has been recently proposed for food applications, and it might provide additional insights into the molecular dynamics of bread. FFC measures longitudinal relaxation times (^{1}H T_1) and molecular dynamics, applying a variable magnetic field to the sample to obtain ^{1}H T_1 at different frequencies and, consequently, widening the range of ^{1}H molecular motions that can be measured. At low frequencies, it is possible to focus on molecular dynamics characterized by very long correlation times, such as molecular surface dynamics and collective effects (Baroni et al. 2008). NMR dispersion profiles [$1/(^1$H $T_1)1/4$ ^{1}H R_1 vs. frequency] are particularly valuable to assess the interactions of water molecules with paramagnetic and large-sized macromolecular systems (Baroni et al. 2008). In particular, the relaxation profile is dominated by the magnetic field dependence of rotationally immobilized protons and dynamically coupled to the spin-lattice relaxation of water protons. The FFC technique has been previously applied to food matrices like cheeses (mozzarella and Gouda; Godefroy et al. 2003), eggs (Laghi et al. 2005), and balsamic vinegar (Baroni et al. 2009) to study the water distribution and interaction with the different food components. In most cases, the phenomenon under study was better resolved at low frequencies. In a recent article, the molecular properties of white bread loaves were characterized by multiple proton NMR techniques (proton FID, T_2 and T_1 relaxation times) over 14 days of storage (Curti et al. 2011). Changes at a molecular level (faster decay of proton FIDs and shifting of proton T_2 relaxation time distributions toward shorter times), indicating a proton mobility reduction of the bread matrix, were observed during storage. Multiple ^{1}H T_2 populations were observed and tentatively associated with water–gluten and water–starch domains. Proton T_1 of bread was, for the first time, measured at variable frequencies (FFC NMR) and was found to be strongly dependent on frequency and to decrease in bread during storage, especially at frequencies lower than or equal to 0.2 MHz. An additional proton T_1 population, relaxing at 2 ms, was detected at 0.52 MHz only at early storage times and tentatively attributed to a water–gluten domain that lost mobility during storage. In particular, MRI has also been applied to other aspects of bread properties. In a study by Goetz et al. (2003), dough fermentation was monitored by MRI and volumetric measurements and the pore structure were observed by MRI. In another study, gas cell formation during the proving of dough was monitored using MRI (van Duynhoven et al. 2003). Various stress effects (kneading, temperature, and molding) were introduced and the dough volume was subsequently investigated by MRI. Using MRI, the largest negative effect could be ascribed to molding. Thus far, only very simple NMR and MRI measurements have been performed to dynamically investigate the dough-to-bread process. Future NMR applications will provide a much more detailed picture of the water migration, intermolecular interactions, and phase transitions occurring during bread baking, allowing for detailed studies of the functionality of additives and ingredients (Viereck et al. 2005).

13.4.7 Dry Process

Dry fractionation technologies allow peripheral tissues to be separated and recovered to efficiently separate valuable from detrimental components (i.e., contaminants, antinutrient compounds, and irritants) to develop nutritionally enhanced ingredients and products. The accessibility of the bioactive compounds with health benefits may also be limited because they are trapped into rigid cell structures. There is therefore a need for a dry grain fractionation technology that can efficiently separate negative and positive elements, and increase the bioavailability of the latter, to produce new flours and ingredients with optimized technofunctional and nutritional attributes. A review by Hemery et al. (2007) provides an overview of the existing processes that can be used for the production of wheat products and fractions with enhanced nutritional interest. The grain composition and properties are briefly introduced with emphasis on nutritionally interesting compounds. Tissue markers and their application in process monitoring are presented, and the physical properties that influence the fractionation properties of grain tissues are developed. The main dry fractionation processes in wheat are then reviewed, including pretreatments, degerming, debranning, and bran fractionations (Hemery et al. 2007). Suggestions for future research to improve dry fractionation technology in wheat to produce higher quality foods and thus answer consumers' demands evidence a move from traditional processing with a high-value product and low-value byproducts toward a true plant refinery processing in which almost every fraction of the raw material could be best exploited for optimal nutritional end use.

13.5   CONCLUSIONS AND FUTURE PROSPECTS

Consumers are increasingly concerned about the contents of the food in their diets and are shifting away from getting nutrients via fortified foods and turning toward products that are naturally high in macronutrients and micronutrients and those that have been blended with other foods to create even higher nutrient levels. Those requirements fully apply to cereal-based goods. From the scientific and technological points of view, cereal trends matching consumers' demands encompass (1) a move from traditional processing with a high-value product and low-value by-products toward processing in which almost every fraction of the raw material could be best exploited for optimal nutritional end use, and (2) an increasing exploration and exploitation of the power of novel technologies, particularly those noninvasive and nondestructive, to create highly nutritious, sensorially accepted, and innovative cereal-based goods. The mild processing is attractive for the consumer and it is expected to have consolidated and increasing applications of the nonthermal technologies in the coming decades—for example, pulsed electric field, high-pressure processing, ultrasound processing, irradiation, ultraviolet, and pulse-light technology—for the development of new cereal-based foods complying with the relevant safety legislations.

REFERENCES

Ahmed, J., Ramaswamy, H.S., Ayad, A., Alli, I., and Alvarez, P. 2007. Effect of high pressure treatment on rheological, thermal and structural changes in basmati rice flour slurry. *Journal of Cereal Science* 46: 148–156.

Allosio, N., Boivin, P., Bertrand, D., and Courcoux, P. 1997. Characterization of barley transformation into malt by three-way factor analysis of near infrared spectra. *Journal of Near Infrared Spectroscopy* 5: 157–166.

Al-Muhtaseb, A.H., Hararah, M.A., Megahey, E.K., McMinn, W.A.M., and Magee, T.R.A. 2010. Dielectric properties of microwave-baked cake and its constituents over a frequency range of 0.915–2.450 GHz. *Journal of Food Engineering* 98: 84–92.

Altamirano-Fortoul, R., Moreno-Terrazas, R., Quezada-Gallo, A., and Rosell, C.M. 2012. Viability of some probiotic coatings in bread and its effect on the crust mechanical properties. *Food Hydrocolloids* in press.

Alvarez-Jubete, L., Arendt, E.K., and Gallagher, E. 2010. Nutritive value of pseudocereals and their increasing use as functional gluten free ingredients. *Trends in Food Science and Technology* 21: 106–113.

Angioloni, A., and Collar, C. 2011a. Nutritional and functional added value of oat, Kamut®, spelt, rye and buckwheat versus common wheat in breadmaking. *Journal of the Science of Food and Agriculture* 91: 1283–1292.

Angioloni, A., and Collar, C. 2011b. Polyphenol composition and "in vitro" antiradical activity of single and multigrain breads. *Journal of Cereal Science* 53: 90–96.

Angioloni, A., and Collar, C. 2011c. Physicochemical and nutritional properties of reduced-caloric density high-fibre breads. *LWT—Food Science and Technology* 44: 747–758.

Angioloni, A., and Collar, C. 2012a. Suitability of oat, millet and sorghum in breadmaking. *Food and Bioprocess Technology* DOI: 10.1007/s11947-012-0786-9.

Angioloni, A., and Collar, C. 2012b. High legume–wheat matrices: an alternative to promote bread nutritional value meeting dough viscoelastic restrictions. *European Food Research and Technology* 234(2): 273–284.

Angioloni, A., Collar, C. 2012c. Promoting dough viscoelastic structure in composite cereal matrices by high hydrostatic pressure. *Journal of Food Engineering* 111: 598–605.

Apichartsrangkoon, A., Ledward, D.A., Bell, A.E., and Brennan, J.G. 1998. Physicochemical properties of high pressure treated wheat samples. *Food Chemistry* 63(2): 215–220.

Arendt, E.A., and Dal Bello, F. 2008. *Gluten-Free Cereal Products and Beverages.* Academic Press, Elsevier. Burlington, MA.

Arvanitoyannis, I.S., and Stratakos, A.Ch. 2010. Potential uses of irradiation. In *Irradiation of Food Commodities: Techniques, Applications, Detection, Legislation, Safety and Consumer Opinion.* Chapter 16. Ed. I.S. Arvanitoyannis. Academic Press, Elsevier. 635–669.

Bárcenas, M.E., Altamirano-Fortoul, R., and Rosell, C.M. 2010. Effect of high pressure processing on wheat dough and bread characteristics. *LWT—Food Science and Technology* 43: 12–19.

Barnard, N.D. 2010. Trends in food availability, 1909–2007. *American Journal of Clinical Nutrition* 91: 1530S–1536S.

Baroni, S., Bubici, S., Ferrante, G., and Aime, S. 2008. Applications of field cycling relaxometry to food characterization. In *Magnetic Resonance in Food Science: Challenges in a Changing World.* Ed. M. Guðjónsdóttir, P.S. Belton, and G.A. Webb. RSC Publishing. UK. 65–72.

Baroni, S., Consonni, R., Ferrante, G., and Aime, S. 2009. Relaxometric studies for food characterization: the case of balsamic and traditional balsamic vinegars. *Journal of Agricultural and Food Chemistry* 57(8): 3028–3032.

Bellido, G.G., and Hatcher, D.W. 2010. Ultrasonic characterization of fresh yellow alkaline noodles. *Food Research International* 43: 701–708.

Brites, C., Trigo, M.J., Santos, C., Collar, C., and Rosell, C.M. 2010. Maize-based gluten-free bread: Influence of processing parameters on sensory and instrumental quality. *Food and Bioprocess Technology* 3: 707–715.

Bro, R. 1996. Multiway calibration, multilinear PLS. *Chemometrics and Intelligent Laboratory Systems* 10(1): 47–61.

Buckow, R., Heinz, V., and Knorr, D. 2007. High pressure phase transition kinetics of maize starch. *Journal of Food Engineering* 81: 469–475.

Cardello, A.V. 2003. Consumer concerns and expectations about novel food processing technologies: Effects on product liking. *Appetite* 40: 217–233.

Chen, P.L., Long, Z., Ruan, R., and Labuza, T.P. 1997. Nuclear magnetic resonance studies of water mobility in bread during storage. *Lebensmittel-Untersuchung und Forschung* 30(2): 178–183.

Chin, N.L., Campbell, G.M., and Thompson, F. 2005. Characterisation of bread doughs with different densities, salt contents and water levels using microwave power transmission measurements. *Journal of Food Engineering* 70: 211–217.

Collar, C. 2008. Novel high fibre and whole grain breads. In: *Technology of Functional Cereal Products*. Ed. B. Hamaker. Woodhead Publishing Limited. Abington Hall, England. Published in North America by CRC Press. Boca Raton, FL. 184–214.

Collar, C. 2011. Revisiting minor cereals and pseudocereals as reinvented healthy and convenient value added goods. *ICC International Conference on Cereals and Cereal Products: Nutrition, Biotechnology and Safety. Keys for Cereal Chain Innovation*. Santiago de Chile, April 10–13, 2011.

Collar, C., and Angioloni, A. 2010. An approach to structure–function relationships of polymeric dietary fibres in foods: Significance in breadmaking applications. In *Dietary Fibre—New Frontiers for Food and Health*. Eds. J.W. van der Kamp, J.M. Jones, B.V. McCleary, and D.L. Topping. Wageningen Academic Publishers. The Netherlands. 91–114.

Collar, C., and Angioloni, A. 2011. Novel high fibre wheat goods from diluted matrices: Visco-elastic network, functional and technological aspects. In *Wheat Science Dynamics: Challenges and Opportunities*. Eds. R.N. Chibbar and J.E. Dexter. Agrobios International. Jodhpur, India. 283–297.

Cui, L., Pan, Z., Yue, T., Atungulu, G.G., and Berrios, J. 2010. Effect of ultrasonic treatment of brown rice at different temperatures on cooking properties and quality. *Cereal Chemistry* 87(5): 403–408.

Curti, E., Bubici, S., Carini, E., Baroni, S., and Vittadini, E. 2011. Water molecular dynamics during bread staling by nuclear magnetic resonance. *LWT—Food Science and Technology* 44: 854–859.

Dessev, T., Jury, V., and Le-Bail, A. 2011. The effect of moisture content on short infrared absorptivity of bread dough. *Journal of Food Engineering* 104: 571–576.

Duodu, K.G., Minnaar, A., and Taylor, J.R.N. 1999. Effect of cooking and irradiation on the labile vitamins and antinutrient content of a traditional African sorghum porridge and spinach relish. *Food Chemistry* 66: 21–27.

Dyson, T. 1999. World food trends and prospects to 2025. Colloquium paper. *Proceedings of the National Academy of Sciences of the United States of America* 96: 5929–5936.

Elmehdi, H.M., Page, J.H., and Scanlon, M.G. 2003. Using ultrasound to investigate the cellular structure of bread crumb. *Journal of Cereal Science* 38: 33–42.

Elmehdi, H.M., Page, J.H., and Scanlon, M.G. 2004. Ultrasonic investigation of the effect of mixing under reduced pressure on the mechanical properties of bread dough. *Cereal Chemistry* 81(4): 504–510.

Engelsen, S.B., Jensen, M.K., Pedersen, H.T., Norgaard, L., and Munck, L. 2001. NMR-baking and multivariate prediction of instrumental texture parameters in bread. *Journal of Cereal Science* 33(1): 59–69.

Estrada-Girón, Y., Swanson, B.G., and Barbosa-Cánovas, G.V. 2005. Advances in the use of highhydrostatic pressure for processing cereal grains and legumes. *Trends in Food Science and Technology* 16: 194–203.

Farhat, I.A., Ottenhof, M.A., Marie, V., and De Bezenac, E. 2003. ^{1}H NMR relaxation of Amylopectin retrogradation. In *Magnetic Resonance in Food Science: Latest Developments*. Eds. P.S. Belton, A.M. Gil, G.A. Webb, and D. Rutledge. RSC. UK. 172–179.

Flander, L., Salmenkallio-Marttila, M., Suortti, T., and Autio, K. 2007. Optimization of ingredients and baking process for improved wholemeal oat bread quality. *LWT—Food Science and Technology* 40: 860–870.

Garcia-Alvarez, J., Alava, J.M., Chavez, J.A., Turo, A., Garcia, M.J., and Salazar, J. 2006. Ultrasonic characterisation of flour–water systems: A new approach to investigate dough properties. *Ultrasonics* 44: e1051–e1055.

Godefroy, S., Korb, J.P., Creamer, L.K., Watkinson, P.J., and Callaghan, P.T. 2003. Probing protein hydration and aging of food materials by the magnetic field dependence of proton spin-lattice relaxation times. *Journal of Colloid and Interface Science* 267: 337–342.

Goetz, J., Gross, D., and Koehler, P. 2003. On-line observation of dough fermentation by magnetic resonance imaging and volumetric measurements. *European Food Research and Technology* 217: 504–511.

Gomes, M.R., Clark, A., and Ledward, D.A. 1998. Effects of high pressure on amylases and starch in wheat and barley flours. *Food Chemistry* 63: 363–372.

Gómez, M., Oliete, B., García-Álvarez, J., Ronda, F., and Salazar, J. 2008. Characterization of cake batters by ultrasound measurements. *Journal of Food Engineering* 89: 408–413.

Hampshire, J. 1998. Composition and nutritional quality of oats. *Enährung/Nutrition* 22: 505–508.

Hemery, Y., Rouau, X., Lullien-Pellerin, V., Barron, C., and Abecassis, J. 2007. Dry processes to develop wheat fractions and products with enhanced nutritional quality. *Journal of Cereal Science* 46: 327–347.

Hüttner, E.K., Dal Bello, F., and Arendt, E.K. 2010. Fundamental study on the effect of hydrostatic pressure treatment on the bread-making performance of oat flour. *European Food Research and Technology* 230: 827–835.

Hüttner, E.K., Dal Bello, F., Poutanen, K., and Arendt, E.K. 2009. Fundamental evaluation of the impact of high hydrostatic pressure on oat batters. *Journal of Cereal Science* 49: 363–370.

Innova Market Insights. 2009. Simplicity Heads Top Ten New Food NPD Trends for 2010—Innova Market Insights. *Food Ingredients First*. Available from: http://www.foodingredientsfirst.com/headlines/Simplicity-Heads-Top-Ten-New-Food-NPD-Trends-for-2010-Innova-Market-Insights.html (accessed November 2011).

Institute of Food Technologists. 2011. Top 10 Food Trends. *Food Technology* magazine 65/4. Available at: http://www.ift.org/food-technology/past-issues/2011/april/features/food-trends.aspx?page=viewall (accessed October 2011).

Jespersen, B.M., and Munck, L. 2009. Cereals and cereal products. In *Infrared Spectroscopy for Food Quality Analysis and Control*. Ed. D.-W. Sun, Chapter 11. Academic Press-Elsevier. 275–319.

Katopo, H., Song, Y., and Jane, J.L. 2002. Effect and mechanism of ultrahigh hydrostatic pressure on the structure and properties of starches. *Carbohydrate Polymers* 47: 233–244.

Kent, M. 2001. Microwave measurements of product variables. In *Instrumentation and Sensors for the Food Industry*. Eds. E. Kress-Rogers and C.J.B. Brimelow. Oxford, UK. Butterworth-Heinemann Ltd. 231–279.

Kent, M., Knochel, R., Daschner, F., and Berger, U. 2000. Composition of foods using microwave dielectric spectra. *European Food Research Technology* 210(5): 359–366.

Keskin, S.O., Sumnu, G., and Sahin, S. 2004. Bread baking in halogen lamp microwave combination oven. *Food Research International* 37(5): 489–495.

Kieffer, R., Schurer, F., Köhler, P., and Wieser, H. 2007. Effect of hydrostatic pressure and temperature on the chemical and functional properties of wheat gluten: Studies on gluten, gliadin and glutenin. *Journal of Cereal Science* 45(3): 285–292.

Laghi, L., Cremonini, M.A., Placucci, G., and Sykora, S. 2005. A proton NMR relaxation study of hen egg quality. *Magnetic Resonance Imaging* 23(3): 501–510.

Lee, A.K. 2007. On analysis in food engineering. In *Near Infrared Spectroscopy in Food Science and Technology*. Eds. Y. Ozaki, W.F. McClure, A.A. Christy A.A. Wiley-Interscience. Hoboken, NJ. 361–378.

Leung, H.K., Magnuson, J.A., and Bruinsma, B.L. 1983. Water binding of wheat flour doughs and breads as studied by deuteron relaxation. *Journal of Food Science* 48: 95–99.

Mehta, K.L., Scanlon, M.G., Sapirstein, H.D., and Page, J.H. 2009. Ultrasonic investigation of the effect of vegetable shortening and mixing time on the mechanical properties of bread dough. *Journal of Food Science* 74(9): 455–461.

Mozhaev, V.V., Heremans, K., Frank, J., Masson, P., and Balny, C. 1996. High pressure effects on protein structure and function. *Proteins* 24: 81–91.

Munck, L., Nørgaard, L., Engelsen, S.B., Bro, R., and Anderson, S.A. 1998. Chemometrics in food science—a demonstration of the feasibility of a highly exploratory, inductive evaluation strategy of fundamental scientific significance. *Journal of Chemometrics and Intelligent Laboratory Systems* 44: 31–60.

Patel, B.K., Waniska, R.D., and Seetharaman, K. 2005. Impact of different baking processes on bread firmness and starch properties in breadcrumb. *Journal of Cereal Science* 42(2): 173–184.

Prepared Foods Network. 2011. *New Product Trends in Cereals and Cereal Bars*. Available at: http://www.preparedfoods.com/articles/109076-new-product-trends-in-cereals-and-cereal-bars (accessed November 2011).

Rosell, C.M., Cortez, G., and Repo-Carrasco, R. 2009. Breadmaking use of Andean crops quinoa, kaniwa, kiwicha, and tarwi. *Cereal Chemistry* 86: 386–392.

Rosell, C.M., and Marco, C. 2008. Rice. In *Gluten Free Cereal Products and Beverages*. Eds. E.K. Arendt and F. dal Bello. Elsevier Science. UK. 81–100.

Salagnac, P., Glouannec, P., and Lecharpentier, D. 2004. Numerical modeling of heat and mass transfer in porous medium during combined hot air, infrared and microwave heating. *International Journal of Heat and Mass Transfer* 47: 4479–4489.

Salovaara, H., Kanerva, P., Kaukinen, K., and Sontag-Strohm, T. 2011. Oats—an overview from coeliac disease point of view. In *The Science of Gluten-Free Foods and Beverages. Proceedings of the First International Conference of Gluten Free Cereal Products and Beverages*. Eds. E.K. Arendt, F. Dal Bello. AACC Press, in press.

Sereno, N.M., Hill, S.E., Mitchell, J.R., Scharf, U., and Farhat, I.A. 2007. Probing water migration and mobility during the aging of bread. In *Magnetic Resonance in Food Science: From Molecules to Man*. Eds. I.A. Farhat, P.S. Belton, and G.A. Webb. RSC Publishing. UK. 89–95.

Shiinoki, Y., Motouri, Y., and Ito, K. 1998. On-line monitoring of moisture and salt contents by the microwave transmission method in a continuous salted butter-making process. *Journal of Food Engineering* 38: 153–167.

Smilde, A., Bro, R., and Geladi, P. 2004. *Multi-Way Analysis*. John Wiley & Sons. Chichester. 381.

Stolt, M., Oinonen, S., and Autio, K. 2001. Effect of high-pressure on the physical properties of barley starch. *Innovative Food Science and Emerging Technologies* 1: 167–175.

Sumnu, G., Datta, A., Sahin, S., Keskin, S., and Rakesh, V. 2007. Transport and related properties of breads baked using various heating modes. *Journal of Food Engineering* 78(4): 1382–387.

Tan, M.C., Chin, N.L., and Yusof, Y.A. 2011. Power ultrasound aided batter mixing for sponge cake batter. *Journal of Food Engineering* 104: 430–437.

Taylor, J.R.N., Schober, T.J., and Bean, S.C. 2006. Novel food and non-food uses for sorghum and millets. *Journal of Cereal Science* 44: 251–271.

Thomas, D.J., and Atwell, W.A. 1997. *Starches*. Eagan Press. St. Paul, MN.

USDA. 2011. Economic Research Service, US Department of Agriculture. Loss adjusted food availability: grains per capita availability adjusted for loss. Available from: http://www.ers.usda.gov/Data/FoodConsumption/FoodGuideSpreadsheets.htm#grain (accessed November 12, 2011).

Vallons, K.J.R., Ryan, L.A.M., and Arendt, E.K. 2011. Promoting structure formation by high pressure in gluten-free flours. *LWT—Food Science and Technology* 44: 1672–1680.

Vallons, K.J.R., Ryan, L.A.M., Koehler, P., and Arendt, E.K. 2010. High pressure–treated sorghum flour as a functional ingredient in the production of sorghum bread. *European Food Research Technology* 231: 711–717.

van Duynhoven, J.P.M., van Kempen, G.M.P., van Sluis, R.R., Rieger, B., Weegels, P., van Vliet, L.J., and Nicolay, K. 2003. Quantitative assessment of gas cell development during the proofing of dough by magnetic resonance imaging and image analysis. *Cereal Chemistry* 80: 390–395.

Viereck, N., Dyrby, M., and Engelsen, S.B. 2005. Monitoring thermal processes by NMR technology. In *Emerging Technologies for Food Processing*. Ed. D.-W. Sun. Academic Press, Incorporated. Amsterdam. 553–575.

Vittadini, E., and Vodovotz, Y. 2003. Changes in the physicochemical properties of wheat-and-soy-containing breads during storage as studied by thermal analyses. *Journal of Food Science* 68(6): 2022–2027.

Vittadini, E., Clubbs, E., Shellhammer, T.H., and Vodovotz, Y. 2004. Effect of high pressure processing and addition of glycerol and salt on the properties of water in corn tortillas. *Journal of Cereal Science* 39: 109–117.

Wang, L., and Sun, D.-W. 2001. Rapid cooling of porous and moisture foods by using vacuum cooling technology. *Trends in Food Science and Technology* 12: 174–184.

Winter, R. 2003. *Advances in High Pressure Bioscience and Biotechnology II*. Springer: Berlin/Heidelberg/New York.

Xue, C., Fukuoka, M., and Sakai, N. 2010. Prediction of the degree of starch gelatinization in wheat flour dough during microwave heating. *Journal of Food Engineering* 97: 40–45.

Zheng, L., and Sun, D.-W. 2005. Vacuum cooling of foods. In *Emerging Technologies for Food Processing*. Ed. D.-W. Sun. Academic Press. Elsevier. 579–602.

14 Effects of Processing on Nutritional and Functional Properties of Cereal Products

Sanaa Ragaee, Koushik Seethraman,
and El-Sayed M. Abdel-Aal

CONTENTS

14.1 INTRODUCTION

Cereal grains such as wheat, rice, and corn are staple foods worldwide. The grains as intact or debranned (pearled) or refined flours are processed into numerous food forms or a grain fraction is incorporated into food recipes to improve the nutritional and functional properties of foods. Cereal grains are an excellent source of carbo-hydrates, dietary fibers (DFs), and proteins (Ragaee et al. 2011, 2012a, 2013, 2006). They are also considered good sources of several vitamins such as B vitamins and vitamin E, and a number of minerals including iron, zinc, magnesium, and phospho-rus. The grain, also known as kernel or caryopsis, is made up of three main parts, endosperm (80%–85% of the kernel), germ (2%–3% of the kernel), and outer layers or bran (13%–17% of the kernel). The endosperm is composed of cells containing starch granules embedded in a protein matrix. The germ is rich in oil and lipid-soluble vitamins. The outer layers contain high concentrations of cellulose, hemi-celluloses, and minerals. The outer layers and germ are considered good sources of many bioactive components such as DFs (β-glucan, lignan, inulin, arabinoxylan, and resistant starch [RS]), phenolic compounds (phenolic acids, alkylresorcinols, and flavonoids), carotenoids (lutein and zanthein), anthocyanins, vitamins, and minerals (Abdel-Aal et al. 2006, 2007; Slavin 2004; Ragaee et al. 2006; Sidhu and Kabir 2007). Many of the bioactive components, particularly phenolic compounds, contribute to the antioxidant properties of cereal grains (Ragaee et al. 2011, 2012a, 2013). In addition, bioactive compounds have also been linked with a reduced risk of cardiovascular disease, cancer, diabetes, obesity, and other chronic diseases (Estruch et al. 2009; Streppel et al. 2008).

Processing of cereal grains has evolved over time to satisfy the needs of today's consumer and also to meet growing market demands, in particular, to provide a variety of cereal foods that are tasty, healthy, and convenient. Cereal processing can be grouped into two categories: primary processing such as milling, pearling, or malting, and others such as fractionation processing in which the grains are broken down or fractionated into flours or components for food and nonfood use. The second category or secondary processing includes a number of processing technologies such as baking, extrusion, puffing, etc., that render cereals into palatable and pleasant foods such as pasta, bread, breakfast cereals, cookies, etc. During these processes, many changes take place that affect the nutritional and functional properties of cereal products. The current chapter focuses on these changes or effects of processing on the compositional and functional properties of cereal products. The emphasis of the chapter is on bioactive compounds due to their important roles in human health and nutrition.

14.2   PRIMARY PROCESSING

14.2.1   MALTING

Malting is the process through which cereal grains are germinated in a controlled manner by soaking in water to boost the activity of enzymes required to modify starch and protein. Germination can be terminated by heat, followed by further heat treatment to "kiln" the grain and produce the required flavor and color. Malted grains are used to make beer, whisky, malted shakes, malt vinegar, confections such as Maltesers, Whoppers, flavored drinks such as Horlicks, Ovaltine, and Milo, and some baked goods such as malt loaf. Ground malted grains are known as "sweet meal." Various cereals such as barley, rye, sorghum, and millet are malted to produce different types of food products (Elkhier and Hamid 2008).

Significant changes in individual and total phenolic compounds and antioxidant capacity were found after the malting of two barley varieties (Lu et al. 2007). Catechin and ferulic acid were the most common phenolic compounds identified in barley. Antioxidant properties such as 2,2-diphenyl-1-picrylhydrazyl radical scavenging capacity, 2,2′-Azino-di-[3-ethylbenzthiazoline] sulfonate radical cation scavenging capacity, and reducing power are strongly correlated with total phenol content and the sum of individual phenolic compounds of malted barley products. Malting was found to slightly reduce tocols, whereas mashing and brewing significantly increased tocol and flavonoid content. Peterson and Qureshi (1993) reported a slight reduction in total tocols in malted barley, and a significant increase (from 57 to 153 mg/kg) in spent grains recovered after the mashing and brewing process. Goupy et al. (1999) reported a significant reduction in flavanols (62%–87%) and flavonols (64%–91%) content in malted barley, whereas phenolic acids were affected to a lesser extent. Other researchers reported a 300% increase in flavanols and flavonols (polyphenols) in five different barley varieties after malting (Maillard et al. 1996). Increasing malting time resulted in increased crude fiber, tannin, and phytic acid levels and in a reduction in carbohydrate and mineral contents; and no significant changes were observed in the protein and oil contents of two sorghum cultivars (Elkhier and Hamid 2008). Malting of red sorghum, millet, and maize was found to improve their nutritional quality resulting in increased protein content and α-amylase activity, as well as decreased lipid, ash, and phytate contents (Traore et al. 2004). The increase in α-amylase activity and cyanide content were more pronounced in red sorghum seeds than in millet and maize seeds, whereas the decrease in phytate concentration was more profound in millet seeds compared with red sorghum and maize seeds. In this study, samples were collected during the entire process, including after soaking, germination, maturation, drying, degerming, and final product. The results suggest the elimination of the maturation step from the malting process due to its adverse effects on sucrose content, whereas it is also recommended that the seeds be degermed after sun drying to achieve a significant reduction in cyanide content. The malted red sorghum or millet flours could be incorporated in infant foods to improve its nutritional properties. In another study, malting and fermentation were found to improve the nutritional quality and functionality of food-grade (Macia and red tannin-containing) sorghum flours (Mella 2011). The malting process resulted in a significant ($p < 0.05$) increase in the amounts of reducing sugars in sorghum flour samples, whereas fermentation caused a significant ($p < 0.05$) increase in the amounts of soluble proteins and free amino acids. When the different sorghum flours were used for the preparation of buns, no effects of malting and fermentation were observed on the textural properties of the buns or on their color and oil uptake. The study demonstrated that sorghum variety had no effect ($p > 0.05$) on the amounts or levels of reducing sugars, soluble proteins and free amino acids, oil uptake, pH, and titratable acidity but had significant effects on the surface color of the buns.

Germination of sorghum grains followed by malting for 5 days increased carbohydrate digestibility and free amino acid content by sixfold and fourfold, respectively (Correia et al. 2008). Fermentation of sorghum was significantly adversely influenced by the tannin concentration in the grains (Elkareem and Taylor 2011), which caused an inhibition of microbial fermentation. It also improved protein quality and physical characteristics of kisra (a naturally lactic acid bacteria-fermented and yeast-fermented sorghum pancake-like flatbread). Germination and fermentation are simple processes and culturally acceptable methods used to enhance nutritional and functional properties of cereal grains and are also used to prepare weaning foods at domestic and industrial levels.

14.2.2 MILLING

In general, milling is the transformation of grains into finer primary products for secondary processing. In cereal grains, there are two types of milling: dry and wet milling. Dry milling is a process that separates the bran and germ from the starchy endosperm to produce refined flours for use in bakery products. Thus, pericarp, testa, and aleurone layers are all removed from the flour fraction. Because the majority of bioactive components are present in the outer layers, dry milling is also used to break down entire kernels into whole-grain flours for the production of whole-grain foods. Wet milling is employed to separate grains into large chemical constituents (starch, protein, fiber, and oil). Pearling is the application of abrasion and friction technologies to effectively remove only the bran layers from the cereal grains, allowing nutritious parts, such as the aleurone layer, to remain in the intact kernels. This procedure, when applied before milling, improves milling yields as it influences the way cereal grains break during milling and as such result in improved flour quality and functionality (Mousia et al. 2004). The application of pearling in the processing of several cereals demonstrated that phenolics and other bioactive components are concentrated in fractions obtained from the outermost layers of the grain as well as the improved antioxidant capacity of these fractions (Zielinski and Kozłowska 2000; Peterson 2001; Skrabanja et al. 2004; Beta et al. 2005; Madhujith et al. 2006; Hung et al. 2009; Holtekjølen et al. 2011).

Dry milling of cereal grains into refined flours considerably affects the health-promoting components present in grains such as minerals, vitamins, fibers, antioxidants, and phytochemicals. For example, the concentration of grain antioxidants was found to be drastically reduced during the refining process. Many studies reported that phenolic and antioxidant compounds are present in the outermost layers of different cereals, and thus, the bran fraction obtained as a milling by-product is a nutritious fraction that could be used as a natural source of antioxidants and value-added products in the preparation of functional food ingredients or to boost DF (Holasova et al. 2002; Awika et al. 2005; Taylor et al. 2006; Zdunczyk et al. 2007; Liyana-Pathirana and Shahidi 2007; Holtekjølen et al. 2006, 2008, 2011; Singh et al. 2006; Zielinski et al. 2007; Hung et al. 2009; Vaher et al. 2010). The effects of milling on phenolic compounds in various cereal grains are summarized in Table 14.1.

Several endogenous enzymes are also present in the aleurone and bran layers of the kernel and in the germ. Therefore, milling will result in the uneven distribution of enzymes in the resultant milling fractions. The lowest endogenous β-glucanase activity was observed in the endosperm fraction, whereas fractions rich in bran and DF had high endogenous enzyme activity levels (Vatandoust et al. 2012). The presence of these enzymes would affect the functionality and health benefits of products fortified with β-glucan (Vatandoust et al. 2012). On the other hand, microbial enzymes have been used in the baking industry for their positive effects on rheological, physical, and sensory characteristics (e.g., color, taste, aroma, crust and crumb texture, crumb softness, freshness, and shelf life) of baked products (Fox and Mulwhill 1982; Stauffer 1987; Hamer 1991; Joye et al. 2009).

Polishing of rice resulted in a marked reduction of several nutrients, that is, a loss of 29% of protein, 79% of fat, and 67% of iron (Abbas et al. 2011). Iqbal et al. (2005)

TABLE 14.1

Effects of Milling and Pearling on Bioactive Compounds in Cereal Grains

Cereal/Conditions	Changes in Phenolic Compounds	References
	Milling	
Wheat	Phenolics concentrated in the outer layers	Barron et al. (2007); Liyana-Pathirana and Shahidi (2006, 2007); Barron et al. (2007)
	The germ fraction possessed the highest total phenolic content, followed by bran, shorts, whole grain, and flour	
	Phytosterols concentrated in the outer layers	
Rye	Phenolics concentrated in the bran fraction	Liukkonen et al. (2003); Glitsø et al. (1999)
Buckwheat	Phenolics mainly located in the outer layers	Hung and Morita (2008); Dietrych-Szostak and Oleszek (1999)
	Pearling	
Wheat	The outer layers contained a significantly higher amount of phenolics and antioxidant capacity than did the whole grain	Beta et al. (2005)
Buckwheat	Phenolics concentrated in the outer layer	Hung et al. (2009); Skrabanja et al. (2004)
Barley	Phenolics concentrated in the outer layer	Madhujith et al. (2006)
Oat	Phenolics concentrated in the outer layer	Peterson (2001)
Wheat	Phenolics concentrated in the outer layer	Liyana-Pathirana and Shahidi (2006, 2007)
Wheat, rye, oat, buckwheat, and barley	Phenolics concentrated in pericarp and testa fractions from wheat, oat, buckwheat, and rye and distributed in evenly in the endosperm and outer layers in barley	Zielinski and Kozłowska (2000)
Sorghum (decorticated)	Phenolics concentrated in the pericarp and testa	Dlamini et al. (2007); Awika et al. (2005)

suggested the use of rice bran in nutraceuticals and functional food industries due to its high tocol and oryzanol content.

In wet milling, soaking is an essential step; however, several nutrients such as proteins and minerals could leach out during this step and could adversely affect the nutritional properties of rice flour. Research suggests that extended soaking times would result in fine flours with less damaged starch and low lipid content that will have high viscosity (Chiang and Yeh 2002).

Wet milling of corn resulted in two types of corn fiber gum: coarse fiber (primarily obtained from pericarp) and fine fiber (obtained from the endosperm; Yadav et al. 2007). Evaluating the nutritional quality of the two fibers showed a high concentration of total phenolic acids in corn fiber gum purified from coarse corn fiber (pericarp fiber) compared with that purified from fine corn fiber (endosperm fiber). The authors reported excellent emulsifying properties of the gum due to the presence of phenolic acids, lipids, and proteins that are strongly associated or bound to the gum. Corn hulls obtained from wet milling are rich sources of ferulic acid (Graf 1992; Rosazza et al. 1995), representing as much as 2% to 4% based on a dry weight of hulls. Corn hulls produced during wet milling could provide more than 1 billion lb. of ferulic acid per year (Hosny and Rosazza 1997). The study also developed a biocatalytic approach to convert ferulic acid into value-added chemical compounds including 4-hydroxy-3-methoxystyrene by enzymatic decarboxylation, vanillic acid by β-oxidative removal of the cinnamoyl side chain, and guaiacol by decarboxylation of vanillic acid.

14.3 THERMAL PROCESSING

Thermal processing of cereals such as baking, roasting, and extrusion causes a number of physical and chemical changes due to starch gelatinization, protein denaturation, component interactions, and Maillard reactions. These changes would result in improved texture and palatability, increased nutrient availability, improved antioxidant properties, and inactivation of heat labile toxic compounds and enzyme inhibitors. Different processing techniques create different effects on the nutritional and functional properties of foods (Ragaee et al. 2013). Traditional cooking of dhal (mix of split pulses) increased RS by 1.6-fold to 9.0-fold, whereas pressure cooking of dhal showed a 2.1-fold to 8.0-fold increase over the uncooked dhal (Mahadevamma and Tharanathan 2004). Blanching and cooking could also result in the loss of several micronutrients to different extents depending on the type of food, processing conditions, and properties of the micronutrient involved. Vaidya and Sheth (2011) investigated the effects of a number of diverse thermal processing methods (boiling, steaming, baking, puffing, roasting, shallow frying, popping, puffing, flaking, and extrusion) on RS in the final product of different cereals (wheat, rice, maize, and pearl millet) and after storage. They reported that roasting, baking, and boiling increased the RS content followed by shallow frying, whereas steaming and frying resulted in reduced RS content. The puffed, flaked, and extruded cereal products also had low RS contents. Storage of different cereal products, up to 12 to 24 h at 4°C significantly increased RS content.

Research on cereal products has shown that thermal processing may assist in releasing bound phenolic acids through the breakdown of cellular constituents and cell walls (Dewanto et al. 2002). In addition, browning reactions occurred during thermal processing resulting in increased total phenolic content and free radical scavenging capacity. This increase could be due to the dissociation of conjugated phenolics during thermal processing followed by some polymerization or oxidation reactions (or both) and the formation of phenolics other than those endogenous in the grains. The unavailable phenolics (stored in the cellular vacuoles) may also be released by thermal processing (Chism and Haard 1996). Other reactions such as Maillard reaction (nonenzymatic browning), caramelization, and chemical oxidation of phenols could also contribute to the increase in total phenol content. Thermal processing may also change the ratio between various phenolic compounds. For example, thermal decomposition of ferulic acid will produce vanillin and vanillic acid (Fiddler et al. 1967; Pisarnitskii et al. 1979), whereas p-hydroxybenzaldehyde can be formed from p-coumaric acid (Pisarnitskii et al. 1979). Furthermore, caffeic acid is heat-sensitive and could be reduced during heat processes, whereas ferulic and p-coumaric acids are susceptible to thermal breakdown (Steinke and Paulson 1964; Pisarnitskii et al. 1979; Huang and Zayas 1991). Heat stress (100°C) could also increase some phenolics such as ferulic, syringic, vanillic, and p-coumaric acids or simple phenolics in wheat flour due to degradation of conjugated polyphenolic compounds such as tannins (Cheng et al. 2006).

Other bioactive compounds such as tocols (tocopherols and tocotrienols) could also be influenced by thermal processing. Moreau and Hicks (2006) reported that thermal treatments of corn germ or other corn oil-containing fractions at high temperatures lead to reductions in γ-tocopherol, γ-tocotrienol, and δ-tocotrienol and also resulted in the production of triacylglycerol oxidation products. Processing of oat groats into rolled oat or flat flakes by using heavy rollers, and then steaming and light toasting, diminish tocotrienols and tocopherols (Bryngelsson et al. 2002). The same authors also reported negative effects on avenanthramides, phenolic compounds found in oats, resulting from heating and drying.

14.3.1 BAKING

Whole-grain bakery products are expected to be better sources of DF, phenolics, and other bioactive compounds compared with refined flour products due to the concentration of DF and phenolic compounds in the outer layers of the wheat kernel. Baking resulted in an increase in phenolic compounds of whole-grain bread regardless of baking time (10, 20, or 35 min; Gelinas and McKinnon 2006). The bread crust contained slightly more phenolic compounds than its crumb. Other studies reported negligible changes in total phenolics caused by baking (Menga et al. 2010). The Maillard reaction is the most common browning reaction occurring as a result of the interaction between reducing sugars and protein during the baking process, which makes lysine unavailable and reduces the quality of the protein, but it was found to produce a dramatic increase in compounds that possess free radical scavenging properties in cookies (Bressa et al. 1996).

The extraction rate or flour yield obtained from cereal grains during milling also has an effect on the changes in phenolic compounds and antioxidant capacity of breads. For instance, the high extraction rate of rye flour enhanced the formation of antioxidant compounds during bread making (Michalska et al. 2008). Cookies baked from 100% extraction rate sorghum flours contained twofold to threefold more total phenolics compared with those made from 70% extraction rate flours. The antioxidant capacity was higher by 22% to 90% depending on the sorghum cultivar (Chiremba et al. 2009). Germination of whole-grain rye before sourdough baking produced bread with increased phenolic compound content (Liukkonen et al. 2003; Michalska et al. 2008). Winata and Lorenz (1997) found that the sourdough baking process decreases alkylresolcinol in rye bread.

Baking was found to have negative effects on some sensitive bioactive compounds such as lutein and zeaxanthin, which undergo oxidation and isomerization when subjected to light, heat, and oxygen. These compounds were found to play significant roles in promoting the health of the eyes and the skin, and in reducing the risk of age-related macular degeneration, cataracts, cancer, and cardiovascular disease. Abdel-Aal et al. (2010) reported that baking resulted in a significant reduction in all-*trans*-lutein and the formation of *cis*-lutein and *cis*-zeaxanthin isomers in naturally high-lutein or lutein-fortified baked products including breads, cookies, and muffins. Subsequent storage of the products at ambient temperature had a slight effect on the content of all-*trans*-lutein. The effect of baking was more pronounced in lutein-fortified products compared with the unfortified products, and the degradation rate of lutein was influenced by lutein concentration and baking recipe. Despite the significant reduction in lutein, the fortified bakery products still possessed reasonable amounts of lutein per serving that would enhance lutein daily intake and consumption of whole-grain foods as reported in the study. Flours obtained from einkorn (ancient wheat) are well suited for soft wheat applications such as cookies and pastries rich in carotenoids, primarily lutein (Borghi et al. 1996; Abdel-Aal et al. 1997, 2002). Storage of einkorn flour or bread for up to 239 days at various temperatures (−20°C, 5°C, 20°C, 30°C, or 38°C) had major effects on carotenoid degradation, and it was influenced by temperature and time following first-order kinetics (Hidalgo and Brandolini 2008). Leenhardt et al. (2006) demonstrated that carotenoids are more susceptible to oxidation by endogenous lipoxygenase than vitamin E during bread making. The study suggested that selecting suitable cereal genotypes rich in carotenoids and a reduction in kneading time and intensity associated with a longer period of dough fermentation could help spare carotenoids and vitamin E by limiting oxygen incorporation. Biscuits made without adding fat and nonfat dry milk, to avoid interferences with the lipophilic oxidation mechanism, had lower levels of carotenoid degradation (31%; Hidalgo et al. 2010). The study also reported carotenoid losses of 21% and 47% for bread crumb and crust, respectively. Bread leavening had almost negligible effects on carotenoid losses, whereas baking resulted in a marked decrease in carotenoids.

β-Glucan is a health-promoting soluble fiber found in cereal grains, particularly in oats and barley, and its cholesterol-lowering effects are linked with molecular weight (MW) and viscosity. A negative correlation between the MW of β-glucan in dough and bread with both mixing time and fermentation time was reported by Moriartey et al. (2010), whereas a positive correlation was observed with bran particle size, which could be due to endogenous enzymes, as reported by Vadantoust et al. (2012). The latter study also reported no effect with oven baking and upon the addition of yeast with regard to β-glucan's MW. MW and the viscosity of β-glucan are significantly correlated with beneficial health effects such as lowering blood cholesterol. On the other hand, longer dough fermentation times and increased baking times or temperature were reported to be considered potential approaches to enhancing total phenolic content and antioxidant properties in whole wheat pizza crust (Moore et al. 2009). The same authors reported no effects of bran particle size on antioxidant capacity or phenolic content.

14.3.2 EXTRUSION COOKING

Extrusion cooking is the simultaneous action of temperature, pressure, and shear force in which their intensities and interactions vary enormously depending on the ingredients, extruder configuration, and desired characteristics of the end products. Nowadays, extrusion is used to produce many food products including pasta, snacks, breakfast cereals, texturized proteins, etc. As a multifunction thermal/mechanical process, extrusion could have beneficial or detrimental effects on the content and bioavailability of nutrients of cereal products. Extrusion would also enhance the stability of foods due to enzyme inhibition (lipases and lipoxidases), increase protein and starch digestibility, and reduce lysine availability due to Maillard reaction, which results in increased antioxidant capacity and phenolic compounds (Cheftel 1986).

Extrusion was found to increase DF content due to the formation of RS. It also affects the redistribution of fiber fractions due to changes in solubility of fibers as reported for corn meal, oatmeal, and wheat (Bjorck et al. 1984; Camire 1988; Camire et al. 1990). These studies showed an increase in the soluble fiber fraction of extruded products. In general, the effect of extrusion cooking on fiber content and composition and physicochemical characteristics depends on the process variables such as temperature, pressure, shear force and screw design, and composition of feed ingredients (Camire 1988; Wang et al. 1993). Changes in physicochemical characteristics such as hydration, cation exchange, adsorption of organic molecules, expansion properties, and particle size would also be expected to cause changes in bacterial degradation of fiber in the human intestine and subsequently changes in physiological effects (Cheftel 1986).

Extrusion processing variables were also found to cause changes in phenolic compounds. For instance, the release of phenolic compounds is highly dependent on moisture content, time, and temperature during extrusion (Dimberg et al. 1996). Little research has focused on the effects of extrusion processes on antioxidant properties and phenolics, and it remains a controversial subject. A study on dark buckwheat flour reported no changes in antioxidant capacity after extrusion at 170°C (Sensoy et al. 2006). Another study showed a significant reduction in both antioxidant capacity (60%–68%) and total phenolics (46%–60%) in barley extrudates compared with that of unprocessed barley flour (Altan et al. 2009). The behavior of phenolic compounds present in selected cereals (wheat, barley, rye, and oat) during extrusion cooking at different temperatures (120°C, 160°C, and 200°C) significantly varied among cereals (Zielinski et al. 2001). Significant increases in free and bound phenolic acids (except for sinapic and caffeic acids) in the extruded grains were found, with rye and oat exhibiting the highest increase. The changes in free phenolic acids were more pronounced when compared with the bound ones. The liberated phenolic acids may contribute to the high antioxidant potential of extrudates when they were considered as a dietary antioxidant. Ferulic acid was found as a predominant compound in raw whole grain as well as in extruded grain.

Research has shown that extrusion affects several other nutrients and phytochemicals in grains. For example, inositol hexaphosphate was reduced from 4% to 50%, tocols from 63% to 94%, glutathione from 20% to 50%, and melatonin from 17% to 63%, depending on the type of cereal (wheat, barley, rye, and oat) and extrusion temperature (120°C, 160°C, and 200°C; Zielinski et al. 2001). The low-temperature extrusion process caused a slight reduction in tocols and oryzanols (~7% and 4%, respectively), whereas the high-temperature process had higher reduction (21% and 8%, respectively). Al-Ruqaie and Lorenz (1992) also reported a 77% reduction in alkylresolcinols during the extrusion of rye and a 53% to 77% reduction in alkyl-resolcinol of extruded wheat bran.

14.3.3 ROASTING

Roasting of grains is a traditional practice in many countries; in particular, it is popular for maize and chick pea to produce a variety of snack foods such as popcorn, aadun, dankuwa, guguru, and elekute (Ayatse et al. 1983; Ihekoronye and Ngoddy 1985). Roasting uses dry heat such as open flame, oven, microwave, or other heat source to cook and convert whole grain into palatable snack foods. Traditional roasting of grains is used primarily to enhance nutritional and sensory (taste and flavor) properties (Huffman and Martin 1994). Significant increase in both antioxidant capacity and total phenols of barley grains was obtained after roasting two layers of grains or 61.5 g in a microwave oven at 600 W power for 8.5 min (Gallegos-Infante et al. 2010; Omwamba and Hu 2010). The increase in phenolic compounds could be attributed to the release of bound phenolics from the breakdown of cellular constituents. On the other hand, roasting resulted in a marked reduction in phenolic content (13.0% and 18.5%) and antioxidant capacity (27.0% and 13.5%) of yellow and white sorghum, respectively (Oboh et al. 2010). A significant decrease in total phenolic content (8.5% to 49.6%) and antioxidant activity (16.8% to 108.2%) was also observed after sand roasting of eight barley varieties (Sharma and Gujral 2011). Roasting of barley at 327°C resulted in a significant reduction in catechin levels (Duh et al. 2001), whereas roasting of dark or white buckwheat at 200°C for 10 min did not affect total phenolic content and caused a slight reduction in antioxidant activity (Sensoy et al. 2006). Because a majority of phenolic compounds in buckwheat are in the free form, roasting would cause a reduction in antioxidant capacity (Zielinski et al. 2009).

14.3.4 CANNING

Canning is the cooking of foods while sealed in an airtight container to increase their shelf life for up to several years. A number of cereal grains, such as corn or sorghum, are thermally processed using canning technology. Canning could cause several changes in nutrient bioavailability and characteristics. It could result in increased solubility of DF that would make it more readily fermentable in the colon, resulting in physiologically active by-products. Ibanoglu and Ainsworth (2004) demonstrated that tarhana (a fermented wheat flour–yoghurt mixture used in soup making) could be produced in a ready-to-eat form using canning. Their results indicated that starch gelatinization was more affected by rotation speed than retorting time or dry solids content of the soup sample during the canning process. They also reported that canning does not cause any significant changes to protein digestibility *in vitro*.

Significant increases in the content of free phenolics, free ferulic acid, and total antioxidant capacity were reported in canned corn heated in a retort at 115°C for 10, 25, or 50 min (Dewanto et al. 2002). Steinke and Paulson (1964) reported that pressure cooking of corn (autoclaved for 40 min at 15 psi) caused a substantial increase in the amounts of free ferulic acid, *p*-coumaric acid, and vanillin.

Studies have shown that heating of corn (canning or cooking) has a slight effect on carotenoids (Kurilich and Juvik 1999; Schlatterer et al. 2006). No significant changes in the contents of lutein and zeaxanthin were observed in white and golden corn after canning in sugar/salt brine solution at 126.7°C for 12 min, but carotenes significantly decreased by approximately 62% (Scott and Eldridge 2005). However, the study did not measure *cis*-isomers of lutein and zeaxanthin, which were found to increase in canned vegetables (Updike and Schwartz 2003). De la Parra et al. (2007) reported that much of the carotenoid loss occurs during the preparation of masa, which is used in making corn chips and tortilla. Various levels of carotenoid bioaccessibility, ranging from 48% for porridge to 63% to 69% for extruded puff and bread, were reported for corn-based food products (Kean et al. 2008). Boiling red sorghum and finger millet at atmospheric pressure resulted in significant reduction in total extractable phenolics, whereas barley exhibited an increase in total phenolic content and antioxidant capacity (Gallegos-Infante et al. 2010). A summary of the changes in bioactive compounds and antioxidant properties during the thermal processing of cereal grains is presented in Table 14.2.

14.4 OTHER PROCESSINGS

14.4.1 Pasta Processing

Pasta is a staple food worldwide, being produced in a large number of appealing shapes and colors. It is made of unleavened dough mostly from durum wheat (whole grain or refined semolina), water, and sometimes eggs. Pasta is produced in two forms, dried or fresh. Dried pasta made without eggs can be stored for up to 2 years under normal conditions, whereas fresh pasta can be stored for a few days in refrigerators. Some whole-grain pasta can provide up to 25% of the daily fiber requirements in every one cup portion. Pasta has low glycemic index, and it can be enriched with various essential nutrients such as iron, folate, thiamine, riboflavin, and niacin. According to the American Pasta Association, enriched pastas provide an excellent source of folic acid and a good source of other essential nutrients. Pasta is generally cooked by boiling, which may influence quality attributes such as water absorption, starch gelatinization, cooking loss (e.g., partial loss of soluble starch and protein, minerals, and water-soluble vitamins), and firmness. Wojtowicz and Moscicki (2009) reported that the moisture content of the raw material and the screw speed of the pasta extruder have significant effects on functionality, microstructure, and sensory characteristic of precooked pasta-like products. They suggested that the most preferable parameters for obtaining good quality products using single screw extrusion cookers are screw speed (80 and 100 rpm) and 30% flour moisture based on excellent properties of precooked pasta products having regular and compact internal structures.

Regarding bioactive compounds, Hidalgo et al. (2010) reported that the longer kneading step had significant effects on carotenoid losses, whereas the drying step did not induce significant changes in carotenoids. New concepts and processing technologies have also been used to enhance the nutritional and functional properties of pasta products. For example, Verardo et al. (2011) used an experimental pasta-making

TABLE 14.2

Effects of Different Thermal Processing on Bioactive Compounds and Antioxidant Properties in Cereal Grains

Cereal	Changes in Phenolic Compounds	References
	Baking	
Wheat	Slight increase in phenolics in bread crust and white bread than whole-grain bread	Gelinas and McKinnon (2006)
Purple wheat bran—or heat-treated purple wheat bran-enriched muffins	Significant reduction in phenolics and ORAC values	Li et al. (2007)
40% barley replacement in wheat bread	Reduction in free phenolics, while increase in bound phenolics	Holtekjølen et al. (2008)
Rye, sourdough	Increased in bound phenolics	Liukkonen et al. (2003)
	Slight or no increase in bound phenolics	Kariluoto et al. (2006)
	Slight reduction in bound phenolic acids while no change in ferulic acid dehydrodimers	Hansen et al. (2002)
Wheat, oat, barley	No changes in phenolics and antioxidant capacity	Menga et al. (2010)
Einkorn wheat	Significant reduction in all-trans lutein in whole-grain flat and pan bread	Abdel-Aal et al. (2010)
	Lutein in whole-grain pan bread dropped to a little extent compared with flat breads	
Einkorn	Marked decrease in carotenoids (21% and 47%) for bread crumb and crust, respectively	Hidalgo et al. (2010)
	Extrusion	
Regular corn flour, corn starch	Reduction in antioxidant capacity	Ozer et al. (2006)
Rye, oat, barley, wheat	Significant decrease in bioactive compounds except for phenolics	Zielinski et al. (2001); Ozer (2006)
Buckwheat at 170°C	No change in wheat	Sensoy et al. (2006)
Barley and barley-fortified products	Reduction in total phenolics	Altan et al. (2009)
Rye at 14% moisture content, 120°C or 180°C	Significant increase in total phenolics Reduction in alkylresolcinols	Gumul and Korus (2006)
Sorghum	Reduction in antioxidant capacity of the products than conventionally cooked porridges	Dlamini et al. (2007)
Barley	Reduction in total phenolics	Altan et al. (2009)

(*continued*)

TABLE 14.2 (Continued)
Effects of Different Thermal Processing on Bioactive Compounds and Antioxidant Properties in Cereal Grains

Cereal	Changes in Phenolic Compounds	References
Einkorn wheat	Significant reduction in carotenoids in pasta	Hidalgo et al. (2010)
Roasting		
Barley, microwave	Increase in total phenolics	Gallegos-Infante et al. (2010);
	Increase in antioxidant capacity and total phenolics	Omwamba and Hu (2010)
	Significant decrease in flavonoids	
Buckwheat at 200°C/ 10 min	Slight reduction in antioxidant capacity	Sensoy et al. (2006); Zhang et al. (2010)
Cooking		
Barley	Increase in total phenolic content	Gallegos-Infante et al. (2010)
Sweet corn	Significant increase in total phenolics	Dewanto et al. (2002)
	No significant change in lutein and zeaxanthin in white and golden corn, but significant reduction (62%) in α-carotene	
	Canning or heating has a slight effect on carotenoids	Scott and Eldridge (2005)
Oat	Autoclaving increase tocols	
Malting		
Barley	Malting slightly reduces tocols, whereas mashing and brewing significantly increases them	Peterson and Qureshi (1993)
	Malting increases phenolics and flavonoids	Maillard et al. (1996); Goupy et al. (1999)
Red sorghum, millet, and maize	Malting improves nutritional quality	Traore et al. (2004)

apparatus composed by a press and a dryer to develop functional pasta from barley flour rich in β-glucan. The developed functional pasta had higher levels of flavan-3-ols and antioxidant capacity as compared with commercial pasta. Another study replaced 50% of durum with β-glucan-enriched barley flour in the presence of 5% vital wheat gluten (85°C drying temperature, 7 h drying cycle) to produce more nutritious pasta (Marconi et al. 2000). The DF (13.1%–16.1% wb) and β-glucan (4.3%–5.0% wb) contents in the barley pastas were much higher than that in the control pasta (4.0% and 0.3% wb, respectively). These levels meet the Food and Drug Administration requirements of 5 g of DF and 0.75 g of β-glucan per serving (56 g in the United States and 80 g in Italy). Tudorica et al. (2002) incorporated different fibers into pasta systems using an experimental pasta-making apparatus composed of a press and a dryer. The fiber-rich pasta characteristics (texture, structure, cooking quality, and nutritional properties) were differently affected depending on the type of fiber. The inclusion of insoluble DF such as pea fiber resulted in the disruption of the protein matrix, whereas incorporating soluble fiber such as guar gum formed the entrapment of starch granules within a viscous protein–fiber–starch network.

14.4.2 HIGH-PRESSURE PROCESSING

Some foods and beverages are exposed to elevated pressure (up to 87,000 psi or approximately 6000 atmospheres) without or with heat to reduce the microbial population of spoilage microorganisms, inactivating unwanted enzymes and compounds, and consequently increasing shelf life and producing safe foods while preserving food quality. High-pressure processing (HPP) was found to minimally affect micronutrients and bioactive compounds in fruits and vegetables, which could be the case if it is applied in processing cereal grains (Estrada-Girón et al. 2005). Various physical and chemical changes result from the application of HPP. In general, HPP has few effects on bioactive compounds, vitamins, and pigments compared with thermal processing.

14.5 BIOFUNCTIONALITY OF CEREAL FOODS AS AFFECTED BY PROCESSING

The biofunctionality of nutrient and bioactive components in cereal foods is influenced by many factors such as food matrix and structure, amount of the component, physical and functional properties of the component, diet composition, and individual responses. For instance, the metabolic effects of DF on glucose and lipid metabolisms are mainly linked to viscosity and ion exchange capacity, which are determined by particle size and bulk volume, surface area, hydration, and rheological properties. But its effects on colonic function are largely associated with its fermentation pattern, bulking effects, and particle size. Therefore, modifying the physicochemical properties of a certain nutrient or component through processing would optimize the biofunctionality and physiological properties of foods (Guillon and Champ 2000). Several studies demonstrated that coarse bran foods exhibit more significant effects on stool weight, speed of intestinal transit, and reduced intraluminal pressure in the colon of patients with diverticular disease compared with foods with fine bran (Brodribb and Groves 1978; McIntyre et al. 1997). Processed whole-grain oat ready-to-eat cereal was found to lower cholesterol in human subjects (Ripsin et al. 1992; Reyna-Villasmil 2007; Maki et al. 2010). In addition, Fraser et al. (1981) reported a reduction of 9 mg/dL in serum cholesterol using a diet containing whole wheat (40 g/day), popcorn and cornmeal plus germ (30 g/day), and oatmeal (30 g/day). Wood (2010) reviewed the health benefits of β-glucans on reducing blood serum cholesterol and regulating blood glucose levels, showing how the amount and MW of solubilized β-glucans in the gastrointestinal tract influence the physiological effects of β-glucans. Other studies have shown that various processing technologies and conditions significantly affect the functionality of β-glucan as a quality attribute of the product or as a biofunctional food ingredient (Beer et al. 1997; Boskov-Hansen et al. 2002; Andersson et al. 2004; Vatandoust et al. 2012). The use of β-glucans exhibiting different MWs in chapattis have demonstrated various levels of starch digestion in vitro and glycemic response *in vivo*; with low MW barley β-glucans not being effective in lowering glycemic response (Thondre and Henry 2009, 2011; Thondre et al. 2010).

Åman et al. (2004) concluded that processing methods such as baking including fermentation, fresh pasta preparation, and processing of fermented soup and pancake batter all resulted in an extensive degradation of the oat β-glucans. Prolonged treatment at low temperature would result in extensive enzymatic degradation and large oat bran particles, whereas short fermentation times would reduce β-glucan degradation during baking of yeast-leavened bread. It was reported that any process that disrupts the physical or botanical structure of grain components would increase the plasma glucose and insulin responses, and thus, it is recommended that the compact structure in common foods be preserved to obtain a low glycemic index (Jarvi et al. 1995; Juntunen et al. 2002; Slavin 2004).

The consumption of whole-grain foods is recommended for healthy eating as well as for reducing the risk of diseases. The consumption of whole-grain wheat breakfast cereals at 48 g per day caused a significant increase in the number of fecal bifidobacteria and lactobacilli (the target genera for prebiotic intake) as compared with wheat bran (Costabile et al. 2008). There were no significant differences in fecal short-chain fatty acids, fasting blood glucose, insulin, total cholesterol, or HDL cholesterol upon ingestion of either whole grain or wheat bran, whereas a significant reduction in total cholesterol was observed in volunteers upon ingestion of either cereal. Jenkins et al. (1986, 1988) suggested that whole-grain foods are more effective in reducing postprandial blood glucose profile in diabetics compared with only bran or fibers. Other components in whole grain that affect gut physiology include oligosaccharides, phytate, lignans, and other phytochemicals (Cummings 1993). The same author reported that wheat bran was very effective in reducing the incident of colon cancer, and each gram of fiber fed as wheat bran increased stool weight by 5.4 g, demonstrating that wheat bran is most effective in increasing stool weight. The effects of barley fiber (a rich source of β-glucan) on reducing postprandial metabolic responses were dependent on product type, for example, cookies or crackers (Casiraghi et al. 2006). The study found that cookies responded better to the addition of barley fiber than crackers. Fortification of granola bars with 15% RS exhibited positive effects on glucose response, which was attenuated after 30 min of consumption of RS-fortified products (Aigster 2009). The properties of starch were found to influence its gelatinization temperature and extent of gelatinization during processing, which in turn may affect its digestibility in humans (Svihus et al. 2005). When starchy foods are heated in a sufficient amount of water, the degree of starch gelatinization becomes greater and, as a result, the susceptibility of starch to digestion in the digestive tract becomes relatively high.

14.6 CONCLUSIONS

The nutritional and functional properties of cereal products are subjected to significant changes during processing, which may or may not be favorable. A number of factors contribute to these changes and they could be adjusted to produce the desired characteristics. Such factors include, but are not limited to, processing technology, whole grain versus refined flours, food recipe and ingredients, food matrix, food type, etc. The availability of more whole grain foods or the incorporation of milling fractions rich in bioactive components such as β-glucan, DF, etc., would improve the nutritional properties and biofunctionality of grain-based foods. In this regard, more research is required to develop innovative processing technologies and grain food recipes that would preserve health-enhancing components during processing and deliver anticipated health benefits and yet are acceptable and tasty.

REFERENCES

Abbas, A., Murtaza, S., Aslam, F., Khawar, A., Rafique, S., and Naheed, S. 2011. Effect of processing on nutritional value of rice (*Oryza sativa*). *World Journal of Medical Sciences* 6: 68–73.

Abdel-Aal, E.-S. M., Hucl, P., Sosulski, F. W., and Bhirud, P. R. 1997. Kernel, milling and baking properties of spring-type spelt and einkorn wheats. *Journal of Cereal Science* 26: 363–370.

Abdel-Aal, E.-S. M., Young, J. C., Wood, P. J., Rabalski, I., Hucl, P., and Frégeau-Reid, J. 2002. Einkorn: A Potential Candidate for Developing High Lutein Wheat. *Cereal Chemistry* 79: 455–457.

Abdel-Aal, E.-S. M., Young, J. C., and Rabalski, I. 2006. Anthocyanin composition in black, blue, purple and red cereal grains. *Journal of Agriculture and Food Chemistry* 54: 4696–4704.

Abdel-Aal, E.-S. M., Young, J. C., Rabalski, I., Frégeau-Reid, J., and Hucl, P. 2007. Identification and quantification of seed carotenoids in selected wheat species. *Journal of Agriculture and Food Chemistry* 55: 787–794.

Abdel-Aal, E.-S. M., Young, J. C., Akhtar, H., and Rabalski, I. 2010. Stability of lutein in wholegrain bakery products naturally high in lutein or fortified with free lutein. *Journal of Agricultural and Food Chemistry* 58: 10109–10117.

Aigster, A. 2009. Physicochemical and Sensory Properties of Resistant Starch-Based Cereal Products and Effects on Glycemic and Oxidative Stress Responses in Hispanic Women. PhD Dissertation. Human Nutrition, Foods, and Exercise. Faculty of the Virginia Polytechnic Institute and State University.

Al-Ruqaie, I., and Lorenz, K. 1992. Alkylresorcinols in extruded cereal brans. *Cereal Chemistry* 69: 472–475.

Altan, A., McCarthy, K. L., and Maskan, M. 2009. Effect of extrusion process on antioxidant activity, total phenolics and β-glucan content of extrudates developed from barley-fruit and vegetable by-products. *International Journal of Food Science and Technology* 44: 1263–1271.

Åman, P., Rimsten, L., and Andersson, R. 2004. Molecular weight distribution of β-glucan in oat-based foods. *Cereal Chemistry* 81: 356–360.

Andersson, A. A. M., Armo, E., Grangeon, E., Fredriksson, H., Andersson, R., and Aman, P. 2004. Molecular weight and structure of (1→3),(1→4)-β-glucans in dough and bread made from hull-less barley milling fractions. *Journal of Cereal Science* 40: 195–204.

Awika, J. M., McDonough, C. M., and Rooney, L. W. 2005. Decorticating sorghum to concentrate healthy phytochemicals. *Journal of Agricultural and Food Chemistry* 53: 6230–6234.

Ayatse, J. O., Eka, O. U., and Ifon, E. T. 1983. Chemical evaluation of the effect of roasting on the nutritive value of maize. *Food Chemistry* 12: 135–147.

Barron, C., Surget, A., and Rouau, X. 2007. Relative amounts of tissues in mature wheat (*Triticum aestivum* L.) grain and their carbohydrate and phenolic acid composition. *Journal of Cereal Science* 45: 88–96.

Beer, M. U., Wood, P. J., Weisz, J., and Fillion, N. 1997. Effect of cooking and storage on the amount and molecular weight of (1→3)(1→4)-β-D-glucan extracted from oat products by an in vitro digestion system. *Cereal Chemistry* 74: 705–709.

Beta, T., Nam, S., Dexter, J. E., and Sapirstein, H. D. 2005. Phenolic content and antioxidant activity of pearled wheat and roller-milled fractions. *Cereal Chemistry* 82: 390–393.

Bjorck, I., Nyman, M., and Asp, N.-G. 1984. Extrusion cooking and dietary fiber—effects on dietary fiber content and on the degradation in the intestinal tract. *Cereal Chemistry* 61: 174–179.

Borghi, B., Castagna, R., Corbellini, M., Heun, M., and Salamini, F. 1996. Breadmaking quality of einkorn wheat (*Titicum monococcum* ssp. *monococcum*). *Cereal Chemistry* 73: 208–214.

Boskov-Hansen, H., Andreasen, M. F., Nielsen, M. M., Larsen, L. M., Bach Knudsen, K. E., Meyer, A. S., Christensen, L. P., and Hansen, A. 2002. Changes in dietary fiber, phenolic and activity of endogenous enzymes during rye bread-making. *European Food Research and Technology* 214: 33–42.

Bressa, F., Tesson, N., Rosa, M. D., Sensidoni, A., and Tubaro, F. 1996. Antioxidant effect of Maillard reaction products: Application to a butter cookie of a competition kinetics analysis. *Journal of Agricultural and Food Chemistry* 44: 692–695.

Brodribb, A. J. M., and Groves, C. 1978. Effect of bran particle size on stool weight. *Gut* 19: 60–63.

Bryngelsson, S. H., Dimberg, L., and Kamal-Eldin, A. 2002. Effects of commercial processing on levels of antioxidants in oats (*Avena sativa* L.). *Journal of Agricultural and Food Chemistry* 50: 1890–1896.

Camire, M. E. 1988. Chemical changes during extrusion cooking. Recent advances. *Advances in Experimental Medicine and Biology* 434: 109–121.

Camire, M. E., Camire, A., and Krumhar, K. 1990. Chemical and nutritional changes in foods during extrusion. *CRC Critical Reviews in Food Science and Nutrition* 29: 35–57.

Casiraghi, M. C., Garsetti, M., Testolin, G., and Brighenti, F. 2006. Postprandial responses to cereal products enriched with barley β-glucan. *Journal of American College of Nutrition* 25: 313–320.

Cheftel, C. 1986. Nutritional effects of extrusion cooking. *Food Chemistry* 20: 263–283.

Cheng, Z., Su, L., Moore, J., Zhou, K., Luther, M., Yin J., and Yu, L. 2006. Effects of post-harvest treatment and heat stress on availability of wheat antioxidants. *Journal of Agricultural and Food Chemistry* 54: 5623–5629.

Chiang, P. Y., and Yeh, A. I. 2002. Effect of soaking on wet-milling of rice. *Journal of Cereal Science* 35: 85–94.

Chiremba, C., Taylor, J. R. N., and Duodu, K. G. 2009. Phenolic content, antioxidant activity and consumer acceptability of sorghum cookies. *Cereal Chemistry* 86: 590–594.

Chism, G. W., and Haard, N. F. 1996. *Characteristics of Edible Plant Tissues in Food Chemistry*, 3rd edition. (Fennema, O. ed.). Marcel Dekker, New York, 944–1011.

Correia, I., Nunes, A., Barros, A. S., and Delgadillo, I. 2008. Protein profile and malt activity during sorghum germination. *Journal of the Science of Food and Agriculture* 88: 2598–2605.

Costabile, A., Klinder, A., Fava, F., Napolitano, A., Fogliano, V., Leonard, C., Gibson, G. R., and Tuohy, K. M. 2008. Whole-grain wheat breakfast cereal has a prebiotic effect on the human gut microbiota: A double-blind, placebo-controlled, crossover study. *British Journal of Nutrition* 99: 110–120.

Cummings, J. H. 1993. The effect of dietary fiber on fecal weight and composition. In *CRC Handbook of Dietary Fiber in Human Nutrition*. (Spiller, G. A. ed.). CRC Press, Boca Raton, FL, 263–333.

De la Parra, C., Saldivar, S. O., and Liu, R. H. 2007. Effect of processing on the phytochemical profiles and antioxidant activity of corn for production of masa, tortillas, and tortilla chips. *Journal of Agricultural and Food Chemistry* 55: 4177–4183.

Dewanto, V., Wu, X., and Liu, R. H. 2002. Processed sweet corn has higher antioxidant activity. *Journal of Agricultural and Food Chemistry* 50: 4959–4964.

Dietrych-Szostak, D., and Oleszek, W. 1999. Effect of processing on the flavonoid content in buckwheat (*Fagopyrum esculentum* Möench) grain. *Journal of Agricultural and Food Chemistry* 47: 4384–4387.

Dimberg, L. H., Molteberg, E. L., Solheim, R., and Frölich, W. 1996. Variation in groats due to variety, storage and heat treatment. I: Phenolic compounds, *Journal of Cereal Science* 24: 263–272.

Dlamini, N. R., Taylor, J. R. N., and Rooney, L. W. 2007. The effect of sorghum type and processing on the antioxidant properties of African sorghum based foods. *Food Chemistry* 105: 1412–1419.

Duh, P. D., Yen, G. C., Yen, W. J., and Chang, L. W. 2001. Antioxidant effects of water extracts from barley (*Hordeum vulgare* L.) prepared under different roasting temperatures. *Journal of Agricultural and Food Chemistry* 50: 1455–1463.

Elkareem, A. M. A., and Taylor, J. R. N. 2011. Protein quality and physical characteristics of kisra (fermented sorghum pancake-like flatbread) made from tannin and non-tannin sorghum cultivars. *Cereal Chemistry* 88: 344–348.

Elkhier, M. K. S., and Hamid, A. O. H. 2008. Effect of malting on the chemical constituents, anti-nutrition factors, and ash composition of two sorghum cultivars (feterita and tabat) grown in Sudan. *Research Journal of Agriculture and Biological Sciences* 4: 500–504.

Estrada-Girón, Y., Swanson, B. G., and Barbosa-Cánovas, G. V. 2005. Advances in the use of high hydrostatic pressure for processing cereal grains and legumes. *Trends in Food Science and Technology* 16: 194–203.

Estruch, R., Martínez-González, M. A., Corella, D., Basora-Gallisá, J., and Ruiz-Gutiérrez, V. 2009. Effects of dietary fibre intake on risk factors for cardiovascular disease in subjects at high risk. *Journal of Epidemiology and Community Health* 63: 582–588.

Fraser, G. E., Jacobs, D. R., Anderson, J. T., Foster, N., Palta, M., and Blackburn, H. 1981. The effect of various vegetable supplements on serum cholesterol. *American Journal of Clinical Nutrition* 34: 1272–1277.

Fiddler, W., Parker, W. E., Wasserman, A. E., and Doerr, R. C. 1967. Thermal decomposition of ferulic acid. *Journal of Agricultural and Food Chemistry* 15: 757–761.

Fox, P., and Mulwhill, D. 1982. Enzymes in wheat, flour and bread. In *Advances in Cereal Science and Technology*, vol. 5. American Association of Cereal Chemists, St Paul, MN, 107–156.

Gallegos-Infante, J. A., Rocha-Guzman, N. E., Gonzalez-Laredo, R. F., and Pulido-Alonso, J. 2010. Effect of processing on the antioxidant properties of extracts from Mexican barley (*Hordeum vulgare*) cultivar. *Food Chemistry* 119: 903–906.

Gelinas, P., and McKinnon, C. 2006. Effect of wheat variety, farming site, and bread-baking on total phenolics. *International Journal of Food Science and Technology* 41: 329–332.

Glitsø, L. V., Gruppen, H., Schols, H. A., Højsgaard, S., Sandström, B., and Knudsen, K. E. B. 1999. Degradation of rye arabinoxylans in the large intestine of pigs. *Journal of the Science of Food and Agriculture* 79: 961–969.

Goupy, P., Hugues, M., Boivin, P., and Amiot, M. J. 1999. Antioxidant composition and activity of barley (*Hordeum vulgare*) and malt extracts and of isolated phenolic compounds. *Journal of the Science of Food and Agriculture* 79: 1625–1634.

Graf, E. 1992. Antioxidant potential of ferulic acid. *Free Radical Biology and Medicine* 13: 435–448.

Guillon, F., and Champ, M. 2000. Structural and physical properties of dietary fibres, and consequences of processing on human physiology. *Food Research International* 33: 233–245.

Gumul, D., and Korus, J. 2006. Polyphenol content and antioxidant activity of rye bran extrudates produced at varying parameters of extrusion process. *Electronic Journal of Polish Agricultural Universities* 9: Issue 4. Available at: http://www.ejpau.media.pl/volume9/issue4/art-11.html.

Hamer, R. J. 1991. Enzymes in the baking industry. In *Enzymes in Food Processing.* (Tucker, C. A., and Woods, L. F. J. eds.), 168–193.

Hansen, H. B., Andreasen, M. F., Nielsen, M. M., Larsen, L. M., Knudsen, K. E. B., Meyer, A. S., Christensen, L. P., and Hansen, A. 2002. Changes in dietary fibre, phenolic acids and activity of endogenous enzymes during rye bread-making. *European Food Research and Technology* 214: 33–42.

Hidalgo, A., and Brandolini, A. 2008. Kinetics of carotenoids degradation during the storage of einkorn (*Triticum monococcum* ssp. *monococcum*). *Journal of Agricultural and Food Chemistry* 56: 11300–11305.

Hidalgo, A., Brandolini, A., and Pompei, C. 2010. Carotenoids evolution during pasta, bread and water biscuit preparation from wheat flours. *Food Chemistry* 121: 746–751.

Holasova, M., Fiedlerova, V., Smrcinova, H., Orsak, M., Lachman, J., and Vavreinova, S. 2002. Buckwheat—the source of antioxidant activity in functional foods. *Food Research International* 35: 207–211.

Holtekjølen, A. K., Kinitz, C., and Knutsen, S. H. 2006. Flavanol and bound phenolic acid contents in different barley varieties. *Journal of Agricultural and Food Chemistry* 54: 2253–2260.

Holtekjølen, A. K., Baevre, A. B., Rodbotten, M., Berg, H., and Knutsen, S. H. 2008. Antioxidant properties and sensory profiles of breads containing barley flour. *Food Chemistry* 110: 414–421.

Holtekjølen, A. K., Sahlstrøm, S., and Knutsen, S. H. 2011. Phenolic contents and antioxidant activities in covered whole grain flours of Norwegian barley varieties and in fractions obtained after pearling. *Acta Agriculture Scandinavica Section B, Soil and Plant Science* 61: 67–74.

Hosny, M., and Rosazza, J. P. N. 1997. Structures of ferulic acid glycoside esters in corn hulls. *Journal of Natural Products* 60: 219–222.

Huang, C. J., and Zayas, J. F. 1991. Phenolic acid contributions to taste characteristics of corn germ protein flour products. *Journal of Food Science* 56: 1308–1315.

Huffman, S. L., and Martin, L. H. 1994. First feedings: optimal feeding of infants and toddlers. *Nutrition Research* 14: 127–159.

Hung, P. V., and Morita, N. 2008. Distribution of phenolic compounds in the graded flours milled from whole buckwheat grains and their antioxidant capacities. *Food Chemistry* 109: 325–331.

Hung, P. V., Maeda, T., Miyatake, K., and Morita, N. 2009. Total phenolic compounds and antioxidant capacity of wheat graded flours by polishing method. *Food Research International* 42: 185–190.

Ibanoglu, S., and Ainsworth, P. 2004. Effect of canning on the starch gelatinization and protein in vitro digestibility of tarhana, a wheat flour-based mixture. *Journal of Food Engineering* 64: 243–247.

Ihekoronye, A. I., and Ngoddy, P. O. 1985. *Cereal Grain Processing in the Tropics. Integrated Food Science and Technology for the Tropics.* Macmillan, London, 241–245.

Iqbal, S., Bhanger, M. I., and Anwar, F. 2005. Antioxidant properties and components of some commercially available varieties of rice bran in Pakistan. *Food Chemistry* 93: 265–272.

Jarvi, A. E., Karlstrom B. E., Granfeldt, Y. E., Bjorck, I. M., and Vessby, B. O. 1995. The influence of food structure on postprandial metabolism in patients with non-insulin-dependent diabetes mellitus. *American Journal of Clinical Nutrition* 61: 837–842.

Jenkins, D. J., Wesson, V., and Wolever, T. M. 1988. Wholemeal versus wholegrain breads: Proportion of whole or cracked grain and the glycaemic response. *British Medical Journal* 297: 958–960.

Jenkins, D. J., Wolever, T. M., and Jenkins, A. L. 1986. Low glycemic response to traditionally processed wheat and rye products: bulgur and pumpernickel bread. *American Journal of Clinical Nutrition* 43: 516–520.

Joye, I. J., Lagrain, B., and Delcour, J. A. 2009. Use of chemical redox agents and exogenous enzymes to modify the protein network during breadmaking: a review. *Journal of Cereal Science* 50: 11–21.

Juntunen, K. S., Niskanen, L. K., and Liukkonen, K. H. 2002. Postprandial glucose, insulin and incretin responses to grain products in healthy subjects. *American Journal of Clinical Nutrition* 75: 254–262.

Kariluoto, S., Liukkonen, K., Myllymäki, O., Vahteristo, L., Kaukovirta-Norja, A., and Piironen V. 2006. Effect of germination and thermal treatments on folates in rye. *Journal of Agricultural and Food Chemistry* 54: 9522–9528.

Kean, E. G., Hamaker B. R., and Ferruzzi, M. G. 2008. Carotenoids bioaccessibility from whole grain and degermed maize meal products. *Journal of Agricultural and Food Chemistry* 56: 9918–9926.

Kurilich, A. C., and Juvik, J. A. 1999. Quantification of carotenoid and tocopherol antioxidants in *Zea mays*. *Journal of Agricultural and Food Chemistry* 47: 1948–1955.

Leenhardt, F., Lyan, B., Rock, E., Boussard, A., Potus, J., Chanliaud, E., and Remesy, C. 2006. Wheat lipoxygenase activity induces greater loss of carotenoids than vitamin E during breadmaking. *Journal of Agricultural and Food Chemistry* 54: 1710–1715.

Li, W., Pickard, M. D., and Beta, T. 2007. Effect of thermal processing on antioxidant properties of purple wheat bran. *Food Chemistry* 104: 1080–1086.

Liukkonen, K. H., Katina, K., Wilhelmsson, A., Myllymaki, O., Lampi, A. M., and Kariluoto, S. 2003. Process-induced changes on bioactive compounds in whole grain rye. *Proceedings of the Nutrition Society* 62: 117–122.

Liyana-Pathirana, C. M., and Shahidi, F. 2006. Antioxidant properties of commercial soft and hard winter wheats (*Triticum aestivum* L.) and their milling fractions. *Journal of the Science of Food and Agriculture* 86: 477–485.

Liyana-Pathirana, C. M., and Shahidi, F. 2007. Antioxidant and free radical scavenging activities of whole wheat and milling fractions. *Food Chemistry* 101: 1151–1157.

Lu, J., Zhao, H., Chen, J., Fan, W., Dong, J., Kong, W., Sun, J., Cao, Y., and Cai, G. 2007. Evolution of phenolic compounds and antioxidant activity during malting. *Journal of Agricultural and Food Chemistry* 55: 10994–11001.

Madhujith, T., Izydorczyk M., and Shahidi, F. 2006. Antioxidant properties of pearled barley fractions. *Journal of Agricultural and Food Chemistry* 54: 3283–3289.

Mahadevamma, S., and Tharanathan, R. N. 2004. Processing of legumes: resistant starch and dietary fiber contents. *Journal of Food Quality* 27: 289–303.

Maillard, M., Boivin, P., and Berset, C. 1996 Antioxidant activity of barley and malt: relationship with phenolic content. *LWT—Food Science and Technology* 29: 238–244.

Maki, K. C., Beiseigel, J. M., Jonnalagadda, S. S., Gugger, C. K., Reeves, M. S., Farmer, M. V., Kaden, V. N., and Rains, T. M. 2010. Whole-grain ready-to-eat oat cereal, as part of a dietary program for weight loss, reduces low-density lipoprotein cholesterol in adults with overweight and obesity more than a dietary program including low-fiber control foods. *Journal of the American Dietetic Association* 110: 205–214.

Marconi, E., Graziano, M., and Cubadda, R. 2000. Composition and utilization of barley pearling by-products for making functional pastas rich in dietary fiber and β-glucans. *Cereal Chemistry* 77: 133–139.

McIntyre, A., Vincent, R. M., Perkins, A. C., and Spiller, R. C. 1997. Effect of bran, ispaghula, and inert plastic particles on gastric emptying and small bowel transit in humans: the role of physical factors. *GUT. An International Journal of Gastroenterology and Hepatology.* 40: 223–227.

Mella, O. N. O. 2011. Effects of Malting and Fermentation on the Composition and Functionality of Sorghum Flour. Dissertations and Theses in Food Science and Technology. University of Nebraska, Lincoln, Nebraska.

Menga, V., Fares, C., Troccoli, A., Cattivelli, L., and Baiano, A. 2010. Effects of genotype, location and baking on the phenolic content and some antioxidant properties of cereal species. *International Journal of Food Science and Technology* 45: 7–16.

Michalska, A., Amigo-Benavent, M., Zielinski, H., and del Castillo, M. D. 2008. Effect of bread making on formation of Maillard reaction products contributing to the overall antioxidant activity of rye bread. *Journal of Cereal Science* 48: 123–132.

Moore, J., Luther, M., Cheng, Z., and Yu, L. 2009. Effects of baking conditions, dough fermentation and bran particle size on antioxidant properties of whole-wheat pizza crusts. *Journal of the American Dietetic Association* 57: 832–839.

Moreau, R. A., and Hicks, K. B. 2006. Reinvestigation of the effect of heat pretreatment of corn fiber and corn germ on the levels of extractable tocopherols and tocotrienols. *Journal of Agricultural and Food Chemistry* 54: 8093–8102.

Moriartey, S., Temelli, F., and Vasanthan, T. 2010. Effect of formulation and processing treatments on viscosity and solubility of extractable barley β-glucan in bread dough evaluated under in vitro conditions. *Cereal Chemistry* 87: 65–72.

Mousia, Z., Edherly, S., Pandiella, S. S., and Webb, C. 2004. Effect of wheat pearling on flour quality. *Food Research International* 37: 449–459.

Oboh, G., Ademiluyi, A. O., and Akindahunsi, A. A. 2010. The effect of roasting on the nutritional and antioxidant properties of yellow and white maize varieties. *International Journal of Food Science and Technology* 45: 1236–1242.

Omwamba, M., and Hu, Q. 2010. Antioxidant activity in barley (*Hordeum vulgare* L.) grains roasted in a microwave oven under conditions optimized using response surface methodology. *Journal of Food Science* 75: 66–73.

Ozer, E. A. 2006. Effect of extrusion process on the antioxidant activity and total phenolics in a nutritious snack food. *International Journal of Food Science and Technology* 41: 289–293.

Peterson, D. M. 2001. Oat antioxidants. *Journal of Cereal Science* 33: 115–129.

Peterson, D. M., and Qureshi, A. A. 1993. Genotype and environment effects on tocols of barley and oats. *Cereal Chemistry* 70: 157–162.

Pisarnitskii, A. F., Egorov, I. A., and Egofarova, R. Kh. 1979. Formation of volatile phenols in cognac alcohols. *Applied Biochemistry and Microbiology* 15: 103–109.

Ragaee, S., Abdel-Aal, E.-S. M., and Noaman, M. 2006. Antioxidant activity and nutrient composition of selected cereals for food use. *Food Chemistry* 98: 32–38.

Ragaee, S., Guzar, I., Dhull, N., and Seetharaman, K. 2011. Effects of Fiber Addition on Antioxidant Capacity and Nutritional Quality of Wheat Bread. *LWT—Food Science and Technology* 44: 2147–2153.

Ragaee, S., Guzar, I., Abdel-Aal, E.-S. M., and Seetharaman, K. 2012a. Bioactive components and antioxidant capacity of Ontario hard and soft wheat varieties. *Canadian Journal of Plant Science* 92: 19–30.

Ragaee, S., Seetharaman, K., and Abdel-Aal, E.-S. M. 2013. Impact of milling and thermal processing on phenolic compounds in cereal grains. *Critical Reviews in Food Science and Nutrition* (in press).

Reyna-Villasmil, N., Bermudez-Pirela, V., and Mengual-Moreno, E. 2007. Oat-derived beta-glucan significantly improves HDLC and diminishes LDLC and non-HDL cholesterol in overweight individuals with mild hypercholesterolemia. *American Journal of Therapeutics* 14: 203–212.

Ripsin, C. M., Keenan, J. M., Jacobs, D. R. Jr., Elmer, P. J., Welch, R. R., Van Horn, L., Liu, K., Turnbull, W. H., Thye, F. W., Kestin, M., Hegsted, M., Davidson, D. M., Davidson, M. H., Dugan, L. D., Demark-Wahnefried, W., and Beling, S. 1992. Oat products and lipid lowering. *The Journal of the American Medical Association* 267: 3317–3325.

Rosazza, J. P. N., Huang, Z., Dostal, L., Volm, T., and Rousseau, B. 1995. Review: Biocatalytic transformations of ferulic acid: an abundant aromatic natural product. *Journal of Industrial Microbiology* 15: 457–471.

Schlatterer, J., Maurer, S., and Breithaupt, D. E. 2006. Quantification of 3R,3-R zeaxanthin in plant derived food by a diastereomeric dilution assay applying chiral high-performance liquid chromatography. *Journal of Chromatography A* 1137: 216–222.

Scott, C. E., and Eldridge, A. L. 2005. Comparison of carotenoid content in fresh, frozen and canned corn. *Journal of Food Composition and Analysis* 18: 551–559.

Sensoy, I., Rosen, R. T., Ho, C. T., and Karwe, M. V. 2006. Effect of processing on buckwheat phenolics and antioxidant activity. *Food Chemistry* 99: 388–393.

Sharma, P., and Gujral, H. S. 2011. Effect of sand roasting and microwave cooking on antioxidant activity of barley. *Food Research International* 44: 235–240.

Sidhu, S. J., and Kabir, Y. 2007. Functional foods from cereal grains. *International Journal of Food Properties* 10: 231–244.

Singh, V., Batie, C. J., Rausch, K. D., and Miller, C. 2006. Wet-milling and dry-milling properties of dent corn with addition of amylase corn. *Cereal Chemistry* 83: 321–323.

Skrabanja, V., Kreft, I., Golob, T., Modic, M., Ikeda, S., and Ikeda, K. 2004. Nutrient content in buckwheat milling fractions. *Cereal Chemistry* 81: 172–176.

Slavin, J. L. 2004. Whole grains and human health. *Nutrition Research Reviews* 17: 99–110.

Stauffer, C. E. 1987. Proteases, peptidases and inhibitors. In *Enzymes and Their Role in Cereal Technology*. Kruger, J. E., Lineback, D., and Stauffer, C. E. (eds.). AACC Int., St. Paul, MN, 166–169.

Steinke, R. D., and Paulson, M. C. 1964. The production of phenolic acids which can influence flavor properties of steam-volatile phenols during the cooking and alcoholic fermentation of grain. *Journal of Agricultural and Food Chemistry* 12: 381–387.

Streppel, M. T., Ocke, M. C., Boshuizen, H. C., and Kok, F. J. 2008. Dietary fiber intake in relation to coronary heart disease and all-cause mortality over 40 y: the Zutphen Study. *American Journal of Clinical Nutrition* 88: 1119–1125.

Svihus, B., Uhlen, A. K., and Harstad, O. M. 2005. Effect of starch granule structure, associated components and processing on nutritive value of cereal starch: a review. *Animal Feed Science and Technology* 122: 303–320.

Taylor, J. R. N., Schober, T. J., and Bean, S. R. 2006. Novel food and non-food uses for sorghum and millets. *Journal of Cereal Science* 44: 252–271.

Thondre, P. S., and Henry, C. J. K. 2009. High-molecular-weight barley β-glucan in chapattis (unleavened Indian flatbread) lowers glycemic index. *Nutrition Research* 29: 480–486.

Thondre, P. S., and Henry, C. J. K. 2011. Effect of a low molecular weight, high-purity β-glucan on in vitro digestion and glycemic response. *International Journal of Food Sciences and Nutrition* 62: 678–684.

Thondre, P. S., Monro, J. A., Mishra, S., and Henry, C. J. K. 2010. High molecular weight barley β-glucan decreases particle breakdown in chapattis (Indian flat breads) during in vitro digestion. *Food Research International* 43: 1476–1481.

Traore, T., Mouquet, C. C., Icard-Verier, C., and Treche, S. 2004. Changes in nutrient composition, phytate and cyanide contents and α-amylase activity during cereal malting in small production units in Ouagadougou (Burkina Faso). *Food Chemistry* 88: 105–114.

Tudorica, C. M., Kuri, V., and Brennan, C. S. B. 2002. Nutritional and physicochemical characteristics of dietary fiber enriched pasta. *Journal of Agricultural and Food Chemistry* 50: 347–356.

Updike, A. A., and Schwartz, S. J. 2003. Thermal processing of vegetables increase cis isomers of lutein and zeaxanthin. *Journal of Agricultural and Food Chemistry* 51: 6184–6190.

Vaher, M., Matso, K., Levandi, T., Helmja, K., and Kaljurand, M. 2010. Phenolic compounds and the antioxidant activity of the bran, flour and whole grain of different wheat varieties. *Procedia Chemistry* 1: 76–82.

Vaidya, R. H., and Sheth, M. K. 2011. Processing and storage of Indian cereal and cereal products alters its resistant starch content. *Journal of Food Science and Technology* 48: 622–627.

Vatandoust, A., Ragaee, S., Wood, P. J. Tosh, S. M., and Seetharaman, K. 2012. Detection, localization and variability of endogenous β-glucanase in wheat kernels. *Cereal Chem. Cereal Chemistry*, 89: 59–64.

Vatandoust, A., Ragaee, S., Tosh, S. M., and Seetharaman, K. 2012. β-Glucan degradation by endogenous enzymes in wheat flour doughs with different moisture contents (in press).

Verardo, V., Gomez-Caravaca, A. M., Messia, M. C., Marconi, E., and Caboni, M. F. 2011. Development of functional spaghetti enriched in bioactive compounds using barley coarse fraction obtained by air classification. *Journal of Agricultural and Food Chemistry* 59: 9127–9134.

Wang, W. H., Klopfenstain, C. R., and Ponte, J. G. 1993. Effect of twin-screw extrusion on the physical properties of dietary fiber and other components of whole wheat and wheat bran on the baking quality of the wheat bran. *Cereal Chemistry* 70: 707–711.

Winata, A., and Lorenz, K. 1997. Effects of fermentation and baking of whole wheat and whole rye sourdough breads on cereal alkylresorcinols. *Cereal Chemistry* 74: 284–287

Wojtowicz, A., and Moscicki, L. 2009. Influence of extrusion-cooking parameters on some quality aspects of precooked pasta-like products. *Journal of Food Science* 74: 226–233.

Wood, P. J. 2010. Review oat and rye β-glucan: Properties and function. *Cereal Chemistry* 87: 315–330.

Yadav, M. P., Moreau, R. A., and Hicks, K. B. 2007. Phenolic acids, lipids, and proteins associated with purified corn fiber arabinoxylans. *Journal of Agricultural and Food Chemistry* 55: 943–947.

Zhang, M., Chen, H., Li, J., Pei, Y., and Liang, Y. 2010. Antioxidant properties of tartary buckwheat extracts as affected by different thermal processing methods. *LWT—Food Science and Technology* 43: 181–185.

Zielinski, H., and Kozłowska, H. 2000. Antioxidant activity and total phenolics in selected cereal grains and their different morphological fractions. *Journal of Agricultural and Food Chemistry* 48: 2008–2016.

Zielinski, H., Ceglinska, A., and Michalska, A. 2007. Antioxidant contents and properties as quality indices of rye cultivars. *Food Chemistry* 104: 980–988.

Zielinski, H., Kozłowska, H., and Lewczuk, B. 2001. Bioactive compounds in the cereal grains before and after hydrothermal processing. *Innovative Food Science and Emerging Technologies* 2: 159–169

Zielinski, H., Michalska, A., Amigo-Benavent, M., Dolores Del Castillo, M., and Piskula, M. K. 2009. Changes in protein quality and antioxidant properties of buckwheat seeds and groats induced by roasting. *Journal of Agricultural and Food Chemistry* 57: 4771–4776.

Index

Page numbers followed by f and t indicate figures and tables, respectively.

Printed and bound by CPI Group (UK) Ltd, Croydon, CR0 4YY
21/10/2024
01777103-0005